SMART CITIES AND FUTURE ARCHITECTURE
智能城市与未来建筑

U0178129

BUILDINGS ALIVE

感应式建筑

物联网时代的建筑

ARCHITECTURE FOR THE INTERNET OF THINGS

鲁道夫·埃尔-库利　克里斯托·马尔科普洛斯　卡罗尔·穆海贝尔　著
魏　秦　张　昕　译
魏　秦　审校

中国建筑工业出版社

著作权合同登记图字：01-2020-7136

图书在版编目（CIP）数据

感应式建筑：物联网时代的建筑：英、汉 /（美）鲁道夫·埃尔–库利（Rodolphe el-Khoury），（美）克里斯托·马尔科普洛斯（Christos Marcopoulos），卡罗尔·穆海贝尔（Carol Moukheiber）著；魏秦，张昕译．—北京：中国建筑工业出版社，2020.7

（智能城市与未来建筑）

书名原文：BUILDINGS ALIVE ARCHITECTURE FOR THE INTERNET OF THINGS

ISBN 978-7-112-25200-8

Ⅰ.①感… Ⅱ.①鲁… ②克… ③卡… ④魏… ⑤张… Ⅲ.①智能化建筑–研究–英、汉 Ⅳ.①TU18

中国版本图书馆CIP数据核字（2020）第091968号

责任编辑：孙书妍　吴　绫
文字编辑：李东禧
版式设计：锋尚设计
责任校对：姜小莲

智能城市与未来建筑
SMART CITIES AND FUTURE ARCHITECTURE
感应式建筑　物联网时代的建筑
BUILDINGS ALIVE ARCHITECTURE FOR THE INTERNET OF THINGS
鲁道夫·埃尔–库利
克里斯托·马尔科普洛斯　著
卡罗尔·穆海贝尔
魏秦　张昕　译
魏秦　审校

*

中国建筑工业出版社出版、发行（北京海淀三里河路9号）
各地新华书店、建筑书店经销
北京锋尚制版有限公司制版
天津图文方嘉印刷有限公司印刷

*

开本：889毫米×1194毫米　1/20　印张：16　字数：581千字
2020年12月第一版　2020年12月第一次印刷
定价：168.00元
ISBN 978-7-112-25200-8
（35942）

教育部人文社会科学研究规划基金项目成果

《大数据时代基于物联网技术的感应式建筑研究》

项目号：16YJAZH059

中英文双语版序
Preface for the Bilingual Edition

As far as I can remember, under the direction of Prof. Rodolphe el-Khoury, Architecture Department and Digital Art Department of Shanghai Academy of Fine Arts once jointly held a significant workshop themed by ‘Digital Space’, during which we probed into the possibility and feasibility of architectural design with the technology of the Internet of Things. Confronted with a fresh challenge, we had to acquire the knowledge alien to us while making the exploration. The students of Digital Art Department are mainly engaged in creating media art works with computer technology, whereas students of Architecture Department knew little about the technology of the Internet of Things. In the context of information age, the students of two different specialties have gained the new technology in architectural design and the Internet of Things when they were making the inter-disciplinary exploration in this design workshop.

As the Internet of Things serves as a link in the process of information age, it frequently brings forth the iterating generations of computer technology, from the popularization of personal computers, LAN architecture, penetration of internet, meg-data to cloud computing. In the course of its development, how far a man masters computer technology decides his proficient use of computer and the achievement. The advent of such tools as 3DMax, CAD and Word are taken as the

在我的记忆中，上海大学美术学院建筑系与数码艺术系的师生曾在鲁道夫 · 埃尔 – 库利教授的指导下举办过一场非常有意义的国际数字空间工作营，用计算机物联网技术探索未来建筑设计的可能性。这是一次全新的挑战，师生们几乎是处于知识盲区的情况下一边学习一边探索的。数码艺术系的同学平时主要运用计算机技术创作媒体艺术作品，而建筑系同学较少接触计算机物联网技术。在信息化时代的背景下，两个不同专业的学生通过这次工作营活动进行了一次跨领域的探索，重新认识了建筑设计与物联网技术。

物联网是信息化时代进程中的一个环节，从 PC 机的普及到局域网的构架，再到互联网的贯通，大数据、云计算迎面而来，计算机技术的迭代加速让人应接不暇，不知所措。在发展过程中，对计算机技术的认知决定了其运用的方式与结果，可谓仁者见仁，智者见智。3DMax 和 CAD、Word

等工具软件的诞生，是将计算机技术看成了工具的结果；而把计算机技术看成媒体，则产生了各类新媒体产业；把计算机技术看成是人机、虚实交互的手段，又产生了人工智能、VR、AR 等技术，实现了“不可能”的现实。如果把计算机技术看成是挑战人类思维和智慧的武器，我们一旦接受挑战，它将帮助人类认知未来，创造更好的未来。

人类的创造价值在不同历史时期反映出的表现形式是不一样的。在工业化时代，人类的创造价值体现在根据需求造物，解决人类生存的物质需求，创造出了物质文明。在信息化时代，计算机技术改变了人们的观念和生活方式，人们对精神的追求大大超出了对物质的需求，人类的创造价值不仅体现在满足人的物质需求和对物质的利用开发上，更反映在为满足人类精神和探索未来需求的思想观念的创新、方式方法的创造以及精神世界的超越上。而这一切变化正是通过信息的传播得到广泛认可，并形成一种社会需求，以此带动社会进步和组织社会生产。其产生的不仅是物质财富，更多的是精神附加值的提升，满足人们精神需求的趋势。数字信息技术造就了信息化时代，它带给人们非线性思维方式，以及共赢、分享的价值取向和民主文明。在新时代，我们应该运用这些新的、独特的思维方式和价值理念去创新和发展，我们将迎来的是真正的信息化社会的生活方式。

互联网、GPS 网、5G 通信网三网合一即将改变人类的认知和世界观，开启人类通向未来世界的大门。5G 带给我们无缝式的时间速度，GPS 技术给了我们可控的空间构架，而云计算技术给人们提供选择的无限可能。当人们掌握了无缝衔接的速度时间、可控的空间构架，加上无限

results of computing technology, while the application of computer technology into media has yielded the industries of various new media. To regard computer technology as the man-machine and virtual-real interactive means generates such technologies as AI, VR and AR, bringing about the ‘impossible’ reality. If we take computer technology as the tool for challenging human thinking and wisdom, we will acquire more knowledge on human recognition to create a better future.

At different historical periods, human beings apply varied means in creating values. During industrialization era, people invent things to meet their material requirements for survival, thus yielding material civilization. Computer technology in the time of information has changed people’s conventional concepts and living styles, during which man is at requesting more spiritual requirement than the material one. At this time, the value created by human beings is shown not only in satisfaction of human material needs, developing and utilizing materials, but also in the innovation of ideas, the creation of ways and methods, and the transcendence of the spiritual world in order to meet the human spiritual needs and explore the future. Through the transmission of the information, the change is accepted by the public and evolves into a social need, which motivates social production and promotes human progress. What it yields include not only material wealth, but also the elevation of spiritual added value so as to meet the increasing spiritual requirement of human beings. Digital information technology has forged the information age, which will offers non-linear mode of thinking, the win-win and shared value orientation and democratic civilization. During the information era, we should make innovation and promote social development with these fresh, unique modes of thinking and value concept. If we succeed, we will usher in the authentic life style of the information society.

The integration of internet, GPS net and 5G net will reverse human recognition of our world, open a way to the future world. 5G net will bring us the seamless time and speed, GPS technology offers us the controlled space, while cloud computing technology provides human beings with infinite options. Once people master the seamless time and speed, the controlled space and never-ending options, their creativity will be inspired as they wish. At that time, we will greet the genuine information era.

This title co-translated by Ms. Wei Qin and Mr. Zhang Xin describes what the architecture in the information age is like and what technical channels we should follow to bring such kind of architecture into fruition. The book introduces the fresh knowledge to China's architectural circles, from which we may consider the re-design of the physical structure from the perspective of information science and learn from the interdisciplinary research mechanism. This bilingual edition of the monograph will be conducive to the architectural design education in China and to the exploration of professional training for the students of information technology in the workshop of Shanghai Academy of Fine Arts.

In this title, the author brings forth such definitions as ‘the integration of network dynamics for architecture and life system’ and ‘the transformation of material entity technology from mechanical paradigm to biological model’ to prescribe the architecture of the Internet of Things. The author no long takes architectural space as a static form, but a random and controlled organism supported by computer technology, in which people can casually change space form, environment and comfort in accordance with man's psychological and mental requirements and the variation of natural environment. In this way, the organism will be the authentic digital space, which frames up a synergistic balancing mechanism driven by internal requirement on the basis of men’ space, material and digital information and which will integrate the advanced technologies, including cloud computing, artificial intelligence, life science and intelligent learning to evolve into a living entity with life consciousness and vitality of organic growth. The future architecture will be the extending part of men's living entity and keep company of human beings forever.

Wang Dawei
Dean of Shanghai Academy of Fine Arts
June, 28th, 2020

可能的选择手段，人的创造力可随心所欲地发挥，真正的信息化时代将要来临。

魏秦老师与张昕先生翻译的《感应式建筑　物联网时代的建筑》一书，给我们描述了信息化时代的建筑是怎样的，以及实现这类建筑的技术路径的多种可能。他们将全新的知识领域介绍给国内建筑设计界，让我们看到了从信息科学的视角重新认识物理结构再设计的可能，书中呈现的跨界研究机制也非常值得我们学习借鉴。该书的翻译出版和上海美术学院跨专业探索性工作营的举办都预示着我国建筑设计教育开启了探索信息化时代专业教育的大门。

本书中，作者提出的“建筑与生命系统的网络动力学相结合”“物质实体技术从机械范式向生物模型的转变”等观点定义了物联网时代的建筑。他不再将建筑空间看作是静止的形态，在计算机技术的支撑下，建筑空间成为随机可控的有机体，可以根据人的生理、心理需求信息和自然环境变化信息随时改变空间形态、环境氛围和舒适度。建筑空间不再是物理空间，而成为名副其实的数字空间，它将人、空间、物质以数字信息的方式构架起一个以需求为内在动力的协同平衡机制，与云计算、人工智能、生命科学、智能学习等先进科学技术相结合，它将成为具有生命意识、有机可生长的生命体。未来的建筑是人生命体的延伸部分，与人类同呼吸、共命运。

汪大伟
2020 年 6 月 28 日于上海

序
Foreword

汤姆 · 万瑞碧斯
Tom Verebes

> **“设计师主导控制系统的建构，因此设计是控制中的核心，换言之，设计师在整个控制系统中起到同样重要的作用，他居于整个系统组织构架中更高的层级。”**
>
> 《控制论的建筑相关性》
> 戈登 · 帕斯克，1969 年
>
> **‘The designer is controlling the construction of control systems, and consequently design is control of control, i.e. the designer does much the same job as his system, but he operates at a higher level in the organisational hierarchy.’**
>
> *The Architectural Relevance of Cybernetics*
> Gordon Pask, 1969

本书记录了一种设计研究的方向，就是将电子传感器、控制体系、制动装置等嵌入建筑的固有空间中，它们反过来会影响人们的行为及与之相互作用的对象。本书作者以教学、研究与实践相结合的路径，陈述了一种有关物质的设计概念。这些物质有着生活中一般物质的特性，但是又有别于传统建筑思维中建筑材料都是不可变的惰性材料。通常的建筑书籍多局限于二维、三维空间与其材质表现，但本书所列举的项目与实践试图为读者展现空间与材料内在的动态与时变属性。

This title documents an arena for design research in which distributed electronic sensors, control systems and actuators are embedded into the hard matter of architectural space, which in turn inflects to the behaviour of its interacting subjects. Through the parallel and intertwined paths of teaching, research and practice, the design work included in this book is concerned with the notion of matter as having lifelike properties, challenging the conventional assumption of material being immutable and inert. The traditional format of a book – which is limited to two dimensional and three dimensional representations and textual description, represents a challenge to capture the inherent dynamic nature of spaces and material assemblages, and their time-varying attributes, which are pursued in each project and studio included in this book.

This body of work forms part of an important arena for experimentation, one which originates with the early work in Cybernetics, and was later propelled by architects' speculations on the relationship between emerging technology and architecture since the 1960s. The cybernetician Gordon Pask wrote in 1969 that architecture need not be inflexible, and its goals have less to do with permanence and endurance, but rather, what he calls a reactive environment in which the inhabitants are in constant dialogue with the mechanisms of design or 'environmental computing machines'. He turns the role of the designer and his / her relation to the subject towards 'the interaction between the designer and the system he designs, rather than the interaction between the system and the people who inhabit it'. In this sense, the designer is no longer conceived as the authoritative controller of the final product, but rather the designer is now the designer of the apparatus that will design the product.

This book raises questions about the relationship between material and electronic technologies, and the ways in which people interact in space. The vehicles of this new paradigm rely on electronic networks and signal processing systems which enable forms of feedback between an architectural assemblage, its host environment, and its users. Through a multitude of designers, these speculative projects are less concerned with the appearance of space than with the dynamic behaviour generated through the interaction of people and the electronic compounding of physical matter. In this light, the projects and studios described in this book can be understood in their totality as experiments in spaces which perform as interfaces of multiple states. It is this heightened potential of interactivity which differentiates this body of work from the representation of conventional spatial or material projects.

Throughout the history of western architecture, the presumption that buildings should be conceived and constructed as enduring and unchanging prevails when the evidence of millennia is that this is an unattainable goal. The suppression of indeterminacy of unwarranted changes misconstrues permanence as the 'dominant cultural expectation that buildings must be built for eternity, when in fact most buildings are built to persist for only a short time.' Design is a process

本书主要记录了一些重要实验性项目的部分内容，这些实验性项目发端于早期的控制论研究。20 世纪 60 年代以来，建筑师通过设计推动了建筑与新兴技术的结合；1969 年，控制论专家戈登·帕斯克提出：建筑不需要是一成不变的，它的目标并不仅仅是持久与耐用；相反他所倡导的是：建筑应该是一个感应式的环境。在这样的环境中，居住者能够与设计机制、“环境计算机设备”不断对话。他将设计师与其设计对象之间的关系聚焦在“设计师与其设计系统的互动上，而非放在设计系统与居住者之间的互动上。”从这个意义上来看，设计师不再被认为是终极产品的权威控制者，而是当下设计产品的装置设计师。

本书提出了一系列材料与电子技术的相关问题，也提出了人与空间互动的方式。这种新范式的载体依赖于电子网络与信号处理系统，通过这一系统在建筑集成、环境与用户之间形成反馈机制。众多设计师在一些前沿性的项目上并非关注空间的表征，而是更关注通过人与电子复合物质之间的交互作用产生的动态行为。从这点来看，本书呈现的项目与工作室总体上可以被理解为复合状态接口的空间表现实验，它有别于传统的空间与材料项目的表现方式，实际上提高了互动性的潜力。

纵观西方建筑史，关于建筑的命题总是认为建筑应被构思与建造成持久不变的构件，而数千年的证据证明这是无法实现的目标。对于无根据变化的不确定性的抑制是将建筑耐久性概念曲解为“建筑通过其存在的永恒性来彰显文化趋势与价值，但事实上大部分建筑都只能维持较短时间”。设计是一个旨在确定其配置与控制机制的过程，同时涌现出“总是处于摆脱控制，使系统处于尽可能自我控制与自我学习的状态中。”这并非就断定物质的短寿是常态，问题是为什么如此多的现代建筑似乎无法适应用户和环境不断变化的需求？设计师如何能将更多的智能技术嵌入设计作品中？

从微观尺度上，尽管速度很慢且常常难以察觉，所有物质都经历着持续的物理和化学变化。除了对建筑材料的风蚀作用，建筑对特定活动的作用方式似乎也缺乏永久的适应性。建筑领域根本性的变革是无法回避的事实，其最终会导致各种各样的功能失调。本书中的项目聚焦于用户生成的可变空间配置的行为活动，而不是设计一个固有的、静态与稳态的建筑空间。这本书远未得出一个综合的结论，它提出的问题比它所回答的要多，它开阔了设计研究的视野，将建筑设计引入未来几年即将到来的智能化与互动性的环境中。随着数字控制系统在建筑和城市空间的全方位普及，设计产品通过内存和评估系统以嵌入式的能量来预测与调整未来的环境状况，具有非常大的智能化前景。我们知道至少在理论上，现在一直处于一种流变的状态，但是鲁道夫 · 埃尔 – 库利、卡罗尔 · 穆海贝尔和克里斯托 · 马尔科普洛斯三人正在为未来而努力实践。

which aims to fix in place its configurations and mechanisms, whilst emergence ‘has always been about giving up control, letting the system govern itself as much as possible, letting it learn from the footprints.’ Far from making claims for a totalising ephemerality of matter, the question is why so much of contemporary architecture is seemingly so unable to adapt to the changing demands of its users and the environment? How can designers embed greater intelligence into design artefacts?

At a microscopic scale, all matter undergoes continual, albeit slow and often imperceptible, physical and chemical changes. Aside from the longer term effects of weathering on material, the ways in which architecture is instrumental to enabling specific activities also seems to lack a permanent fitness. Programmatic change is an inevitable fact which can eventually render any form dysfunctional. The projects in this book focus on the activity of users generating variable spatial configurations rather than the design of fixed, static and homeostatic architectural spaces. Far from reaching a synthesised conclusion, this book asks more questions than it answers, and thus opens a rich field of design research into intelligent and interactive environments for years to come. Given the proliferation of digital control systems into all aspects of architectural and urban space, there is the very real prospect of greater intelligence in design products with an embedded capacity to anticipate and adjust to future conditions, through memory and evaluation systems. We know the present is always, at least theoretically, in a state of flux, but Rodolphe el-Khoury, Carol Moukheiber, and Christos Marcopoulos are working on the future.

目 录
CONTENTS

中英文双语版序
Preface for the Bilingual Edition 004
序 Foreword 007
导言 Introduction 012
RAD 简介 Introduction: RAD 018

1 人工自然
ARTIFICIAL NATURE

1.1 过滤块 Filtration Block 040
1.2 深圳当代艺术与城市规划馆 MOCAPE 046
1.3 光合作用立面 Photosynthetic Facade 053
1.4 氧气窗帘 Oxygen Curtain 057
1.5 合成景观 Synthetic Landscapes 060
1.6 室外的房间 The Out-House 067
1.7 Rgb 花园 RGB garden 071

2 沉浸式空间
IMMERSIVE SPACES

2.1 天空之宅 Sky House 076
2.2 全视之屋 All Seeing House 081
2.3 可调声云 Tunable Sound Cloud 086
2.4 微风穿过的墙 Breeze-Thru Wall 088
2.5 忙着游戏 Play on the Fly 091
2.6 看不见的住宅 Invisible House 097
2.7 范斯沃斯墙 Farnsworth Wall 102
2.8 范斯沃斯帘幕 Farnsworth Curtain 105
2.9 双重状态房间 Dual State Room 108
2.10 西安气味花园 Xi'an Scent Garden 113
2.11 纪念碑 + 比特 Monuments + Bits 117
2.12 闪烁 Blink 122
2.13 光之隧道 Tunnel of Light 126
2.14 规则 Rule 130
2.15 自然 2.0 Nature 2.0 133
2.16 雾宅 Fog House 137
2.17 同步实例化 Simultaneous Instantiations 142

3 能动构件 KINETIC COMPONENTS

3.1 可变隔热体 Variable Insulation — 148
3.2 新风窗户 Fresh-Air Window — 153
3.3 供热与制冷玻璃 Heating and Cooling Glass — 156
3.4 可变几何桁架 Variable Geometry Truss — 160
3.5 寻光砖 Light-Seeking Brick — 165
3.6 智能墙体 Wallbot — 169
3.7 窗帘 Curtain — 173
3.8 感应窗帘 Sentient Curtain — 178

4 可变覆盖面 VARIABLE CLADDING

4.1 群落瓷砖 Swarm Tile — 186
4.2 充气围护结构 Pneumatic Envelope — 194
4.3 聚光幕墙 Poly-Glazed Curtain Wall — 198
4.4 修道院的微气候 Iviron Microclimate — 201
4.5 可调节的天气 Mediating Weathers — 206
4.6 日光建筑 The Sunlight House — 211

5 表面作为接口 SURFACE AS INTERFACE

5.1 IM 感应毯 IM Blanky — 218
5.2 发光混凝土 Light-Emitting Concrete — 225
5.3 环境壁纸 Ambient Wallpaper — 230
5.4 数字化窗户 Digital Window — 236
5.5 窗户 2.0 Window 2.0 — 239
5.6 我的城市 My City — 243
5.7 无穷集合 Aleph TO — 246
5.8 可连接的遮阳 Linking Shade — 248
5.9 斯特拉特福德广场 Stratford Square — 251
5.10 超级块 Superblock — 256
5.11 隐形屋 Invisible Room — 260
5.12 微环境装置 Micro-environment Gear — 267
5.13 数字时代的涂鸦和建筑 Graffiti and Architecture in the Digital Age — 271

6 个体感应 INDIVIDUATED RESPONSE

6.1 胡氏人体模型 Hu-mannequins — 278
6.2 实时遮阳 Spot Shade — 281
6.3 光晕室 Auratic Chamber — 285
6.4 气味筛选器 Scentisizer — 289
6.5 变色龙休息厅 Chameleon Lounge — 293
6.6 空气中的建筑 The Architecture of Air — 296

作者简介 About the Authors — 304
作品目录 List of Projects — 306
致谢 Acknowledgments — 314
译后记 Postscript — 316

导言
Introduction

图 0-1

图 0-1 微芯片
Microchip

The integration of information technology into objects, buildings and cities is radically transforming the built environment. This book suggests a proactive role for architecture in shaping this new reality: a world where every object is in a network of communication and interaction. The migration of computing from dedicated appliances to physical environments, thanks to increasingly proliferating microchips and ever-expanding information networks, directly empowers architecture as a transformative agent. The fact that objects can now sense, think, act and communicate with the help of embedded technology is opening up potential for an architecture that is more closely aligned with the networked dynamics of living systems – a sentient architecture. The technological enhancement of physical matter charts a movement away from a mechanical paradigm towards a biological model. This shift manifests itself on several levels: from the micro scale in the form of new composite or ‘smart’ materials capable of responding to external stimuli, to larger network formations between people, objects, spaces, and landscapes. Radical artifice here serves to imitate nature, enmeshing built environments in a complex web of interactions whose emergent properties approximate the resiliency of natural ecologies (figure 0-1).

信息技术在物体、建筑与城市中的集成正在从根本上改变着建成环境。本书提出了建筑在塑造新的现实中起到的前瞻性作用，而这个现实就是世界上每一个物体都融入一个交流的互动网络中。归因于越来越多的微芯片和不断扩展的信息网络，计算从专有设备转移到物理环境，直接赋予建筑成为可变计算实体的能力。在嵌入式技术的支撑下，物体可感知、思考、行动与交流，并不断激发潜能，它与建筑同生命系统的网络动力学结合更紧密，从而实现一种可感知的建筑。物质实体的技术增强描绘了从机械范式向生物模型的转变，这种转变体现在几个层面上：从新的复合材料或能应对外部刺激的“智能”材料的微观尺度，到人、物体、空间和景观之间更大的网络结构。这里的基本技巧是模仿自然，将建成环境融入一个复杂的交互网络中，其涌现出的性质近似于自然生态的弹性（图 0-1）。

正是这种与生命系统的紧密同步，使得这些新兴技术在处理身体康复、调节全球与气候能量系统等广泛问题上都非常契合。本书介绍的项目特点在于：通过工作原型，展示了综合感知在环境智能、响应与交互系统、增强现实、嵌入式 / 定位技术、移动计算和定位媒体等广泛研究领域的建筑应用。本书所进行的研究是基于这样一个理念：每个建筑或景观构件都可以具备运算能力、无线通信或至少一个 IP 地址。本书记录了由多伦多大学、麻省理工学院和香港大学的研究者发起和指导的多项研究路径。项目建构了一个由艺术家、设计师、科学家和工程师参与的跨学科平台，以及与生命系统动力学相协调的、能够解决空间难题的方法。我们的共同目标是通过协作实验开发一个数字增强的架构，能够很好地应对摆在眼前和接踵而至的挑战。

It is precisely this close synchronisation with life which has made these emerging technologies pertinent in dealing with a wide range of issues from therapeutic benefits to the body, to the mediation of global and climatic energy systems. The projects featured in this book demonstrate, through working prototypes, architectural applications of synthetic sentience in the broad research area of ambient intelligence, responsive and interactive systems, augmented reality, embedded / situated technology, mobile computing and locative media. The research presented in this book is premised on the notion that every building or landscape component can be equipped with computational power, wireless communication or at least an IP address. It documents multiple research trajectories launched and guided by the authors at the University of Toronto, MIT and the University of Hong Kong. The projects have established an interdisciplinary platform involving artists, designers, scientists and engineers and an approach to spatial problems that is attuned to the dynamics of living systems. The collective aim is to develop a digitally enhanced architecture from the collaborative experiments that is well equipped to handle persistent and emerging challenges.

普适计算和综合感知

Ubiquitous computing and synthetic sentience

“普适计算基本上与虚拟现实相反。虚拟现实将人们置身于计算机生成的世界中，而普适计算则使计算机与人共同生活在这个世界中。”——马克 · 威瑟

‘*Ubiquitous computing is roughly the opposite of virtual reality. Where virtual reality puts people inside a computer generated world, ubiquitous computing forces the computer to live out here in the world with people*.’—Mark Weiser

这个故事现在已经广为人知：“嵌入式”“感应式”“自适应”“定位的”“环境式”和“交互式”已经或多或少地成为描述“普适计算”条件

The story is by now well known: ‘embedded’, ‘responsive’, ‘adaptive’, ‘situated’, ‘ambient’, and ‘interactive’ have become more or less synonymous terms to describe a condition of ‘ubiquitous computing’. This term was coined by Mark Weiser at Xerox PARC in the 1980s to describe the migration of computing power from personal computers into the physical environment – a development made possible by the evolution and proliferation of small, cheap sensors and microcontrollers. Weiser described this new emerging reality of pervasive technology as the third wave in computing, where the first wave brought forth the development of the mainframe computer, and the second the personal computer. The process of continuous miniaturisation and dispersion of ever-growing computing power is becoming increasingly evident in the objects that surround us. From the autonomous robot vacuum that cleans the house when no one is home, to the coffee shop that can track a customer's location in the city through a phone's embedded GPS sensors, new affordable technologies endow previously inert objects with the capacity to sense, process and act upon incoming information.

Ubiquitous computing presents physical objects or artifacts in the real world with the possibility of gaining agent characteristics, exhibiting therefore a certain level of autonomy, agency, or synthetic sentience. Object-agents in that sense are seen as variations of physical robots. This augmented condition shares some critical overlaps with Artificial Intelligence and its preoccupation with the concept of embodiment which views the body of an agent, in this case an artifact – in its interaction with the world — as the primary shaper of intelligence. In other words the brain is not the only computational system; computation is manifested in or distributed into the morphology and materiality of the body. The question regarding object intelligence is not just 'what' does the object do, but 'how' does it do it.

的同义词。这个术语是由施乐帕克研究中心[1]的马克・威瑟在 20 世纪 80 年代创造的，用来描述计算能力从个人电脑向物理环境的迁移——这一发展是由小型廉价的传感器和微控制器的发展和普及而实现的。威瑟将这种新兴的普及技术描述为计算机领域的第三次浪潮（第一次浪潮催生了大型计算机的发展，第二次浪潮是个人计算机的发展）。不断增长的计算能力的持续小型化和分散过程在我们周围的物体中变得越来越凸显。从无人在家时自主机器人真空清洁房间，到通过在手机中嵌入全球定位系统传感器，可以在城市中跟踪客户位置的咖啡馆。新型的经济适用技术赋予了传统惰性物体以感知、处理和依据收到的信息采取行动的能力。

普适计算为现实世界中的物质对象或人工制品提供了获得代理特征的可能性，因此表现出一定程度的自治、代理或综合感知能力。在这个意义上，对象代理就被看作是物理机器人的变体。这种增强条件与人工智能有一些重要的重叠，其专注于将代理主体视为具象概念。在这种情况下，人工物在与世界交互的过程中成为智能的主要塑造者。换句话说，大脑并非是唯一的计算系统，计算表现或分布在实体形态和物质中。关于对象智能的问题不仅是对象“做什么”，而且是对象“如何做”。

---

[1] 译者注：Xerox PARC（Xerox Palo Alto Research Center，简称 Xerox PARC）即施乐帕克研究中心，是施乐公司所成立的最重要的研究机构，成立于 1970 年。施乐帕克研究中心是许多现代计算机技术的诞生地，他们创造性的研发成果包括：个人电脑、激光打印机、鼠标、以太网图形用户界面、Smalltalk、页面描述语言 Interpress（PostScript 的先驱）、图标和下拉菜单、“所见即所得”文本编辑器、语音压缩技术等。

20 世纪 80 年代，随着罗德尼 · 布鲁克斯的机器人实验，人工智能中的具象概念出现了，即智能需要实体。人工智能领域（在 1956 年左右）在理解有关自然智能的基本问题方面遇到了障碍。经典的人工智能是在思想和身体分离的笛卡尔逻辑下运行的，即思想（软件）控制身体（硬件）。智能被理解和追求为计算，或以算法和程序形式存在的抽象的符号控制。大脑运行各种程序，身体作为硬件是无关紧要的。然而，随着人工智能未能超越单纯的数字运算和持续的问题解决，这种追求的局限性就变得显而易见。那些以试图通过自主解决问题来模仿或模拟人类复杂思维为目标的任务（如“专家系统”）未能像他们试图复制的专家大脑那样发挥作用。智能不能简化为抽象计算。罗德尼 · 布鲁克斯提出，将人工智能的关注点从高层次的认知，即一种基于大脑内省的固有偏见，转移到与现实世界的互动中。当一个人可以利用真实世界的时候，为什么要模拟这个世界呢？著名的六脚昆虫机器人“根吉斯”（1989 年）是布鲁克斯第一次试图脱离大脑，将之作为认知和自主行为的中心，关注物质实体与环境的互动。他探索昆虫世界的倾向源于这样一个事实：从无机物到昆虫（历经 30 亿年）的自然进化时间是从昆虫到人类（历经 5 亿年）的 6 倍，从物质到昆虫的飞跃比从昆虫到人类的飞跃更为复杂。

布鲁克斯指出控制论是灵感的源泉。对动物和智能的理解是由诺伯特 · 维纳领导的控制论的多学科领域（约 20 世纪 40 年代）的核心，该领域与人工智能平行，但采用了完全不同的方法。他们的信息模型将动物 / 有机体耦合起来，将其建模成一台机器，与环境形成一个紧密相连的回路。在两

The concept of embodiment in AI - the notion that intelligence requires a body — emerged in the 1980s with Rodney Brooks's experiments in robotics. The field of Artificial Intelligence (b. 1956) had hit a roadblock in its progress towards understanding fundamental questions regarding natural intelligence. Classical AI had operated under the Cartesian logic of the separation of mind and body, whereby the mind (software) exercised control over the body (hardware). Intelligence was understood and pursued as computation, or abstract symbol manipulation in the form of an algorithm, or a program. The brain runs the program; the hardware, the body is irrelevant. The limits of this pursuit became evident however with the failure of AI to go beyond pure number crunching and sequential problem solving. Tasks that aimed to mimic or model complex human thinking through autonomous problem solving, such as 'expert systems' failed to function as the expert's brain they were attempting to replicate. Intelligence could not be reduced to abstract computation. Rodney Brooks proposed shifting AI's focus from high-level cognition, which was inherently biased in its brain-based introspection, and instead focus on interacting with the real world. Why try to model a representation of the world when one can use the real world. 'Ghengis' (1989), the famous insect like six-legged robot was Brooks' first attempt at leaving the brain as the seat of cognition and autonomous behavior and instead investing in the body's materiality in its interaction with the environment. His inclination for exploring the insect world stemmed from the fact that it took natural evolution approximately six times longer to go from inorganic matter to insects (3 billion years) than from insects to humans (0.5 billion years). The leap from matter to insect was more complex than from insect to human.

Brooks points to Cybernetics as a source of inspiration. Understanding animals and intelligence was central to the multidisciplinary field of Cybernetics (b. 1940s) led by Norbert Wiener and running parallel to AI, but with an entirely different approach. Their information models coupled the animal / organism, modeled as a machine with its environment as a tightly linked circuit. The system of exchange or communication, feedback, between the two constituted essentially continuous and dynamic behavioral adaptation of organism to environment. Early analog circuit robots based on cybernetic principles exhibited goal-seeking behaviors such as moving towards a light source. The physically embodied machines of W. Grey Walter and Braitenberg's imaginary ones, influenced Brook's insistence on embodiment in its ability to ground an organism in the real world.

The projects presented here all seek this situated, intimate physical interaction or dynamic engagement between bodies, as agents, and their ever changing environment. Through their distribution of sensors materials and particular morphologies, the projects explore the opportunities for an architecture that exploits its ecological niche towards greater sensitivity and performance. This necessitates a close communion or meditation with between architecture, the bodies it envelops, and those that envelop it in turn. In doing so the projects seek to test computing's augmentation of the traditional role of objects to scaffold our environment — in effect extending us through our increased interaction with, and dependency on our natural and synthetic environment (figure 0-2).

者之间的交流或沟通、反馈系统，基本上构成了有机体对环境的连续动态的行为适应。早期基于控制论原理的模拟电路机器人表现出目标寻求行为，类似向光源移动。格雷·沃尔特和布瑞滕贝格虚构的物质具象化的机器，影响了布鲁克斯对具象化的坚持，因为其能够让有机体在现实世界中落地。

这里呈现的项目都在作为代理人的机体和不断变化的环境间寻求情境化的、紧密的物理互动或动态参与。通过传感器材料和特定形态的分布，这些项目尝试以建筑学的视角，探索利用生态位实现更高的敏感度和性能。这需要在建筑、建筑所包裹的形体以及那些反过来包裹建筑的形体之间进行密切的交流或思考。在此过程中，这些项目试图尝试对支撑环境的对象的固有效能进行数据增强——实际上是通过增加自然和合成环境的互动和依赖性来拓展自身（图 0-2）。

图 0-2

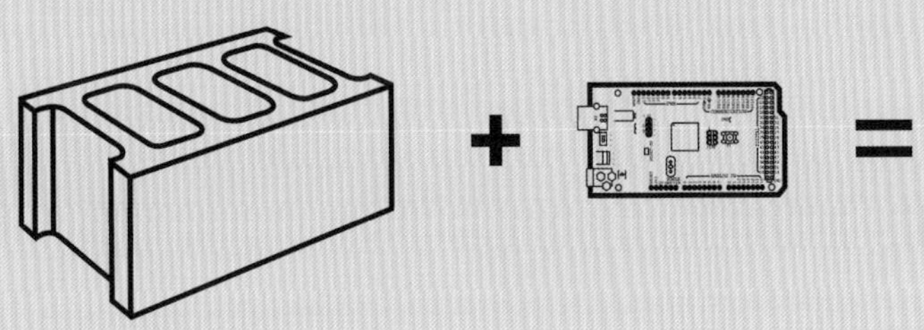

图 0-2　砖 + 技术
Brick + tech

Prototype as Ideas

The projects ask basic questions: What happens when we add computation and communication in the form of a microcontroller

作为理念的原型

这些项目提出了一些基本问题：当我们以微控制器及其相关微技术的形式将计算和通信添加

到建筑模块中时，会发生什么？建筑构件如何与它们所调节的物理和社会环境建立更紧密的关系？当我们提高一个材料的整合多种信息的能力时会发生什么？他们试图一砖一瓦、一件一件地质疑数字增强的物理现实含义和机遇。研究围绕着新获得的对象代理的形式、体验和生态含义，这些新获得的感知能力如何反过来影响对象的形态演变？对象的社会能力是什么？它的关系网是由什么构成的？它如何通过社会互动与其他事物、人、空间和环境产生新知识？它们能带来了什么新的类型和体验？换句话说，什么是新型微技术植入所带来的本质性转变：从“固件＋电脑”的模式转变为固件本身就是电脑，或者从嵌入式技术到技术物化（图 0-3）。

这些样本项目表达了 RAD 研究的性质、视野和雄心。该实验室是由多伦多大学的学者们建立和启动的，他们通过大量的实验测试普适计算呈现建筑的机遇。他们将研究聚焦在物质的数字增强，挖掘更具有感应性和适应性的形式，在此通过迭代生产和实验的“制造”被视为获取和拓展知识的核心。本书确定了进化和开放式的生产研究链：人工自然、沉浸式空间、能动构件、可变覆盖层、表面作为接口、个体感应，这些依赖于项目本身来整合新形态和应用。

图 0-3

and its related micro technologies to a building module? How can architectural components engage in a more intimate relationship with the environments (physical and social) they mediate? What happens when we enhance a material's ability to integrate a manifold of information? They attempt to question brick by brick, object by object the implications and opportunities of this digitally enhanced physical reality. The investigations revolve around the formal, experiential, and ecological implications of this newly gained object agency. How do these newly acquired sentient capacities inform in turn the evolution of an object's form? What are the object's social capacities? What constitutes its network of relations? How does it connect with other things, people, spaces, and environments to produce new knowledge through its social interactions? What new typologies and experiences do they enable? In other words, what are the qualitative transformations made possible by the addition of new micro technologies: moving from 'object + computer' to object as computer, or from embedded technology to embodied technology (figure 0-3).

The sampled projects represent the nature, scope and ambition of research at RAD. The lab was set up and launched by the authors at the University of Toronto as a way to fully experiment with and test the opportunities ubiquitous computing presents architecture. The work focuses on the digital enhancement of matter, examining the potential for more responsive and adaptive forms, where 'making' through iterative production and experimentation is seen as central to the acquisition and extension of knowledge. The book identifies evolving and open ended strands or categories for of productive investigations: artificial nature, immersive spaces, kinetic components, variable cladding, surface as interface, and individuated response, relying on the projects themselves to articulate novel morphologies and applications.

图 0-3 砖 +IP
Brick + IP

RAD 简介
Introduction: RAD

鲁道夫 · 埃尔 – 库利、卡罗尔 · 穆海贝尔
Rodolphe el-Khoury and Carol Moukheiber

This book documents the work of RAD and is premised on the notion that every building, city and landscape component can be – and will be – equipped with communicative and computational capacities. The migration of computing from dedicated appliances to physical environments directly empowers architecture as a transformative agent. The fact that objects can now sense, think, act and communicate with the help of embedded technology is opening up the potential for an architecture that is more closely aligned with the networked dynamics of living systems – a sentient architecture. Our projects establish an interdisciplinary platform involving artists, designers, scientists and engineers //spanning different institutions and continents in a technological approach to spatial problems that is attuned to the dynamics of living systems. The collective aim is to develop from the collaborative experiments to a digitally enhanced architecture that is well equipped to handle persistent and emerging challenges in building a healthy and sustainable environment.

For designers, the ongoing merging of artifacts, information technology and data requires new tools and methodologies. How do we design for the Internet of Things? How do we embed technology in

本书记载了 RAD 的工作实录，提出了所有的建筑、城市与景观构件等能够且将会具备交流和计算的能力。这种从专业设备到物理环境的计算方式的转变使建筑成了一种转换媒介。在嵌入式技术的帮助下，物体可以感知、思考、表现以及交流，这一事实挖掘了建筑的潜能，使其成为与生活系统的网络动力更加紧密关联的一种建筑——感应式建筑。我们的项目建立了一个跨学科的平台，包括艺术家、设计师、科学家与工程师，跨越不同的机构和地区，以一种与生命系统动态协调的方式来解决空间问题。我们的共同目标是将合作研究发展成为一个数字增强的研究架构，能够更好地应对在构建健康与可持续发展的环境中不断出现的挑战。

对于设计师而言，这种不断融合的构件、信息技术以及数据集合需要新的工具和方法。我们如何为物联网设计？我们又如何在日常生活中嵌

入技术？在解决这些问题时，我们不可避免地要解决更基本的问题：当一个对象连接到互联网时会发生什么？传感器和处理器将如何开展工作？它们应该做什么与如何做？当这些对象具备新功能时又将会发生什么？最终所有问题都归结为一个问题：什么才是对象？

物联网是一项高端技术，其价值 15 万亿美元的业务是机器智能化在我们生活的环境中空前拓展的一部分。伴随全球近 35 亿个传感器的应运而生，马克 · 威瑟在 1991 年设想出普适计算的概念。在嵌入式虚拟技术的影响下，一架包含了传感器和制动器的小型计算元素的飞行器突破了台式电脑外壳的限制进入物理环境本身，这种虚拟性已经成为现实。

正如 N. 凯瑟琳 · 海尔斯所描述的，这种现象扩展了控制论思想的体系[1]，并形成了四阶控制论或海尔斯所说的计算体系。第一阶控制论（1945 ~ 1960 年）：在自主与自发、主体和环境之间建立一个反馈机制，使系统达到稳态；第二阶控制论（1960 ~ 1985 年）催生了一种自生系统的概念，在这个封闭的自我复制系统中，主体不再是一个界限分明的观察者，而是与环境诱因交织在一起，而环境诱因具有反射和循环的信息，能够催生更为复杂的反馈；第三阶控制论（1985 ~ 1995 年）：随着以虚拟自适应显现的人工

[1] N. 凯瑟琳 · 海尔斯（2013）："未完成的工作：从半机械人到认知空间"，载于《环境建筑理论：死后的领土》，阿丽亚娜 · 劳里 · 哈里森（编）。纽约：劳特利奇出版社，第 39 页。

objects of everyday life? In tackling these questions we inevitably end up with more basic ones: what happens when an object gets connected to the internet? How does sensing and computing expand its agency? What should it do? How should it do it? What are the consequences of these new capacities on the object, and finally the most fundamental of all questions: what is an object?

The IoT is an advanced technological development, a $15 trillion business that is part of an unprecedented expansion of machine intelligence in our environment. With already 3.5 billion of sensors in the world, Mark Weiser's concept of ubiquitous computing conceived in 1991, with its implications of an embodied virtuality – a flight of miniaturized computational elements including sensors and actuators from the confines of a desktop computer's shell into the physical environment itself – has become a reality.

As described by N. Katherine Hayles, this phenomenon extends a genealogy of cybernetic thought and ushers fourth-order cybernetics[1], or what Hayles calls the Regime of Computation. First-order cybernetics (1945-1960) posited a feedback loop towards a homeostatic condition between an autonomous, self-directed, subject and the environment; second-order cybernetics (1960-1985) ushered in the notion of autopoesis, whereby the subject, no longer a clearly delimited observer, became intertwined with environmental triggers in a closed self-reproducing system with reflexive and recursive information that enables more complex feedback; third-order cybernetics (1985-1995) saw the concept of emergence, order out of chaos, with the rise of artificial life in the form of virtual self-evolving programs no longer concerned with simply self-maintenance and equilibrium; and now fourth-order cybernetics

[1] N. Katherine Hayles (2013) 'Unfinished Work: From Cyborg to Cognisphere', Ariane Lourie Harrison (ed.) Architectural Theories of the Environment: Posthuman Territory. New York: Routledge, p. 39.

with the pervasive penetration of computational processes into, networked cognitive systems.

Accordingly this process is reshaping us into posthuman subjects, as we become more deeply intertwined with other kinds of cognitive systems, contributing to a dispersed sense of self. Here the liberal humanist subject, the autonomous master, ceases to exist as such, and instead joins an intricate ecology of natural and artificial distributed cognition systems.

While Hayes shows how this computational regime and its algorithmic logic leads to a disembodied view of the world — rationalizing the exchange and processing of information and separating it from the physicality of matter within which or even through which information is processed — the possibility of ubiquitous computing, in its connection to the physical environment invites an opportunity to activate, and deeply engage the body, or the physical object in which this virtual computing capacity is activated.

In other words matter here really matters. In this sense, multi-scalar everyday objects from appliances to buildings, landscapes and cities can be understood as having a newly gained virtual space of possibilities, what Weiser coined 'embodied virtuality'. With the more recent development of the IoT, this virtual capacity is further tied to the vastness and incomprehensible scale of the Cognisphere.[1] This massive global network of interconnected objects consisting of ever more intelligent cognitive agents in which humans are embedded is an opportunity as Hayles sees it to re-evaluate 'human agency, rationality, and affective capacities' and has the potential to initiate new relations between a myriad of distributed and diverse cognitive actors, human and non-human, animal, machine, and inert artifact.

[1] N. Katherine Hayles uses this borrowed term to describe not just the Internet but all data networks and programmable systems across the electromagnetic spectrum which feed into the Internet.

生命学科的兴起，产生了“从混沌到有序”的概念，不再单纯关注自我维持和平衡态；而当下的第四阶控制论：在大规模分布的网络认知系统中，普适计算广泛地渗透到生活的方方面面。

因此，这个过程正在将我们重塑为新兴人类，随着我们与其他类型认知系统的交织关系越来越紧密，有助于形成分散的自我意识。在这里，自由人文主义与自我控制都不复存在，而是进入一种错综复杂的自然生态与人工分布认知系统。

虽然海尔斯阐明了这种计算规则和算法逻辑是如何带领我们进入一种无形的世界观——合理化的信息交换和处理，并且通过信息处理把他们从物质的物理属性中分离出来——这种普适计算的可能性在与物理环境的连接中加入了可激活的机会，并且深入地接触身体或激活具有虚拟计算能力的物理对象。

换句话说，物质才是最重要的。从这个意义上说，从电器到建筑、景观和城市的多标量日常物品都能被理解为开始获得充满可能性的虚拟空间，这就是威瑟所创造的“实体虚拟性”。随着物联网技术的最新发展，这种虚拟能力进一步与认知层的巨大和难以想象的规模相关联。[1] 这个由相互关联的对象所组成的巨大全球网络包含着更为智能化的、被人工嵌入的认知主体，这就是海尔斯所认为的机遇，它能够重新评估“能动性、理性和情感能力”。而且这种关联对象具有在众多分散而多样的认知行为体之间、人类和非人类之间，以及动物、

[1] N. 凯瑟琳 · 海尔斯不仅用这个借来的术语描述互联网，还用它描述传输到互联网的跨电磁波谱的全部数据网络和可编程系统。

机器、惰性物质之间创造新型关系的潜能。

在建筑设计领域，从一阶到四阶控制论以镜像序列的方式运行，从一种系统扩展回归到尼古拉斯 · 舍弗在 20 世纪 50 年代做的著名的控制性试验中。这个实验是基于第一阶控制论中对于自我平衡的反馈和循环的因果关系概念；塞德里克 · 普莱斯和戈登 · 帕斯克紧随其后，在 20 世纪 60 年代，通过用户参与方式，基于第二阶控制论自生系统的概念开展了更为复杂的实验活动；戈登 · 帕斯克的对话理论概括了循环复杂性阶段，创建了一种更有趣的人机交互模式。

这些范例仍然引导着当前交互建筑的模式，其中交互主要围绕用户及其输入进行。20 世纪 80、90 年代，约翰 · 弗雷泽在美国空军航空局开展了大量的自下向上系统和人工生命原型的实验，交互建筑模式开始转向基于抽象系统设计的演化模型。

这种感应性是基于外部环境条件的动态变化，它能够根据气候变化最大限度地优化建筑性能。这些项目的范围从纯粹的机械反应到形而上的创新实验。从联接对象的物联网到巨大的信息和通信网络，都驱使着对象进入极端的社会混沌状态，用相关性、无迹可寻的连接和不可预测的模式来取代直接的因果关系。我们感兴趣的是这种状态及其通过设计对物质产生的影响。

更准确地说，正是从用户主导（从主体到对象）的交互到物联网所要求的广义的物物对象的概念转变，激发了对建筑设计对象的工作内容和审视。在物联网背景下，建筑设计的对象变成了什么？

In architecture, first to fourth order cybernetic movements played out in a mirroring sequence, with a lineage extending back to the notable cybernetic experiments of Nicholas Schoffer (1950s) based on first order cybernetic notions of feedback and circular causality towards homeostasis, followed by Cedric Price and Gordon Pask (1960s) pursuing more complex behavior based on second order notions of autopoesis through user based participation. Gordon Pask's conversation theory epitomizes this phase of recursive complexity, creating a more engaging interaction between user and machine.

These paradigms still guide current models of interactive architecture where interactivity revolves largely around the user and their inputs. Experiments with more bottom up systems and artificial life prototypes were carried out at the AA by John Frazer in the 80s and 90s began to move towards an evolutionary model based on abstract systems design.

More recently this responsiveness has engaged with the dynamics of external environmental conditions in an effort to maximize a building's performance with respect to climatic changes. These projects range from purely mechanistic responses to more formally innovative experiments. The IoT capacity for linking objects to a vast information and communication network propels the object into situations of extreme social promiscuity, replacing direct causality with correlation, untraceable connections, and unpredictable patterns. We are interested in this condition and its repercussions on matter by means of design.

More precisely, it is the shift from a user dominant (subject-object) interaction to broader notions of object-to-object considerations demanded by the IoT that motivates the work and its interrogation of the architectural object. In the context of the IoT, what becomes of the architectural object?

What is RAD?

RAD stands for 'Responsive Architecture at Daniels'. RAD is literally a space: a workshop with rapid prototyping equipment situated right off the Bay Street Station at the heart of Toronto's subway system. It is an outpost of the University of Toronto's campus that is well poised for partnership with the private sector, in full public view through a storefront window. Most importantly, RAD is a research platform that provides a home for projects we develop with our research and professional teams and with an interdisciplinary network of collaborators across the university. The projects fall under the thematic umbrella of the following terms: *Internet of Things, Embedded Technology, UbiComp, Physical Computation, and Every ware.*

The themes or technologies defined by these terms all point from different disciplinary and discursive perspectives to the tendency for computation to migrate from its familiar dedicated appliances towards a seamless and ubiquitous integration into the built environment, into the very substance of our homes, cities and landscapes. RAD is premised on a mandate for architecture to take an active role in shaping this emerging informational landscape as it becomes literally spatial and material.

Since RAD has a pedagogical ambition we use a graphic motto to diagram and clarify our process. Most of the work at RAD can be reduced to this formula that also comes in different iterations to cover several variation on a common theme. They envision possibilities for embedding different forms of technology into a brick. They show it incorporating computational power, harvesting energy, or communicating online with a dedicated IP address. The brick, the most basic and universal building block stands by extension for all architectural elements and the diagram for their capacity or necessity to assimilate emerging technology (figures 0-4~0-7).

什么是 RAD？

RAD 就是“丹尼尔斯的感应式建筑”（Responsive Architecture at Daniels）的缩写。RAD 实际上是一个空间：一个拥有快速原型设备的工作室，工作室位于多伦多地铁系统中心的海湾街站附近。它是多伦多大学校园的一个前哨站，随时准备与私营部门建立合作关系，通过店面的窗户能看到整个公共性景观。最重要的是，RAD 是一个研究平台，为我们的研究和专业团队，以及通过校际跨学科协作网络共同开发的项目提供一个平台。这些项目包括以下主题范畴：物联网、嵌入式技术、普适计算与按照自然法则计算。

这些术语所定义的主题或者技术，都从不同学科的论证观点转化为计算，即从熟悉的专用设备融入无缝联结的、无处不在的建成环境、住宅、城市以及景观实体中。RAD 的宗旨是要建筑在塑造新兴的智能化景观方面发挥积极作用，因为它已经成为名副其实的空间和材料。

由于 RAD 有自己的教学目标：就是利用图形化的范式去描绘和阐述我们的研究过程。大多数 RAD 的工作都被简化为这种形式，即通过不同的迭代法来涵盖同一主题的几个变量。他们设想在砖块中嵌入不同技术所产生的各种可能性，他们表现其合并计算能力、获取能源与专用 IP 地址在线交流的能力。砖是最基本最普遍的建筑构件，为了适应新兴科技的容量和需要，其作为建筑构件的必需元素与图表被推广（图 0-4～图 0-7）。

图 0-4

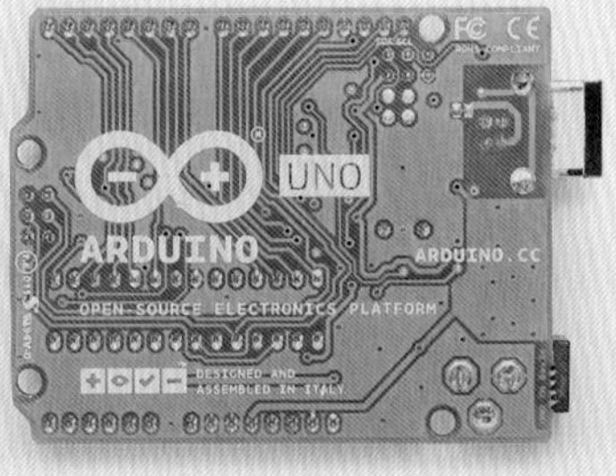

图 0-5

图 0-6

图 0-7

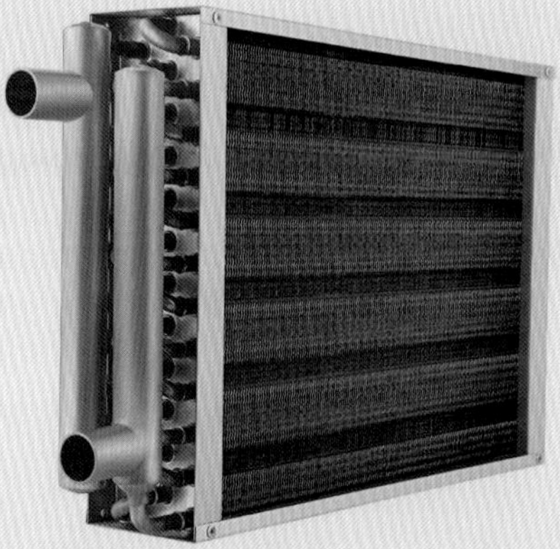

图 0-4 开源硬件
Arduino

图 0-5 柔性太阳能板
flexible solar panel

图 0-6 风力涡轮机
wind turbine

图 0-7 热交换机
heat exchanger

Three years of RAD activity inspired by these diagrams are documented here in a variety of approaches while demonstrating eight common defining premises.

RAD 三年来的工作受到以上图表的启发，并在此以不同的方法进行记录，演示了八个常见的定义前提。

RAD premises

1. Make it

We are committed to prototyping the ideas we develop in our lab. Drawing and diagraming concepts falls short in demonstrating the relevance of the emerging technology and its potential for transforming our environments. We take our projects through all stages of development from initial concepts to fabrication and prototyping (figure 0-8). The primary mission of RAD — and its well-equipped rapid prototyping facilities — is to manifest ideas in concrete functioning objects. Our approach is bottom-up, open source, DIY and opportunistic; it aligns the ethos — technological and entrepreneurial — of the Maker's movement with established forms of funded university-based research.

2. Ask basic questions

We start with basic questions. For instance, we explore what happens when we substitute an LCD screen and a camera for a windowpane. This is a startlingly simple operation. We think of it as the 'degree zero' of electronic mediation. Yet it has significant consequences in the potential of reconstituting reality by means of digital signal processing. The possibilities are endless once you start to imagine how a window is not simply an aperture that gives visual access to the world outside but also an active agent with a capacity to transform this visual experience with spatial and temporal manipulation.

With 'degree-zero' Digital Window (figure 0-9) we reimagined the visual function of the window. In the HVAC Window series (figure 0-10) we tackle the environmental role of the window as a thermal and atmospheric mediator. In one instance a pair of

RAD 的前提

1. 努力实现

我们致力于在实验室中研发设计构思的原型。绘图与图表概念并不适合展现新兴科技的相关性以及表达转换环境潜能的标准。我们的项目经历了从最初概念、制造到原型制作的所有研发阶段（图 0-8）。RAD 的最初任务就是以精良的快速原型设备，通过具象且具有功能性的实体去表达设计创意。我们的方法是自下而上、开源、DIY 和投机取巧，运用创客理论在技术与企业、高校资助之间建立紧密的协作研究。

2. 提出基本问题

我们从基本问题谈起，比如：我们探索用窗格替代液晶屏和相机时会发生什么，这是一个相当简单的操作，我们把它看作“零度”电子媒介。然而，它在通过数字信号处理重建现实的潜力方面产生了显著的效果。一旦你开始想象窗户不仅仅是一个能让你看到外面世界的洞口，而是一个能够通过时空性操作来转换视觉体验的主动装置，那么这种可能性就是无限的。

我们使用这种“零度”数字窗户（图 0-9），重新设计了窗户的视觉功能。在暖通空调窗系列（图 0-10）中，我们将窗的环境角色功能定义为

图 0-8

图 0-9

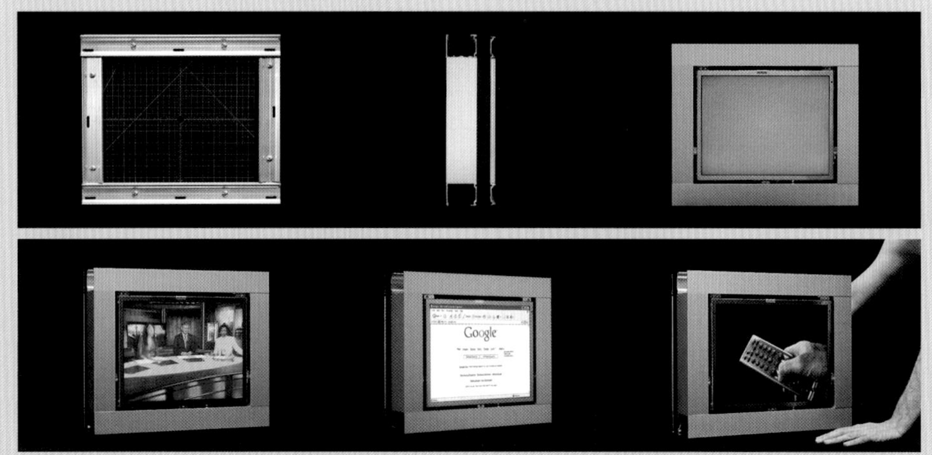

图 0-8　动态照明系统（来自“看不见的住宅”项目，详见 97 页）：当人行走时地板和顶棚的原型能动态地照亮；顶棚的 LED 是由与地砖连接的红外光束的干扰而触发的（NMinusOne 工作室、卡罗尔·穆海贝尔、克里斯托·马尔科普洛斯）
Dynamic Illumination System (from the ‘Invisible House’): Prototype of floor and ceiling tiles that dynamically illuminate a person walking; ceiling tile LEDs are triggered by an interruption in the infrared beams aligned with the floor tile (Studio NMinusOne, Carol Moukheiber and Christos Marcopoulos)

图 0-9　数字化窗户（详见 236 页项目）：通过连接一侧的摄像头和另一侧的 LCD 屏幕，可以实现数字化的调节透明度（NMinusOne 工作室、卡罗尔·穆海贝尔、克里斯托·马尔科普洛斯）
Digital Window: a digitally mediated transparency is achieved by connecting a camera on one side and an LCD screen on the other (NMinusOne, Carol Moukheiber and Christos Marcopoulos)

图 0-10　暖通空调窗（详见 156 页项目）：典型的幕墙玻璃系统，具有背对背的中央处理冷却单元充当热回收通风器（瓦伦蒂娜·梅勒、卡罗尔·穆海贝尔、克里斯托·马尔科普洛斯）
HVAC Window: A typical curtain wall glazing system with back-to-back CPU cooling units act as a Heat Recovery Ventilator (Valentina Mele, Carol Moukheiber and Christos Marcopoulos)

图 0-10

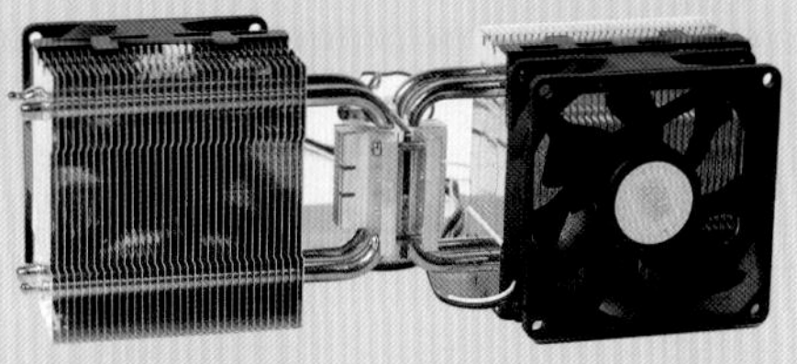

modified CPU cooling units (heat sinks and fans typically used in cooling a computer's CPU) was integrated into the window pane. Back-to-back CPU cooling units act as a Heat Recovery Ventilator conditioning incoming fresh air with the energy of outgoing stale air. In another prototype, a Peltier device❶ inserted between the cooling units enhances the conditioning effect.

Subsequent attempts rethink the aluminum frame itself. One instance has the Peltier module replacing the thermal break to function as a thermal pump that transforms the window frame into a heating and cooling surface (figure 0-11). The HVAC Window series thus aims to decentralize infrastructure by means of embedded technology. Such experiments replace heavy HVAC machinery with distributed and

热量与空气的介质，比如有一组改良的 CPU 制冷单元（散热器和风扇通常用于冷却计算机的 CPU）被集成到窗口窗格中；背对背的 CPU 制冷单元作为一个热回收通风系统，用排出的废气的能量来调节进入的新风。在另一个原型中，我们在制冷单元里置入珀耳帖效应设备❶以提高调节效率。

随后的各种尝试让我们对铝框架本身进行了重新考量。例如用珀耳帖效应模块替代断热层，相当于热力泵的功效，将窗框转换成可加热与制冷的面板（图 0-11）。因此，暖通空调窗系列旨

❶ A Peltier device is thermoelectric heat pump that is customarily used for cooling CPUs but could also be used for heating. This is a solid state module with no moving parts. Its technology is evolving rapidly and is expected to gain in efficiency.

❶ 珀耳帖效应设备是热电热泵，通常用于冷却 CPU，但也能用于制热。这是一个没有运动部件的固态模块。它的技术正在迅速发展，预计将提高效率。

❷ 译者注：珀耳帖效应是指当有电流通过不同的导体组成的回路时，除产生不可逆的焦耳热外，在不同导体的接头处随着电流方向的不同会分别出现吸热、放热现象。

图 0-11

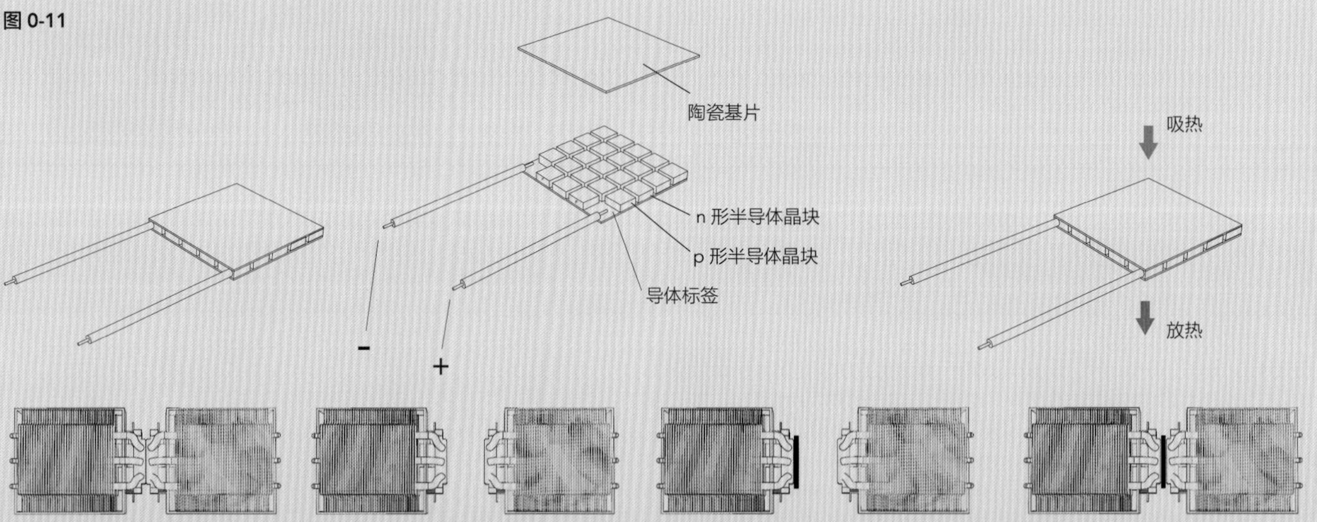

图 0-11 暖通幕墙框架（详见 156 页项目）：采用珀耳帖效应❷设备替代断热层，该框架的表面积倍增，以最大限度地实现对流（NMinusOne 工作室、卡罗尔 · 穆海贝尔、克里斯托 · 马尔科普洛斯）

HVAC Curtain Wall Frame: A Peltier device replaces the thermal break; the frame's surface area is multiplied to maximize surface to air exchange (NMinusOne, Carol Moukheiber and Christos Marcopoulos)

在通过嵌入式技术分散基础设备，这些实验是采用分散联网的小型设备集成到建筑构件中，以替代重型暖通空调设备，他们在数字技术的辅助下能够调节建筑室内的微气候。

3. 走向市场

我们为美国家得宝公司的货架设计了原型产品。这不仅是由于 RAD 活力化经营的企业精神。这其实是一种概念化的基础，因为我们知道这种技术对于解决建筑业的现实问题所产生的必要性。为了让世界知道我们，我们的项目设计必须融入当前制造、分销和建造的模式中去。

我们的产品之一是一种操作简便的集光砖，即插即用，能像乐高积木一样适用于更大的区域，它能通过光纤将光收集并引导到建筑物的更深处。它不仅可以被单独使用，也可以大量使用并构建成墙体。另一个例子是数字地砖，这是一种嵌入液晶屏、传感器以及探测器的混凝土砖（图 0-12），用于为公共空间的大面积表面配备数字广播传送功能。我们认为数字地砖是设计师在选择用于公共广场或人行道中的材料时可以考虑的选择之一，而这种选择无需因不同预算或建造技术而改变设计方案。

4. 自下而上的思考

这种自下而上的组织偏好是为了确保产品能够应用简便，满足即插即用的产品承诺。我们意图将小单元组合成能够产生巨大功效的大构件。因此，我们采用不需要集成操控或中枢控制的自下而上的装置，该策略支持利用无线电通信和分

networked miniaturized devices integrated into building components. They enable responsive local interior micro-climates with the assistance of a digital technology.

3. Put it on the shelf

We prototype products for the shelves of Home Depot. This is not only due to the entrepreneurial spirit that animates much of the activity at RAD. This desire has a conceptual basis in the necessity we see for this technology to address the realities of the construction industry. In order to have any traction in this world, our projects would have to be designed with a capacity to integrate into the current modes of fabrication, distribution and construction.

A light harvesting brick, which can be aggregated like Lego blocks into larger fields, to gather light and channel it by means of fiber optics into the deeper recesses of a building is designed for plug-and-play installation. It could be bought of the shelf as a single item or in great quantities for large wall sections. Another example is a Digital Paver, a concrete tile with embedded LEDs, sensors and transducers (figure 0-12) that is designed for use on large surfaces to equip public spaces with digital broadcasting capabilities. We imagine Digital Paver as one of the many options that a designer could chose when selecting a material for a public square or a side-walk, a that this choice wouldn't shift the project into a different category of budget and building technology.

4. Think bottom-up

The preference for bottom-up organization is a consequence of the commitment to off-the-shelf and plug-and-play products. We are invested in the possibility of aggregating small units into large configurations that would have a greater impact. We hence explore bottom-up setups that do not require centralized wiring and command. This strategy enables the deployment of technology with wireless communication and distributed light infrastructure. It is particularly suited for retrofitting existing structure while forgoing prohibitive heavy

图 0-12

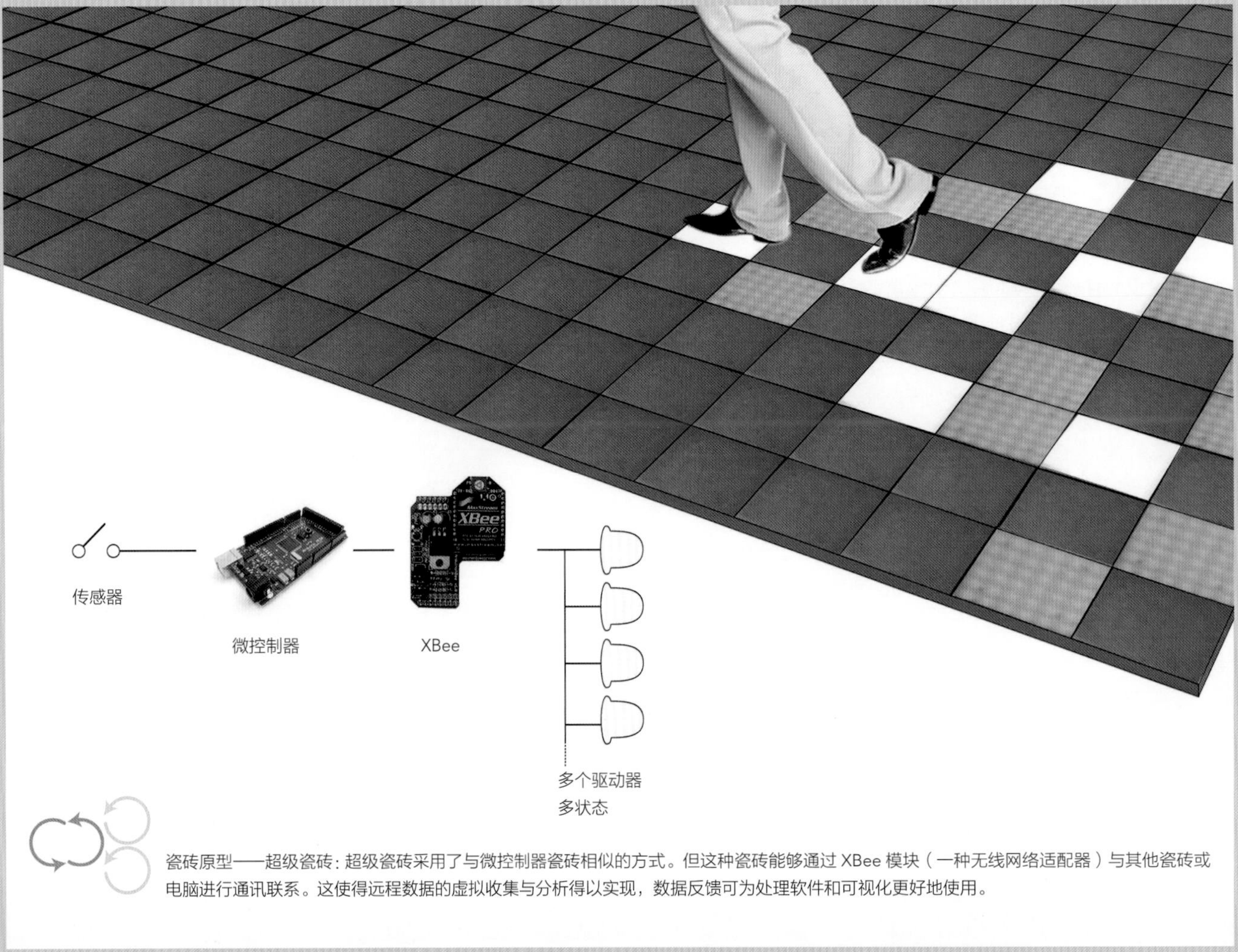

传感器
微控制器
XBee
多个驱动器
多状态
瓷砖原型——超级瓷砖：超级瓷砖采用了与微控制器瓷砖相似的方式。但这种瓷砖能够通过 XBee 模块（一种无线网络适配器）与其他瓷砖或电脑进行通讯联系。这使得远程数据的虚拟收集与分析得以实现，数据反馈可为处理软件和可视化更好地使用。

图 0-12 数字地砖（详见 186 页项目）：预制混凝土地砖被嵌入传感器、LED 驱动器和扬声器（克里斯多夫·陈、鲁道夫·埃尔 – 库利）
Digital Paver: a precast concrete paver is embedded with sensors and actuators – LEDs and speakers (Christopher Chung, Rodolphe el-Khoury)

布式的光基础设施做技术部署，当我们放弃高昂的前期投资时，它尤其适用于改造现有的结构。

例如，有一种可响应活性荷载的可变几何桁架结构，每一个结构构件都能独立响应局部荷载，桁架的整体形状是由单个部件在群动力作用下集合而成（图 0-13），它不需要被输入中枢计算命令与指令图形。智能墙体（图 0-14）是一种自我推移和分控引导的系统，它可以重新配置你的空间，以适应你的使用偏好、环境或是各种设定的场景。想

upfront investment.

For instance, a structural truss with a responsive variable geometry (figure 0-13) that reacts to live loads is designed with each structural member responding independently and individually to local loads. The overall shape of the truss emerges from the sum of individual parts reacting in a swarm dynamic. There is no input from a central command computing and dictating the geometry. WallBot (figure 0-14) is a system of moving self-propelling and guiding partitions which will reconfigure the plan of your loft to suit your use habits or following environmental inputs and various programmable scenarios. Imagine the floor plan

图 0-13

图 0-14

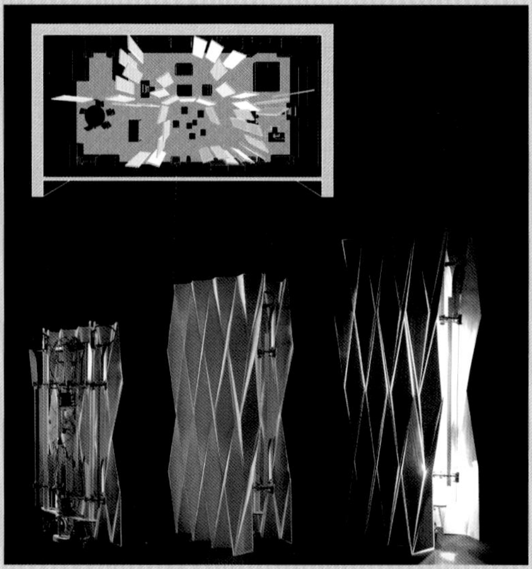

图 0-13 可变几何桁架（详见 160 页项目）：设计使用了“群”逻辑，允许每个驱动桁架杆件能够连续和独立地适应由传感器阵列测量和传递的局部压力（戴维 · 朗、鲁道夫 · 埃尔 – 库利）
Variable Geometry Truss: the design uses a swarm logic allowing each actuated truss member to continuously and independently adapt in response to local pressures measure and relayed by an array of sensors (David Long, Rodolphe el-Khoury)

图 0-14 智能墙体（详见 169 页项目）：它的原型通过扩展折纸表皮和能动构件，可以从 1 米扩展到 1.5 米（奥托 · 吴、鲁道夫 · 埃尔 – 库利、纳希德 · 纳比安）
Wallbot: the prototype stretches from 1m to 1.5m by expanding its origami skin and kinetic skeleton (Otto Ng, Rodolphe el-Khoury and Nashid Nabian)

reconstituting continuously to adapt to changing conditions. Each partition is acting individually but in communication with others as they swarm into various configurations throughout the day.

象一下你的房间平面能够不断地重组以适应环境的改变。房间每一分区都能独立使用，但又相互紧密联系，因为它们时刻被整合在不同的场景配置中。

5. Work with the flow

Responsive technologies are very well equipped for dealing with the dynamic aspects of the environment. The projects use the technology no so much with the aim of achieving homeostasis, but rather for their potential in domesticating turbulence. 'Breeze-Thru Wall' is a project where micro heat pumps integrated into building blocks condition the air as it traverses the building's entire envelope (figure 0-15). A gentle breeze would literally flow through the building just as if in an outdoor space with no obstacle in its path. Embedded responsive technologies here work in sync with the dynamic flows of the environment rather than struggle against them.

In another instance a wall features individually motorized and sensor-equipped bricks that react independently to programmable environmental and use-related scenarios. Imagine the ensuing flux of continually adjusting patterns adapting in real time to anything from a shadow of a passing cloud outside to the presence of an occupant in need of privacy inside. In Pneumatic Envelope, a system of inflatable pillows (figure 0-16) embedded within a standard wall adjusts in real time the insulating capacity of the building envelope in reaction to environmental dynamics and in response to user input. The possibility of modulating insulation vertically by differentiating between lower and higher sections of the wall would also aid in the circulation of the air within the room.

5. 流动式操作

感应式技术非常适合解决动态环境问题，使用这项技术并不是为了达成某种稳态，而是为了操控干扰的能力。“微风穿过的墙”这一项目是将微型加热泵集成到建筑构件中，当空气穿过建筑的整个围护结构时，墙体能够对空气进行调节（图 0–15）。微风可以自由地在建筑中流动，就像在室外空间一样毫无阻碍，这样嵌入式的感应技术可以与环境动态流同步，而非单纯地抵御环境。

另外一个案例的墙体特征是其可以独立制动且装备有传感器的砖，它们可独立对设定的环境和使用相关场景做出反应。想象一下，这种随后持续而来的动态可调节模式，可以实时地调节任何事物，包括从室外飘过一片乌云到居住者所需要的室内隐私空间。在充气围护结构中，嵌入标准墙体内的充气枕头系统可根据环境动力学和用户输入实时调整围护结构的隔热性能，它能通过区分墙体截面的高低来垂直调节隔热的可能性，也有助于室内空气的循环（图 0–16）。

6. Construct atmosphere

We have a growing interest in the possibility of using embedded technology to precisely mobilize non-substantial means for the orchestration of ambience and atmosphere. Tunable Cloud is a digitally augmented acoustical tile (figure 0-17) that modulates the acoustical signature of a room in real time in response to stage events,

6. 建构氛围

我们对于使用嵌入式技术精确地调动非实质手段，来建构环境与氛围的可能性有着浓厚的兴趣。可调声云是一种数字增强声学片（图 0–17），它能够根据舞台事件、表演类型，或其他输入和

图 0-15

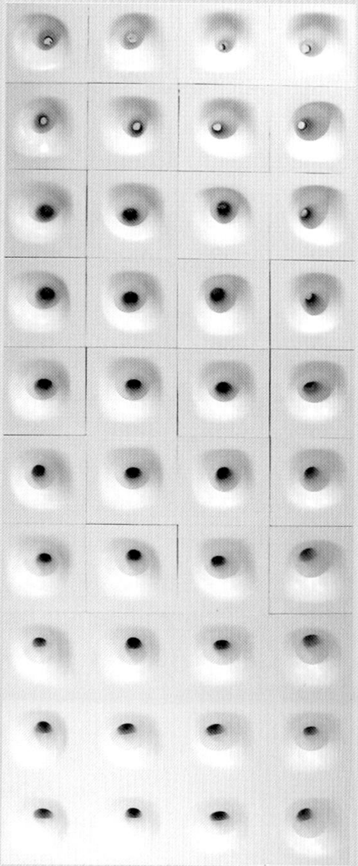

图 0-16

图 0-17

图 0-15 微风穿过的墙（详见 88 页项目）：在交换循环中，冷空气通过墙进入室内，并被墙体内的暖空气加热；在回收循环中，室内暖空气通过墙体排出，并加热了墙体内循环的空气（亚瑟 · 曾、卡罗尔 · 穆海贝尔、克里斯托 · 马尔科普洛斯）

Breeze-Thru Wall: in the exchange cycle incoming cool air blows through the wall, where it is heated by warm air housed within the wall cavity; in the recovery cycle interior warm air exits through the wall where it heats up the air circulating within the wall cavity (Arthur Tseng, Carol Moukheiber and Christos Marcopoulos)

图 0-16 充气围护结构（详见 194 页项目）：每个“枕头”充气或放气以提供局部的热舒适度。墙体垂直截面隔热水平的变化使得空间内产生空气对流，以减轻垂直热滞留问题（瑞克 · 索莱、卡罗尔 · 穆海贝尔、克里斯托 · 马尔科普洛斯）

Pneumatic Envelope: each ‘pillow’ inflates or deflates as necessary to provide local thermal comfort; a variation in insulation levels across the wall’s vertical section allows for the generation of convective flows within a space, alleviating the problem of vertical heat entrapment (Rick Sole, Carol Moukheiber and Christos Marcopoulos)

图 0-17 可调声云（详见 86 页项目）：应用程序能手动调控定制空间矩，或通过编程让软件通过一套反馈机制来自动校准系统，可调声云的模块化设计允许大规模地推广与现场定点安装（曼妮 · 曼妮、卡罗尔 · 穆海贝尔、克里斯托 · 马尔科普洛斯）

Tunable Cloud: the application can be manually adjusted for customized spatial moments, or programmed to allow the software to autonomously calibrate the system through a series of feedback loops; the modular design allows for large scale expansions and site specific installations (Mani Mani, Carol Moukheiber and Christos Marcopoulos) performance genres, or other inputs and criteria. Scent Garden (figure 0-18) was commissioned by the Xi'an International Horticultural Exhibition. It features scent diffusers doubling as light fixtures. The diffusers build on experiments with Scentisizer(figure 0-19), a device that is designed to digitize scents and perform the digital files of recorded olfactory compositions. The diffusers of Scent Garden overlay a sensor-controlled olfactory image on top of the landscape to complement natural fragrances with a curated artifice.

7. Calibrate experience

The possibility of precisely customizing environments for an individual's use is often touted as a promising potential of digital embedded technology. We are not so much interested in optimizing the delivery of services to individual preferences. We are more invested in the possibility of a sensuous and affective resonance between people and spaces. Chameleon Lounge features stools that imitate the color of people's clothing. While 'Auratic Chamber' (figure 0-20) responds to the mood of individuals, it does so by reading the brainwaves with a brainwave controller — a device that has had limited applications in the gaming industry.

8. From embedded to embodied

Our intervention upon familiar objects of everyday life is not merely an additive operation but also a qualitatively transformational one. Technology is not simply layered onto an indifferent object. The object, through its reinvented morphology and materiality, becomes the sensual embodiment of the technology.

IM BLANKY (figure 0-21) harnesses the intricacies of traditional ornamental patterns for the sake of functionality, enabling the blanket to become a soft computational surface. Sensitized with an array of soft tilt sensors (figure 0-22), the blanket is a digitally empowered interface (figure 0-23) capable of registering and relaying in real-time its own position and shape in time and space (figure 0-24).

标准实时调节房间的声学特征。气味花园在西安国际园艺博览会上参展，它的特点是气味扩散器兼作照明装置（图 0-18）。扩散器是建立在气味筛选器实验的基础上，气味筛选器是一种用于将气味数字化，并执行数字文件记录嗅觉成分的设备（图 0-19）。气味花园的发散器在景观顶部覆盖了一个由传感器控制的嗅觉图像，通过精巧的人工技术来补充一些自然香氛。

7. 优化体验

为客户个人精确定制环境的可能性常常被热捧为数字嵌入技术大有潜力的方面。我们对根据个人喜好优化服务并不感兴趣。我们更注重人与空间之间感官和情感共鸣的可能性。变色龙休息厅的特点是凳子模仿人们的衣服颜色。虽然“光晕室”（图 0-20）项目通过使用脑电波控制器读取脑电波可以对个体情绪做出反应，但脑电波控制器在游戏行业的应用非常有限。

8. 从嵌入式到呈现式

我们对日常生活中熟悉事物的干预，不仅是一种加法操作，而且是一种质的转换操作。技术不是简单地分层到一个无关紧要的对象，对象通过其形态和物质性的再造，成为技术的感性体现。

例如，出于功能的考虑，IM感应毯（图0-21）利用了传统装饰图案的复杂性，使毯子成为一个柔软的计算机接口。毯子被一系列柔软翘起的传感器激活（图 0-22），成为一个数字增强的接口（图 0-23），能够在不同的时空中实时记录和传送自己的方位和形态（图 0-24）。

图 0-18

图 0-19

图 0-20

图 0-18 气味花园（详见 113 页项目）：气味柱可以兼作照明装置，它们由不锈钢管、集成的气味盒和 LED 灯组成（KLF 事务所）

Scent Garden: the scent poles double as light fixtures; they consist of stainless steel tubes with integrated scent cartridges and LED lights (Khoury Levit Fong)

图 0-19 气味筛选器（详见 289 页项目）：原型由气味分配器的驱动矩阵组成。分配给每个分配器的气味是根据香水行业开发的一种分类方法的变体绘制的，该分类方法将气味划分为不同但相关的类别：花香、清新、东方调、木香和馥奇香调（鲁道夫 · 埃尔 – 库利、纳希德 · 纳比安）

Scentisizer: the prototype consists of an actuated matrix of scent dispensers. The scents assigned to each dispenser are mapped according to a variation on a taxonomy developed by the fragrance industry to classify scents in distinct but related families: Floral, Fresh, Oriental, Woody, and Aromatic Fougere (Rodolphe el-Khoury and Nashid Nabian)

图 0-20 光晕室（详见 285 页项目）：由室内空间表面排列在一起的联锁砖构成。环境利用光和色彩成为一个人心理状态的延伸（吉米 · 陈、克里斯托 · 马尔科普洛斯）

Auratic Chamber: the chamber consists of interlocking tiles that line the interior surfaces of a space; the environment uses light and color to become an extension of one's psychological state (Jimmy Tran, Christos Marcopoulos)

图 0-21

图 0-21 IM 感应毯（详见 218 页项目）：尺寸为 230cmx127cm；材料是绿色塔夫绸、镀镍银布、导电珠、玻璃珠、导电线、电阻；组件是开源硬件、模块与多路复用器（卡罗尔·穆海贝尔、克里斯托·马尔科普洛斯、鲁道夫·埃尔－库利）

IM BLANKY: dimensions: 230 centimetersx127 centimeters centimeters; materials: green taffeta, nickel coated silver taffeta, conductive beads, glass beads, conductive thread, and resistors; components: Arduino LilyPad, XBee, Multiplexers (Carol Moukheiber, Christos Marcopoulos and Rodolphe el-Khoury)

图 0-22

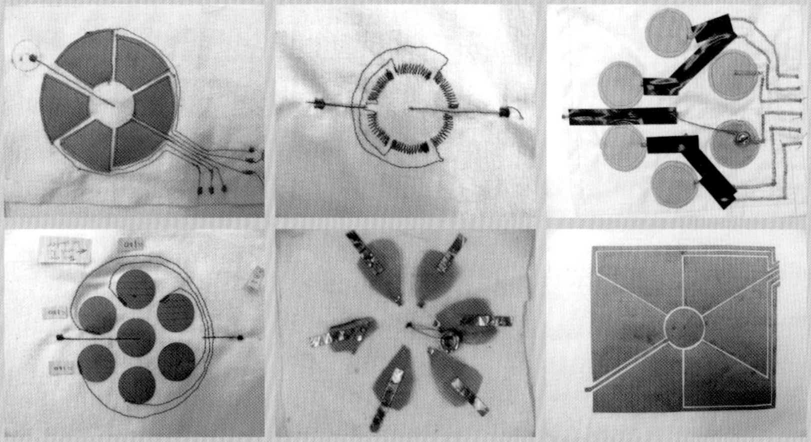

图 0-22 IM 感应毯（详见 218 页项目）：地毯传感器阵列由 104 个倾斜传感器组成，呈六边形网格状排列，均匀分布于整个感应毯

IM BLANKY: sensor array with 104 tilt sensors, arrayed in a hexa-grid formation and distributed uniformly over the entire field

图 0-23

图 0-24

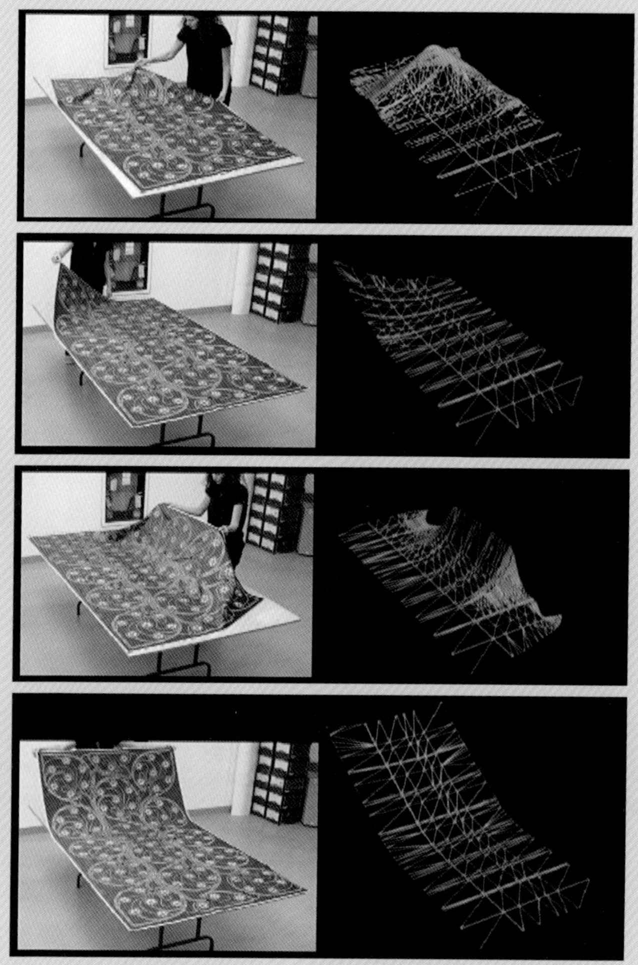

图 0-23 IM 感应毯的电路原理图（详见 218 页项目）：16 个多路复用器从塔斯勒接点各接收 7 个输入；输入被重新配置到 2 个主多路复用器，并依次传递到开源硬件板
IM BLANKY: Circuit schematic;16 Multiplexers receive seven inputs each from the tassle's contact; the inputs are relayed to 2 master Multiplexers and in turn to the Lilypad

图 0-24 IM 感应毯（详见 218 页项目）：描述现实活动的实时数字复制的视频截图；毯子体现了实体和数字化可以同时存在
IM BLANKY: stills from a video depict real time digital replication of actual movement; the blanket embodies simultaneously a physical and digital presence

The blanket's technical and aesthetic innovation is based on the functional and formal convergence of electronic circuit and plant morphology, complete with branching patterns, stems, petals, and pistils (figure 0-25). Using conductive fabric and beads, the tilt sensors and their attendant circuitry are constituted by the very material and ornamental techniques of textile itself (figure 0-26). The blanket relies on the movement of the body along with gravity for its own sensory stimulation, turning movement directly into usable information. The sensing components and their organization are integral to and not external to its form. In other words, IM BLANKY's body is also an information processing system.

The project began with an invitation to re-think the craft of stitching and embroidery in the 21st century, bringing digital processes to bear on issues of craft and ornament.[1] Capitalizing on potential healthcare applications, next generation IM BLANKY senses additional biometric data such as breathing. It belongs to an eco-system of domestic objects that capture a variety of vital signs, forming over time a dynamic evolving profile of the embodying subject.

What next?

Recent projects at RAD recognize how environments designed with densely woven informational links and feedback loops are suited to interface with natural systems. They hybridize living organic material with building components to capitalize on their environmental benefits and responsiveness. A series of projects explore the ceiling as an untapped site for interior landscapes. It is an undisturbed zone that can be reclaimed by nature and linked to resilient outdoor ecosystem by means of networked embedded technology. In a departure from its focus of the object RAD is also currently invested in the development of a user interfaces and communication platform for the deployment and navigation of the Internet of Things.

---

❶ IM BLANKY was commissioned in 2011 by Workshop Gallery, Toronto, for 'Re-Stitching', an exhibition of traditional Chinese embroidery and new works that project the tradition into the 21st Century.

毯子的技术和美学创新是基于电子线路和植物形态在功能和形式上的融合，包括分支图案、茎、花瓣和雌蕊（图 0-25）。利用导电织物、挂珠、倾斜传感器和四周的线圈构成富有质感和装饰性的织物工艺（图 0-26）。毯子依赖身体的运动和重力来产生感官刺激，将运动直接转化为可用信息。传感器构件及其组织是其形式的整体，而非形式的外在。换言之，IM 感应毯整体也是一个信息处理系统。

该项目开始重新思考 21 世纪的缝纫和刺绣工艺，将工艺、装饰问题与数字化应用紧密结合。[1]有效利用潜在的医疗保健应用，下一代的 IM 感应毯可以感知更多类似呼吸之类的生物指征数据，它可以被看作是一个能捕捉各种生命迹象的家庭生态系统，形成一个能够随时间体现主体动态演变的轮廓。

接下来是什么?

在 RAD 最近的项目中，人们认识到，由信息链接和反馈回路紧密交织而成的环境设计如何适应自然系统的界面。他们将活的有机材料与建筑构件混合，发挥它们的环境效益和响应能力。一系列项目探索了顶棚作为室内景观的潜在领域。它是一个自然可再生的不受干扰的区域，并通过网络嵌入式技术与弹性的户外生态系统相联系。RAD 已经不再只关注个别物体，也开始投资研发一个用于部署和导航物联网的用户界面和通信平台。

---

❶ 2011 年，多伦多的工作室画廊以“重新缝合”为主题举办展览，展示中国传统刺绣和新技术的结合，IM 感应毯作为将传统投射到 21 世纪的新作品在展览中展出。

图 0-25

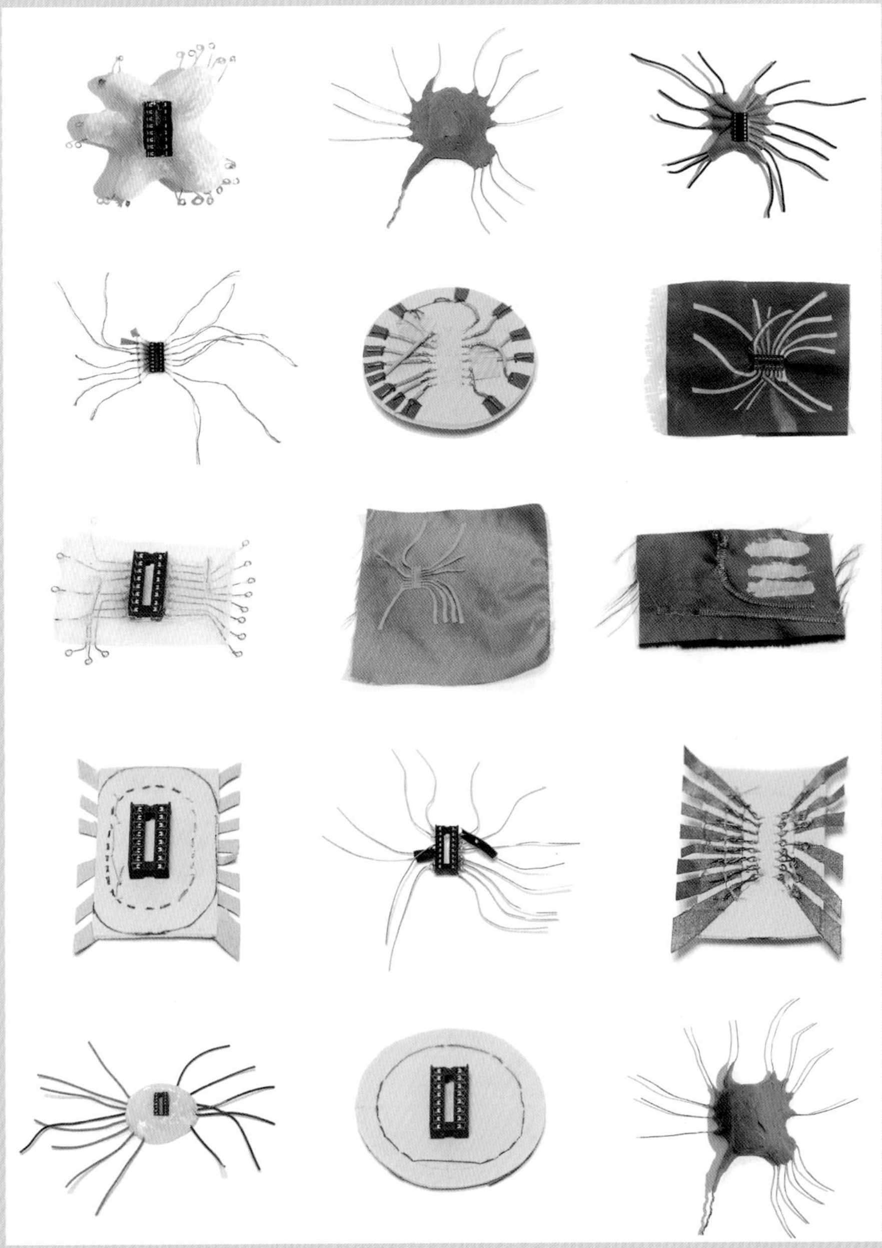

图 0-25 IM 感应毯（详见 218 页项目）：多路复用集线器突破迭代；测试了各种铸造和成型技术，以确保传入的倾斜传感器输入 / 连接到多路复用集线器
IM BLANKY: multiplexer hub break-out iterations; various casting and molding techniques were tested to secure the incoming tilt sensor input / connections to the Multiplexer hub

图 0-26

图 0-26 IM 感应毯细部
IM BLANKY: Close-up view

植物物种高度进化的生物化学过程被纳入环境控制机制。有机物质融入机械系统，并与建筑建造的模块化机制相结合，从而产生了一种人工合成物质。一个新的物种由此产生：植物机械体——部分活体、部分技术支撑，它通过化学反应在环境输入、环境输出与居住者之间建立共生的关系。这些项目通过将自然纳入建筑运营，能够产生能源、净化空气、吸收二氧化碳和修复生态景观。正如在生态系统中，弹性修复被定义为物种的异质性，能够在具有生产效能的生态系统中正常运行。

The highly evolved bio-chemical processes of the vegetal species are enlisted into the mechanism of environmental control. The incorporation of organic matter into mechanical systems and its integration with the modular logic of construction produces a hybrid artefact. A new species emerges: a plant cyborg – part living, part technological – working symbiotically on a chemical level with outputs and inputs from the environment and the user. By putting nature to work, these projects produce energy, purify air, sequester carbon, and remediate landscapes. As in biological systems, resilience here is defined by a heterogeneity of species working in a productive ecosystem.

1

人工自然

ARTIFICIAL NATURE

1.1 过滤块 Filtration Block

作为室内空气过滤系统的模块化结构单元。❶

伊莲 · 童

A modular structural unit that performs as an indoor air filtration system.❶

Elaine Tong

'Filtration Block' is a modular structural unit that performs as an indoor air filtration system to create healthier air by absorbing common indoor air toxins. The Filtration Block module utilises common indoor plants as filters.

'The plants' roots are highly efficient at absorbing an array of indoor chemicals, such as Formaldehyde and Benzene. The module structure is based on the Weaire-Phelan geometry for maximum spatial flexibility. The modules lock together to form a structural wall or ceiling system. Each module is glazed for light exposure, and acts as a micro-greenhouse. The plants are sustained by a water misting infrastructure that calibrates the precise delivery of water through the use of humidity sensors and atomisers.

“过滤块”是一个模块化的结构单元，能够作为室内空气过滤系统吸收空气中常见的空气毒素，创造健康的空气质量。过滤模块使用常见的室内植物作为过滤装置。

这种植物的根能高效吸收大量的室内化学物质，如甲醛和苯。该模块结构基于维埃尔 – 费伦几何结构❷，以获得最大的空间灵活性。这些模块连接在一起构成一个结构墙体或屋顶系统，每个模块都暴露于光照下，成为一个微型阳光间。植物通过水雾喷射设施来维持生命，并通过湿度传感器与雾化器来校准精确的水量传送。

❶ Guided by Christos Marcopoulos and Carol Moukheiber.

❶ 由克里斯托 · 马尔科普洛斯与卡罗尔 · 穆海贝尔指导。
❷ 译者注：1993 年两位爱尔兰物理学家丹尼斯 · 维埃尔与罗伯特 · 费伦依靠计算机技术模拟出最优的泡沫结构，这种十四面体结构由 2 个正六边形和 12 个正五边形构成，被称为维埃尔 - 费伦结构。

图 1.1-1

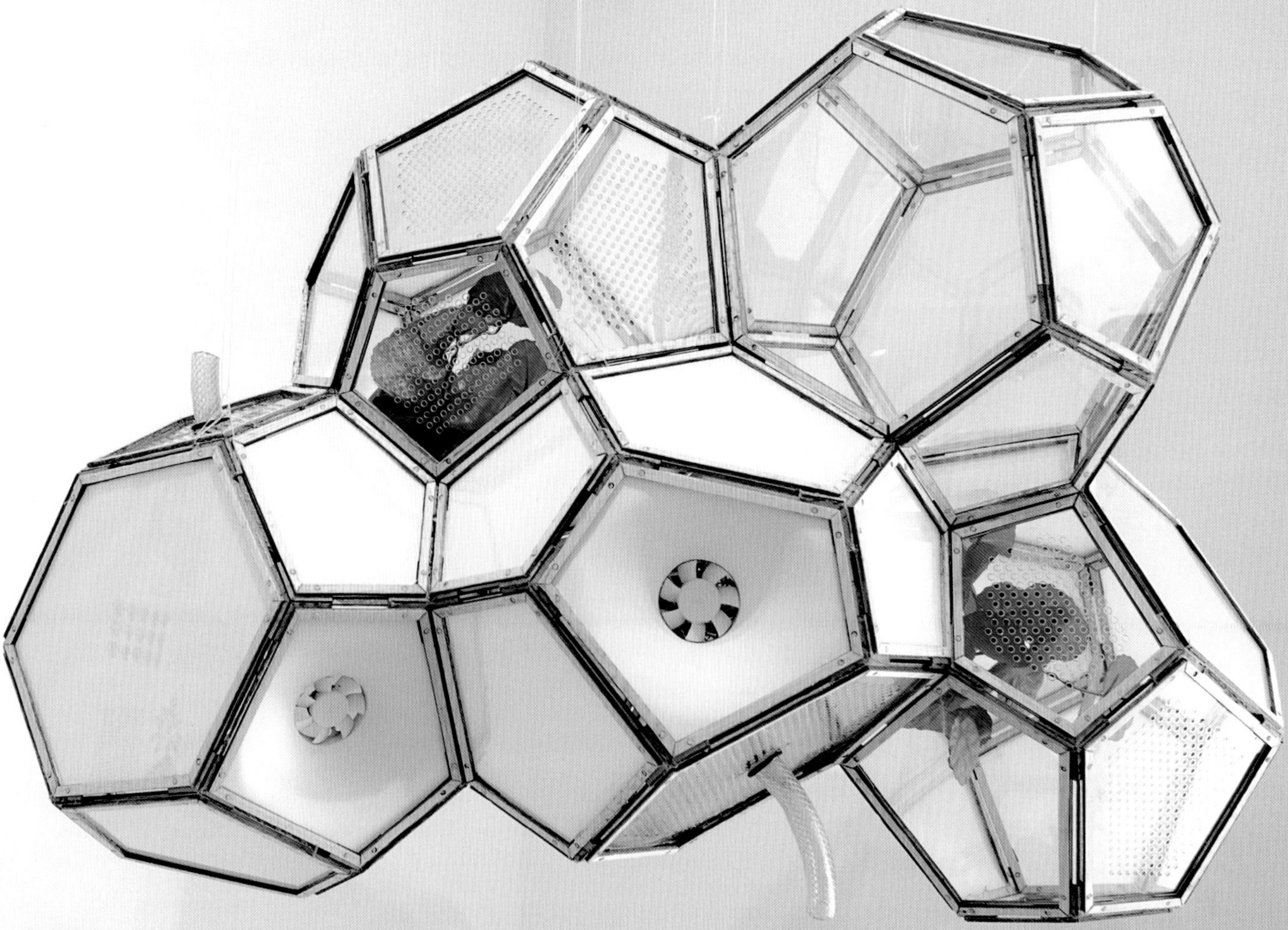

图 1.1-1 项目原型表现了一个模块集合体，其中不透明模块将植物的根封装起来，并内置空气进气口与雾化装置，透明模块将植物具有装饰性的部分包围起来
Prototype showing an aggregation of modules. The opaque modules encapsulate the roots, carrying air intake and misting infrastructure. The transparent modules enclose the ornamental aspect of the plant

图 1.1-2

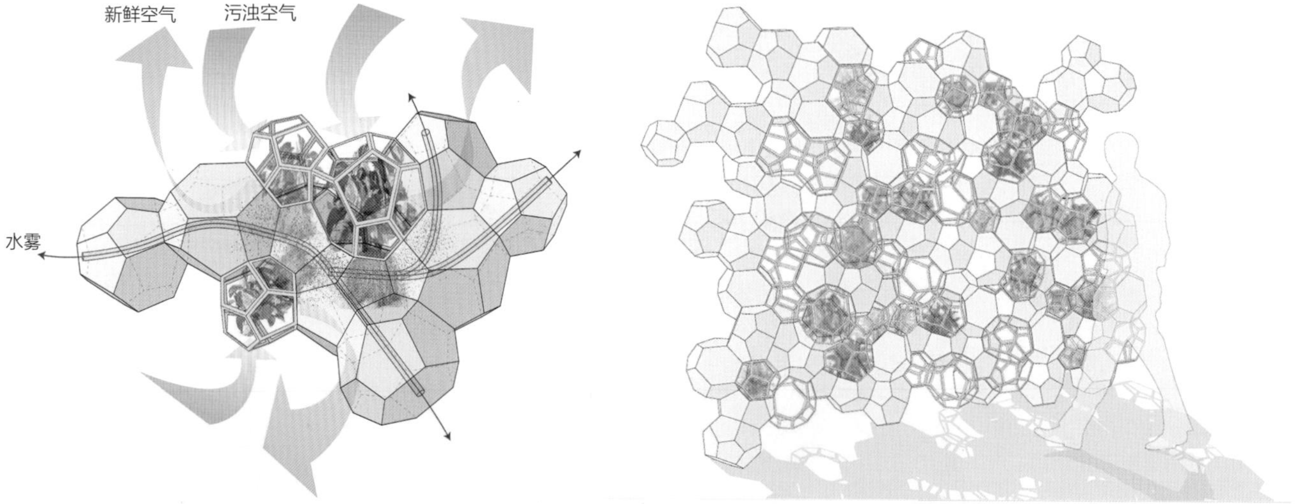

图 1.1-2 室内被污染的空气被装载植物根的模块吸收进去，并被根过滤处理，再将净化的空气释放回空气中
Polluted indoor air is pulled in through the root-carrying modules. The air is filtered by the roots, and then released back into the atmosphere

图 1.1-3

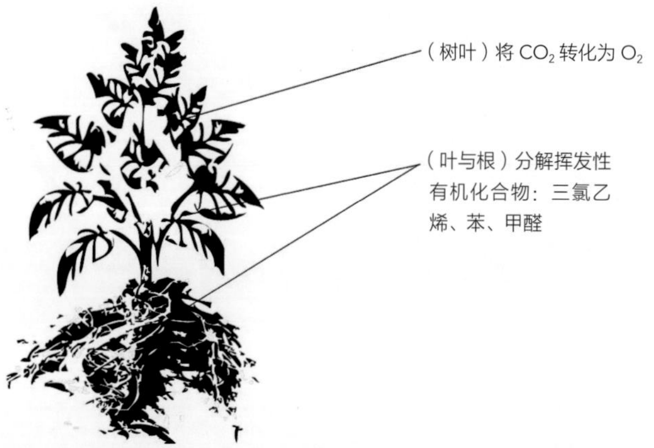

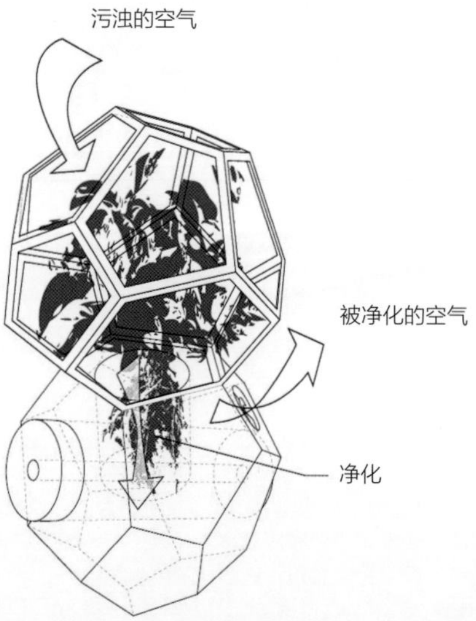

图 1.1-3 叶与根对比
Leaves vs. roots

图 1.1-4

图 1.1-4 不同植物被用于分解一些特定的毒素，一些植物适合过滤一些化学物质，而其他植物可以过滤几种常见的室内毒素
Various plants are used for their ability to break down specific toxins. Some plants are optimal for the filtration of one chemical, while others filter several common indoor toxins

图 1.1-5

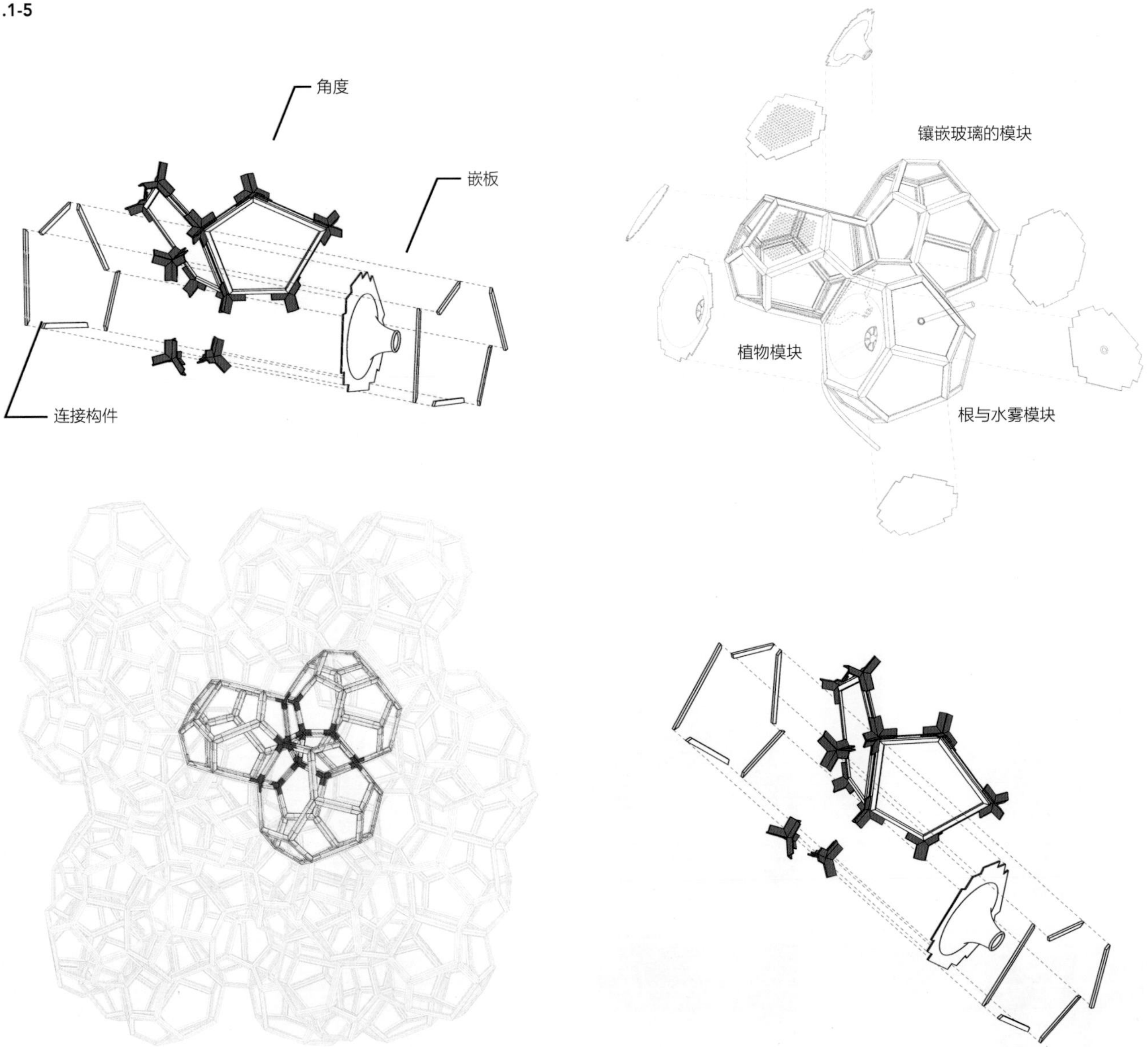

图 1.1-5 模块分解图。该模块是一套可以通过连接器系统组装的组件。模块可以锁在一起形成不同的配置，以确保结构的完整性
Anatomy of a module. The module comes as a kit of parts, and can be assembled through a system of connectors. The modules can lock together into different configurations, ensuring structural integrity

图 1.1-6

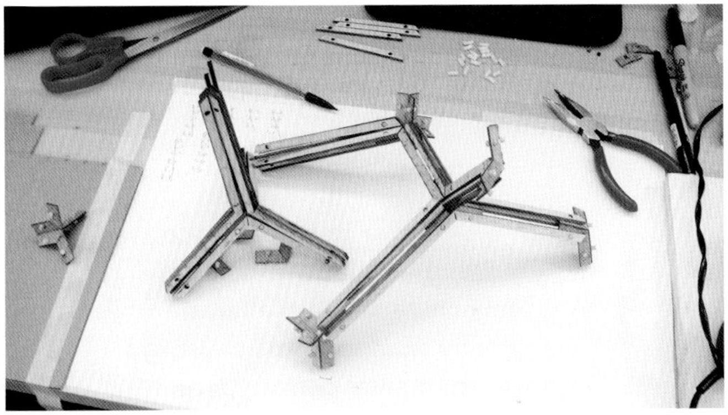

图 1.1-7

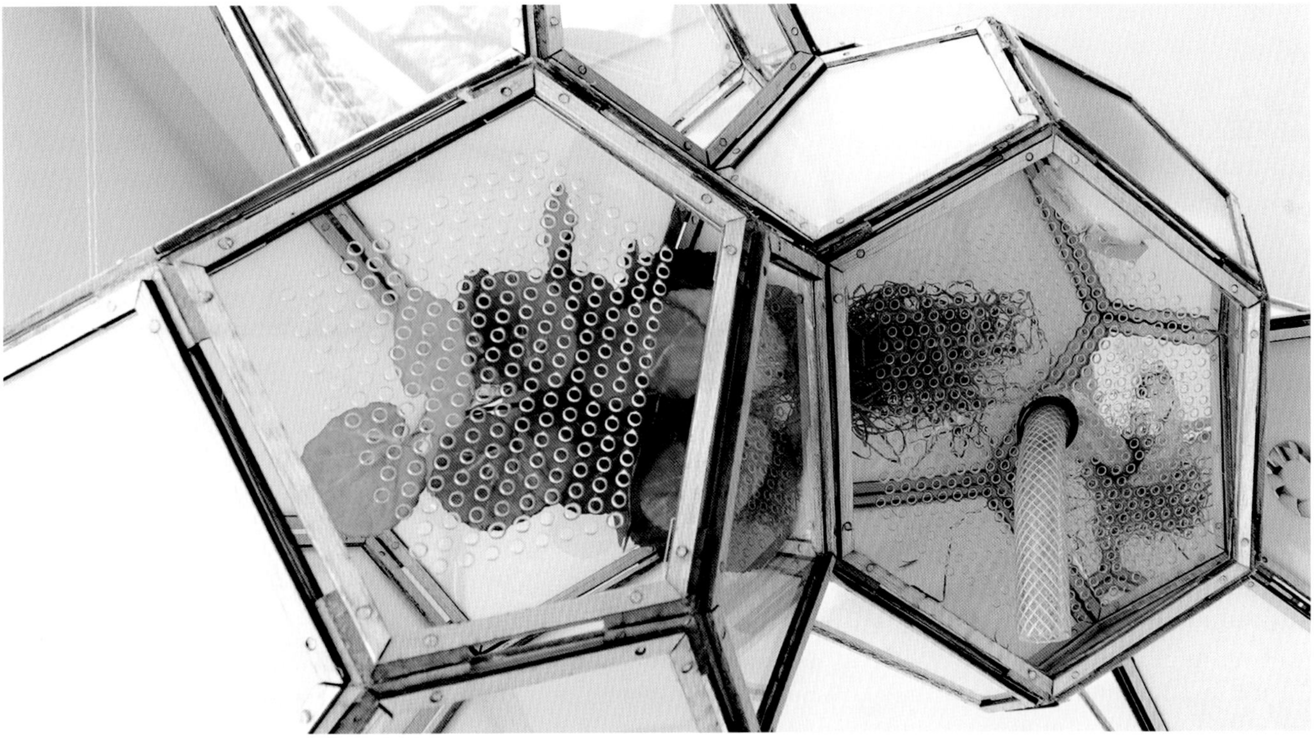

图 1.1-6 模块框架棱边相互咬合，将不透明和半透明面板系统固定在适当位置
The ridged edges of the frames snap onto one another and hold the opaque and translucent panel system in place

图 1.1-7 最终模型
The final modules

1.2 深圳当代艺术与城市规划馆 MOCAPE

将一个生物反应器集成到建筑围护结构中，给立面带来了引人注目的绿叶外形，并大大提升建筑的环境性能。

KLF 事务所

A bio-reactor integrated into the building envelope gives a striking leafy character to the facade while greatly enhancing the building's environmental performance.

Khoury Levit Fong

This project for a 75,000 square metre museum of contemporary art and exhibition hall for the city of Shenzhen was one of four finalists in a combined open and invited international competition.

The project is simultaneously monumental and small in scale. It establishes a clear and elemental figure at the scale of the civic mall that it faces, while the interior breaks down into numerous complementary museum volumes and outdoor courtyards. These courtyards are open and accessible to the surrounding city and shelter a finer tuned public scale that is an alternative to the grandeur of the mall.

The exterior of the building combines two languages: mineral and vegetal. The crystals are enclosed in dark glass of varying transparency and set within a structural frame of angled members. Materially and geometrically the surface and structure of the crystals reinforce its cool mineral identity. The 'bridges' which join the crystals and close the perimeter figure of the building are covered in curved algae-filled tubes and supported by a branching pattern of columns. The 'bridges' are arboreal in nature and create thresholds onto the landscaped courtyards within the building. The algae tubes

该项目为占地 7.5 万平方米的深圳当代艺术与城市规划馆，是一场公开邀请的国际设计竞赛，方案最终进入了设计竞赛的前四名。

项目在尺度上既可以认为是纪念碑式的，也可以是小巧玲珑的，它在正对的市民广场的尺度上建立了一个清晰而基本的形象，然而建筑室内却分解成许多形状互补的博物馆单元与室外庭院。这些庭院是开敞的，可达城市周边，并提供了更为精致和适宜的公共空间尺度，可替代宏大的城市购物中心。

建筑物外观结合了两种设计语汇：矿物质与植物。晶体被包裹在不同透明度的黑色玻璃中，放置在倾斜的结构框架内。晶体表面、结构材料与几何形态强化了矿物质降温的特性，连接各晶体结构与包围建筑界面的“桥梁”被填充了藻类植物的弯管所覆盖，并由柱状分支支撑。这些“桥”实质上是栖于树上，并在建筑内部设置了通往景观庭院的门槛。藻类弯管是一个创新生态系统，能够调节建筑的空气质量，并产生一些生物质能作为替代性能源。建筑立面上的藻类弯管形态形成了一种来自生物机械的植物图景。穿过从庭院和人行道边长出的茂密竹林，我们可以看到这些弯管，它构建了一个迷人的城市情感花园。

are an innovative ecological system that conditions the ‘building’s air and produce biomass used as an alternative energy source. The patterning of the algae tubes on the façades creates a vegetal iconography from biological machinery. These tubes, seen through the leafy bamboo that grows from the courtyards and sidewalks, produce an alluring affective garden in the city.

图 1.2-1

图 1.2-1 深圳当代艺术与城市规划馆鸟瞰
Bird's eye view of MOCAPE

图 1.2-2

图 1.2-2 生物反应器与立面细部
Detail of bio-reactor and elevation

图 1.2-3

图 1.2-3 生物反应器中的藻类养殖
Algae culture in bio-reactors

图 1.2-4

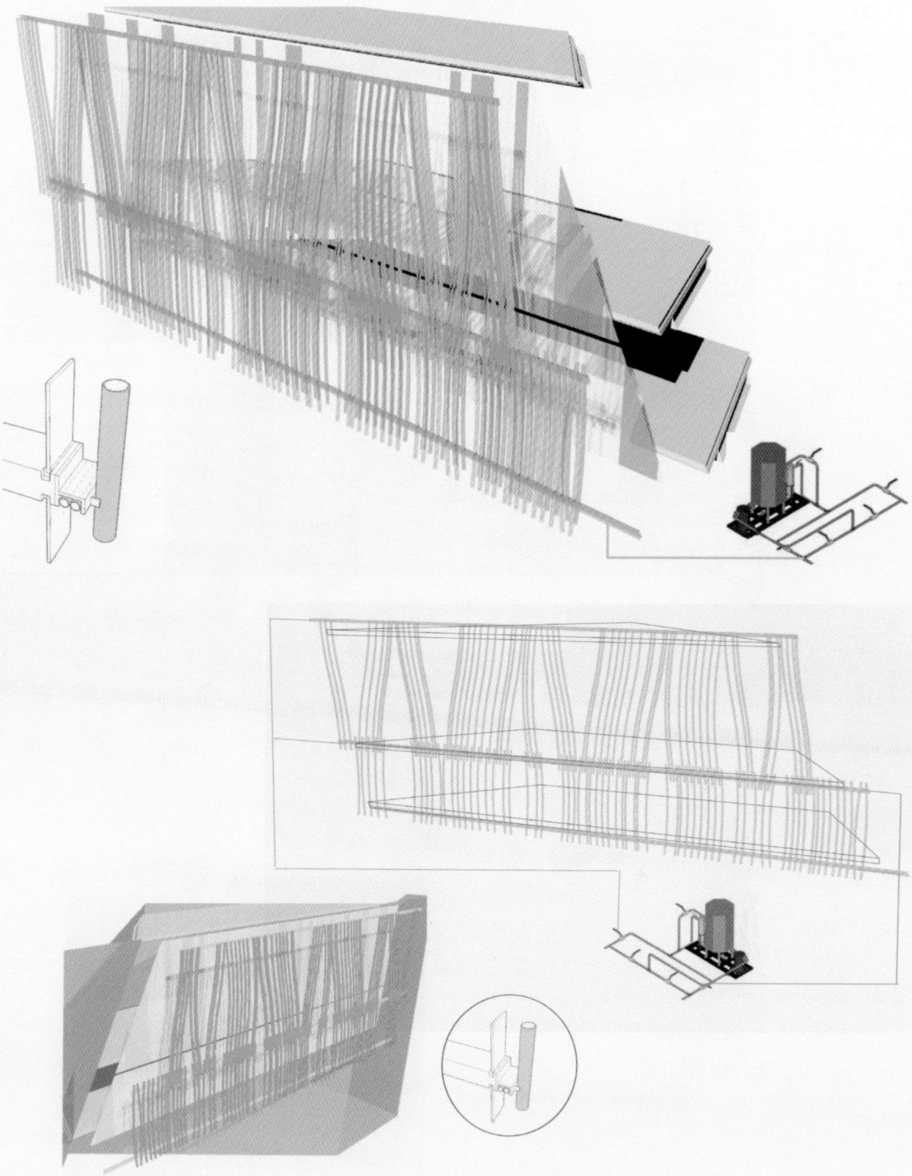

图 1.2-4 生物反应器图解与细部
Diagram and detail of bio-reactor

图 1.2-5

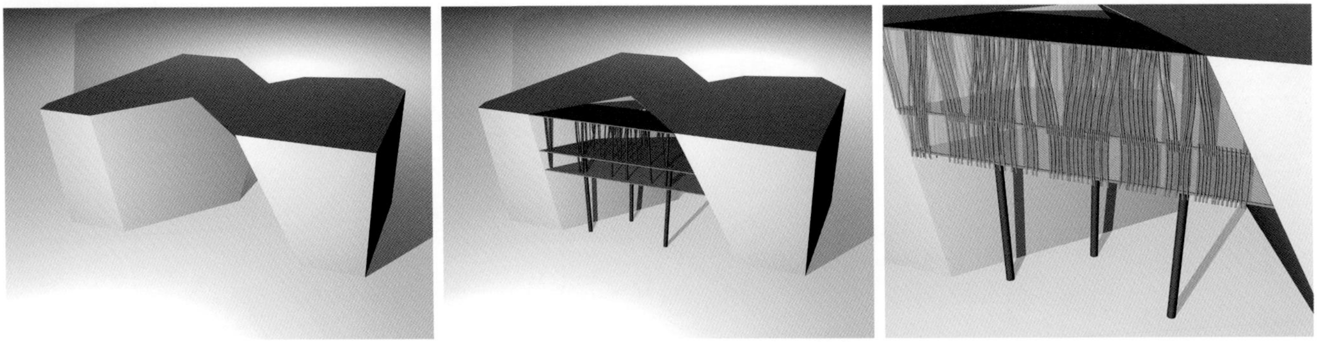

图 1.2-6

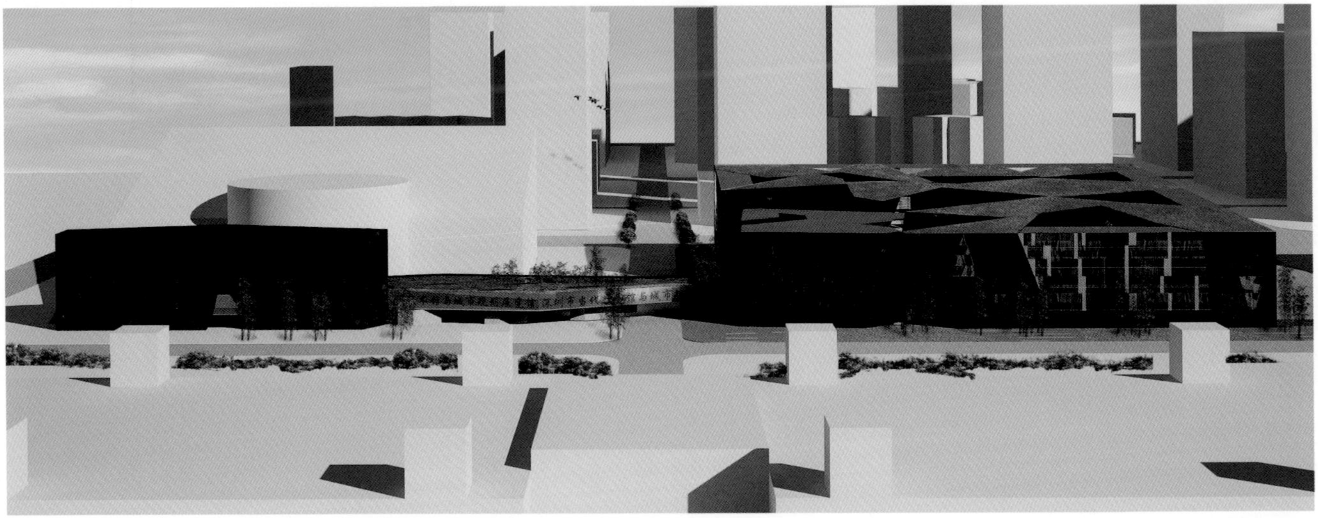

图 1.2-5 树形桥接块与叶状生物反应器
Tree-like bridging block with leaf-like bio-reactors

图 1.2-6 从市民广场鸟瞰建筑立面
Bird's eye view of elevation along the civic mall

图 1.2-7

图 1.2-7 通过生物反应器过滤光的研究
Study of light as filtered through the bio-reactors

图 1.2-8

图 1.2-8 建筑全景
Architectural panorama

1.3 光合作用立面 Photosynthetic Facade

在建筑表皮植入藻类植物生产场，建筑围护结构就成为能源净生产者。❶

霍利 · 乔丹

Injecting the building's skin with an algae production farm, a building envelope aims to become a net producer of energy.❶

Holly Jordan

光合作用立面是一个能够应对温度波动地区高舒适度要求的围护结构系统，它采用嵌入式的藻类植物生产器，创造一个具有最小净热损失与得热高效能的围护结构系统。

设计重点在于发掘了蕴藏在建筑表皮与其整套独特的能效标准中的巨大潜力，建筑围护结构系统能够根据居住者的体验与外部环境因素调节自身。可持续发展的目标就是减少能量使用与能量消耗，以使高效能的建筑围护结构可以满足更高的需求。总之，我们期望身处能够自我调节舒适度的室内空间中，同时还能够获得最大限度的室外景观，而在实现它的同时要尽可能的经济。

Photosynthetic Facade is an envelope system suited to the high demands of fluctuating climatic regions. It uses embedded algae production to create a high performance envelope system with minimal net heat loss and gain.

This proposal focuses on the potential found within building skins and their unique set of functional criteria. The envelope system modulates an occupant's experience and exposure to exterior elements. Sustainability objectives focus on reducing energy use (and energy waste), thus placing even greater demand on high performance building envelopes. In short, we want maximum views to the outside while we sit in the modulated comfort of an interior space, paying as little as possible to achieve this.

❶ 由鲁道夫 · 埃尔 – 库利监制。

❶ Supervised by Rodolphe el-Khoury.

For as long as our internal environmental temperatures differ from external temperatures, net heat loss or gain is inevitable. We therefore need to develop alternate means to offset that loss or gain. By thickening the building skin and injecting it with an algae production farm, this building envelope aims to become a net producer of energy as opposed to the 'standard' mediated heat loss / gain system. A network of flexible, ETFE tubing houses the algae, suspended in water and placed inside a double skin facade system. The internal air naturally stratifies and exhausts at the ceiling of each floor into the double skin cavity. The CO_2 laden return air feeds a series of bubbles which aerate the algae tube system. In conjunction with its exposure to the sunlight, the algae feeds on the CO_2 and therefore grows. The water solution within the tubes also acts as a heat sink to absorb heat from sunlight and exhaust air. The heat energy can be redistributed to other zones of the building by passing air through the double skin cavity. Over time, the algae solution intensifies in colour as it develops, until it is harvested through a main drainage system. The cycle is completed once the tubes are refilled with fresh, young algae solution. In response to occupant use, activity, and outward expression, the tubular system can be manipulated and controlled by a system of muscle wire. 'Muscle wire' responds to charges sent through it by elongating or shrinking. The resulting network is a curtain of tubes surrounding the building.

只要我们的内部环境温度与外部环境温度不同，热损失与得热就不可避免，因而我们需要开发替代性的方式来补偿那些热损失与得热。通过增加建筑表皮的厚度与植入藻类生产器，使建筑围护结构成为一个有别于标准的得热与失热调节系统的净能源生产者。藻类植物被放进一个柔性的氟塑料（ETFE）[1]制成的管状网络，悬浮于水中，并放进一个双层皮的立面系统中。室内空气自然分层，并从每层顶棚将气体排入双层皮的空腔中。二氧化碳气体包裹着回风形成许多气泡，使藻类的管状系统充气。在阳光照射下，藻类植物吸收了二氧化碳并生长，管内的水溶液也起到吸收与存储阳光热量的作用，并释放出空气。同时，热量又随着空气流动进入双层皮的空腔而被带到建筑的其他区域。随着时间的推移，藻类溶液的颜色会随着它的生长而变深，直到它通过一个主要的排水系统被收集。一旦管内重新被鲜嫩的藻类溶液充满，这个循环周期就结束了。为了回应居住者的使用、活动与相关外显性行为，管状系统可由类似肌肉一样的丝线操控，"肌丝"能够通过伸长与收缩来响应它的控制，由此产生的网状物质就是在建筑表皮围合而成的管状帷幕。

[1] 译者注：ETFE 的英文为 ethylene-tetra-fluoro-ethylene，中文全称为乙烯 - 四氟乙烯共聚物，俗称 F-40。ETFE 是最强韧的氟塑料，它在保持了聚四氟乙烯（PTFE）良好的耐热、耐化学性能和电绝缘性能的同时，耐辐射和机械性能有很大程度的改善，拉伸强度可达到 50MPa，接近聚四氟乙烯的 2 倍。

图 1.3-1

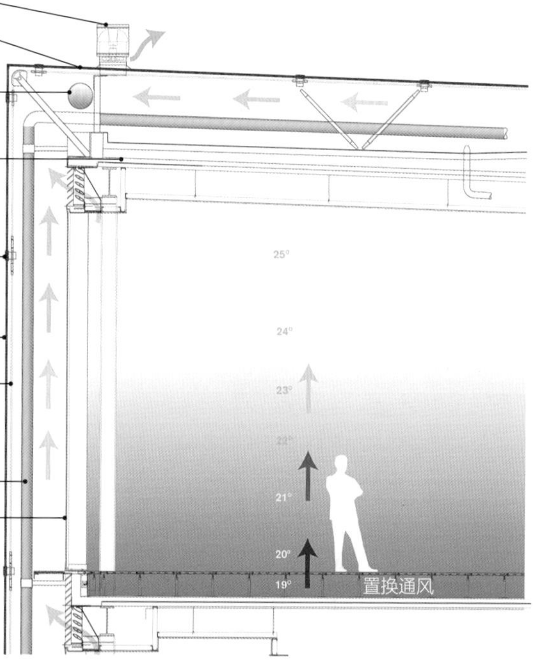

图 1.3-2

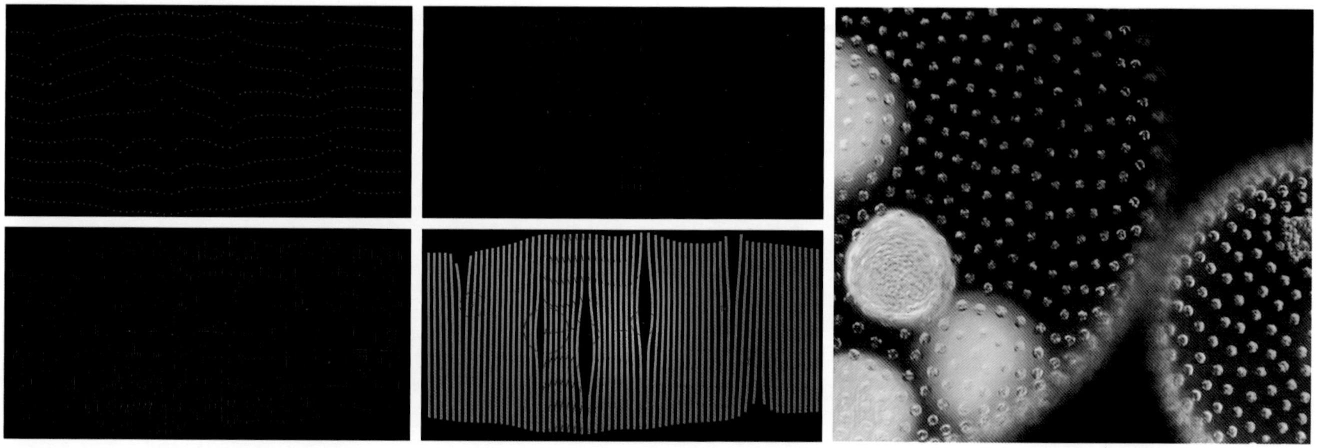

图 1.3-1 综合生物反应器的立面与剖面
Elevation and section of facade integrated bioreactors

图 1.3-2 塑料管通过肌丝发生变形，以应对气候变化与观景需要，建筑表皮材料的厚度增加可以遮挡阳光直射，或者表皮材料变松以制造景观
Plastic tubes deform by means of muscle wires in response to climate and the presence of viewers. The field thickens to shield from direct sunlight or loosens up to yield the view

图 1.3-3

图 1.3-3 用来培育藻类的塑料管顺着屋顶延伸，可以提供屋顶遮阳与最大效能地获得太阳能
The plastic tubes of algae culture extend along the roof to provide shade and most efficiently capture the energy of the sun

图 1.3-4 建筑全景
Architectural panorama

图 1.3-4

1.4 氧气窗帘 Oxygen Curtain

藻类光合作用气体交换器是采用窗帘的形式，在调节光照的同时，为室内空气提供氧气。[1]

曼 · 沙巴

An algae photosynthetic gas exchanger takes the form of a curtain, oxygenating indoor air while modulating light.[1]

Mae Shaban

“氧气窗帘”是将高效的有机活性炭融汇于室内空间，能主动感应室内二氧化碳的波动情况。

“氧气窗帘”产生的氧气量相当于一株成熟阔叶树的氧气产出量——一种释放氧气能力非常强的室内植物。窗帘由一系列藻类生物反应器构成，一个由室内空气、电力与养分供给链构成的网络将生物反应器编织成一个单一膜结构。营养物质由建筑内的废水提供，窗帘由居住者呼出的二氧化碳支撑，它能够直接感应居住者与环境的变化。每个模块都自主运行，当空间中二氧化碳的浓度变化水平达到适宜的浓度时，传感器将会激活被选择的模块，这些模块会随着循环空气膨胀和收缩，显现出能够持续不断净化空气的机械有机体。

The ‘Oxygen Curtain’ integrates an efficient organic living carbon sink into an interior space, responding automatically to fluctuating CO_2 levels.

The ‘Oxygen Curtain’ produces an amount of oxygen equivalent to a mature broad-leaved tree — it is a dramatically enhanced house plant. The curtain is composed of an array of algae bioreactors. A network of indoor air, power and nutrient supply lines weave the bioreactors into a single membrane. The nutrients are supplied by the building’s waste water. The curtain is nourished by the CO_2 exhaled by the inhabitants. It is directly responsive to the users and the environment; each module operates autonomously and sensors activate select modules as appropriate to the changing levels of CO_2 within a space. The modules then expand and contract with circulating air revealing a mechanic-organic organism that is continuously refreshing the air.

❶ 由克里斯托 · 马尔科普洛斯与卡罗尔 · 穆海贝尔指导。

❶ Guided by Christos Marcopoulos and Carol Moukheiber.

图 1.4-1

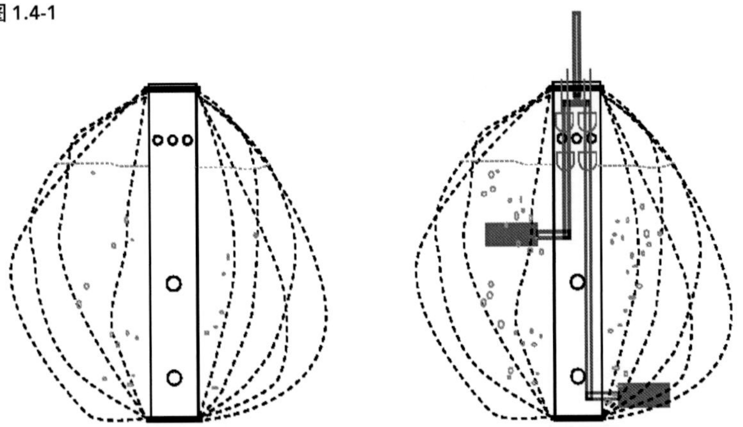

图 1.4-2

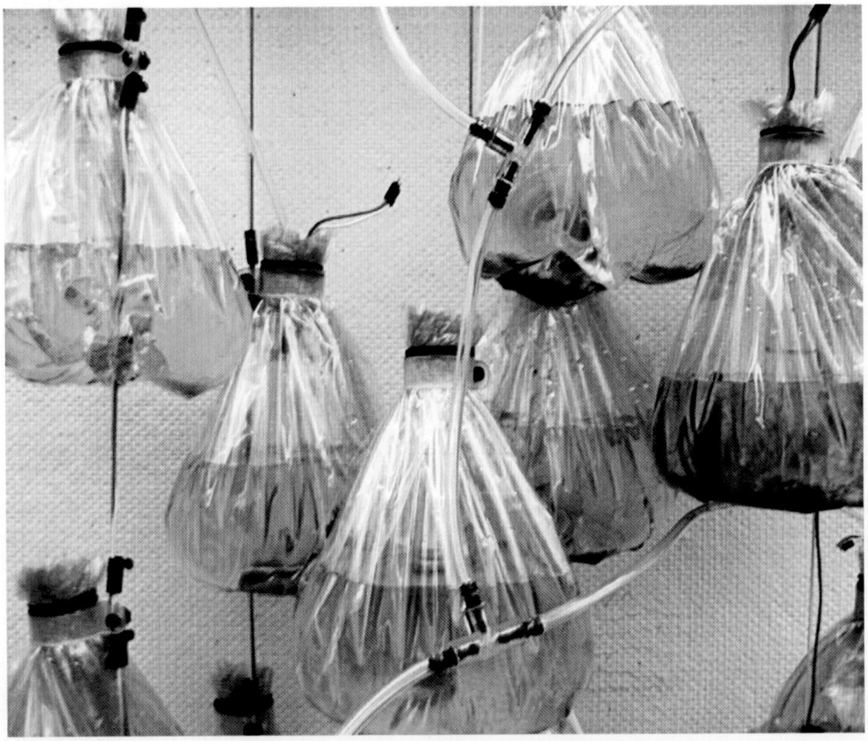

图 1.4-1 窗帘包括两种不同的生物反应器：主动式生长模块配备有通气装置与能够加速藻类生长周期的 LED 灯；被动式生长模块只依赖空间中的环境光线。整个窗帘是由主动式与被动式模块混合而成的

The curtain contains two different types of bioreactors: the active growth modules are equipped with aerators and LED lights that speed up the growth cycle of the algae. The passive growth modules rely solely on ambient light within a space. The curtain exists in mixtures of active and passive modules

图 1.4-2 提供电力与通风装置的供给管道形成藻类袋的结构基质

Electrical and aeration infrastructure supply tubes form the structural matrix for the algae bags

图 1.4-3

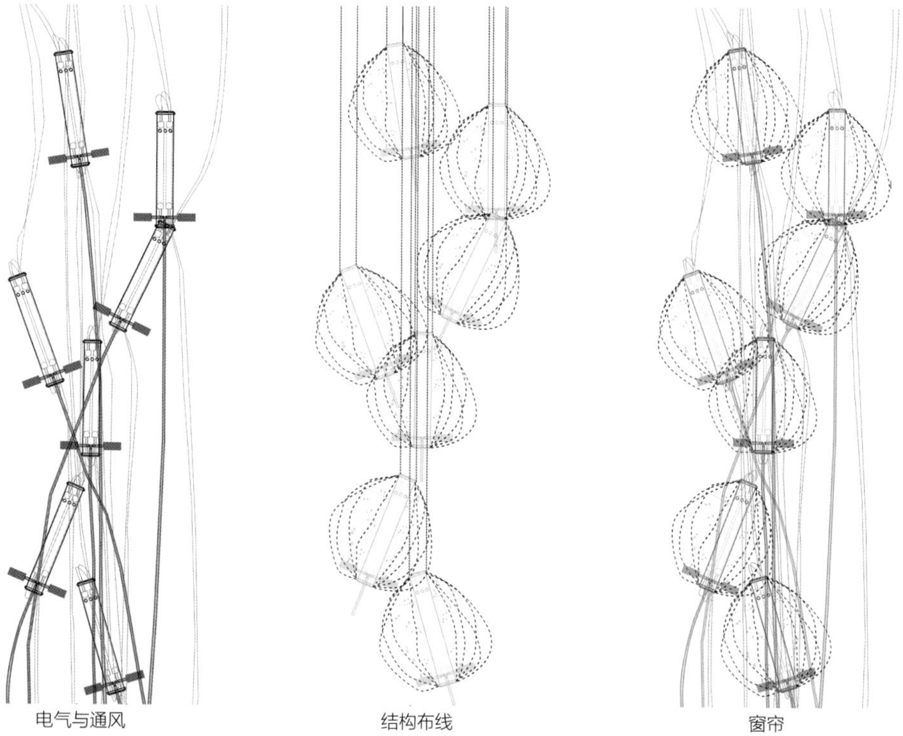

建构海藻窗帘

为了响应有毒气体排放到环境中的问题，窗帘将高效的有机碳汇集到家庭环境中。幕墙的建造和装配使用了一个悬挂系统，其中电气基础设施作为结构支撑。同样的，主动式和被动式模块的布置决定了空气循环网络的横向连接。

虽然下面的结构是一个原型，这个窗帘的概念框架体现一个控制的、关闭的模块系统。在此意义上，窗帘将会更直接地响应用户和环境；所有的模块将是可激活的，并且二氧化碳感应器将激活选择模块以适应空间中不断变化的二氧化碳水平。这些模块随着被引入的循环空气扩大和收缩，而呈现为机械的有机生物体。

图 1.4-3 每个袋内海藻的生长速度略有不同，形成袋内不同的藻类密度，而自然光的调节起到改变藻类密度的作用
Each algae bag grows at a slightly different rate, creating different algae densities within the bags. The modulation of natural light becomes a function of changing algae density

1.5 合成景观 Synthetic Landscapes

在魁北克城的加斯佩矿山的景观实验项目案例 [1]

玛雅 · 德赛

A case for the implementation of a Beta landscape at the Gaspé Mines, Québec [1]

Maya Desai

Challenging the notion of ‘naturalness’ and presenting a new means of remediating irreversibly degraded land areas, Synthetic Landscapes deploys a range of cybernetic mechanisms to create an alternative ecosystem that mimics the performative and experiential qualities of its natural counterparts.

Situated at the Gaspé Mines near Murdochville, the proposed project is a synthetic provincial ecotourism park. Drawing on research from across various fields of study, from synthetic biology to robotics, this project brings together seemingly opposite notions of nature and machinery. Based on automated systems, the implementation of synthetic and interdependent hydrological and energy cycles creates the rudimentary ecosystem for the Synthetic Landscape, while designed elements of performance, interactivity, and programmed environmental cues create the experience of nature. Tourists and researchers alike are invited to camp in the park, experience their surroundings, and conduct their own experiments at various scales, thus contributing to a constantly evolving, synthetic nature.

合成景观是对“自然性”的挑战，它提出了一种新的能够修复不可逆土地退化区域的方法，并实施了一系列的控制机制来建立一个替代性的生态系统，这种生态系统可以模拟真实自然的运行与体验特质。

该项目位于默多克维尔附近的加斯佩矿山，是一个省级综合生态旅游公园。该项目借鉴了从合成生物学到机器人等跨学科的不同研究领域成果，将看似对立的自然与机械两个概念融为一体。基于自动化系统，合成技术的实现与紧密契合的水文与能量循环为合成景观创造了基本的生态系统，而性能、交互性和可定制环境等设计元素创造了自然的体验。游客和研究人员都被邀请到公园里露营，体验周围的环境，并在不同的尺度上进行自己的实验研究，从而促成了一个不断发展的合成自然。

---

1. Supervised by Rodolphe el-Khoury .

1. 由鲁道夫 · 埃尔 – 库利监制。

四种技术干预措施构成了合成景观的基本骨架：带有嵌入式集群逻辑的污染物自动收集器不断地在站点中导航，创造一个动态的智能维护系统；网络昆虫是一种飞行机器人，能够监测环境中的有毒气体浓度，为游客提供露营环境安全区域的视觉线索。同时通过无线网络将这些信息传递给集中收集器；光动力树能够模拟自然中树的特性，同时可充当自然能源的收集器，并从现场收集的水中制造出氢气；雨水收集池能保留与净化雨水，并用于野营与制造氢气。

Four classes of technological interventions constitute the backbone of the proposed synthetic landscape: automated pollutant collectors with embedded swarming logic constantly navigate the site, creating a dynamic and intelligent maintenance system. Cyber bugs — flying robots that monitor the landscape's concentration of toxic gases — provide visitors with visual clues about environmentally safe zones for camping, simultaneously communicating this information wirelessly to the swarming collectors. The photokinetic trees mimic the experiential aspect of natural trees while functioning as collectors of natural energy and producing hydrogen from the water harvested onsite. Rainwater catchment ponds retain and remediate the rainwater used onsite both for camping and hydrogen production.

图 1.5-1

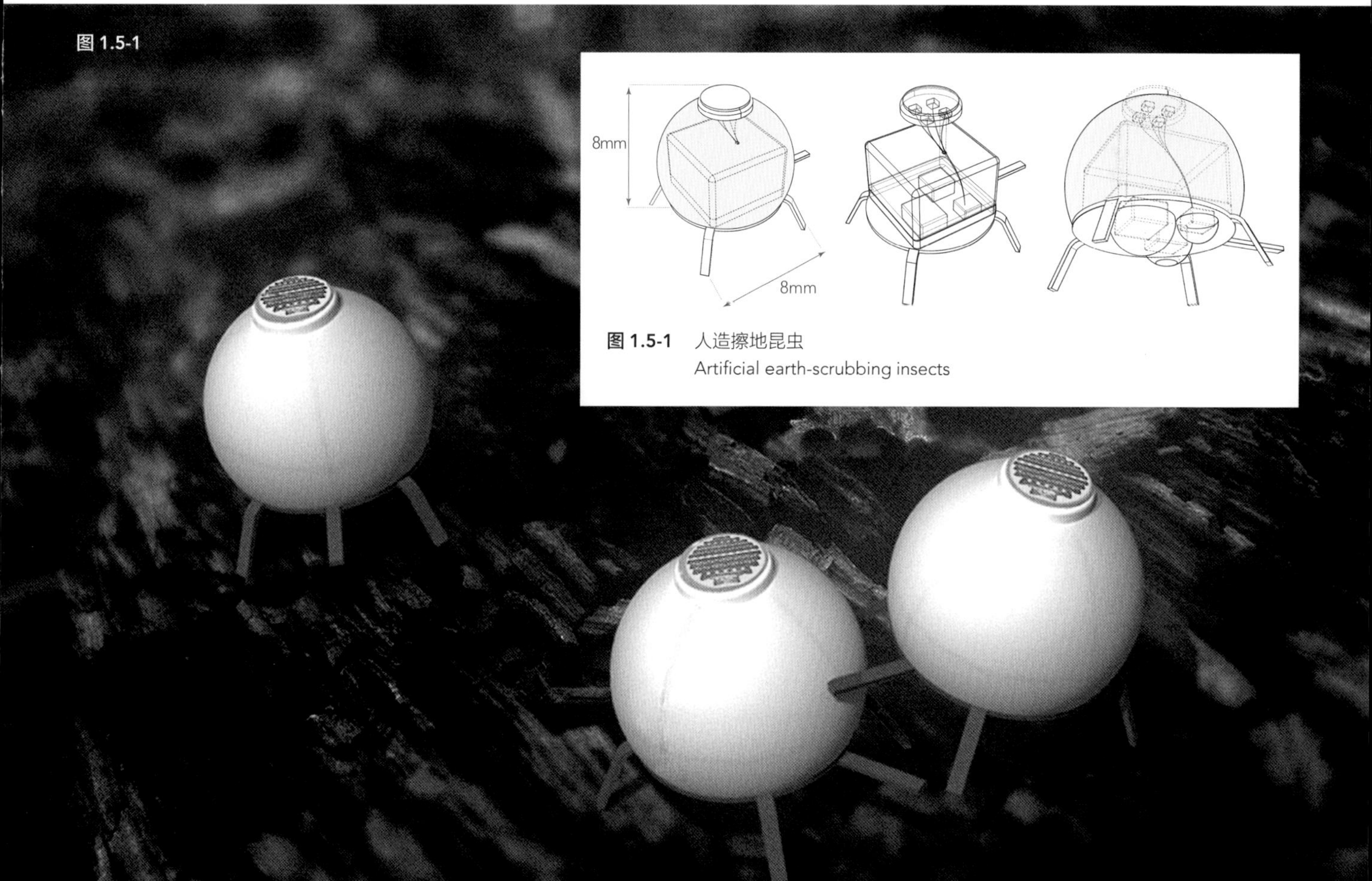

图 1.5-1 人造擦地昆虫
Artificial earth-scrubbing insects

图 1.5-2

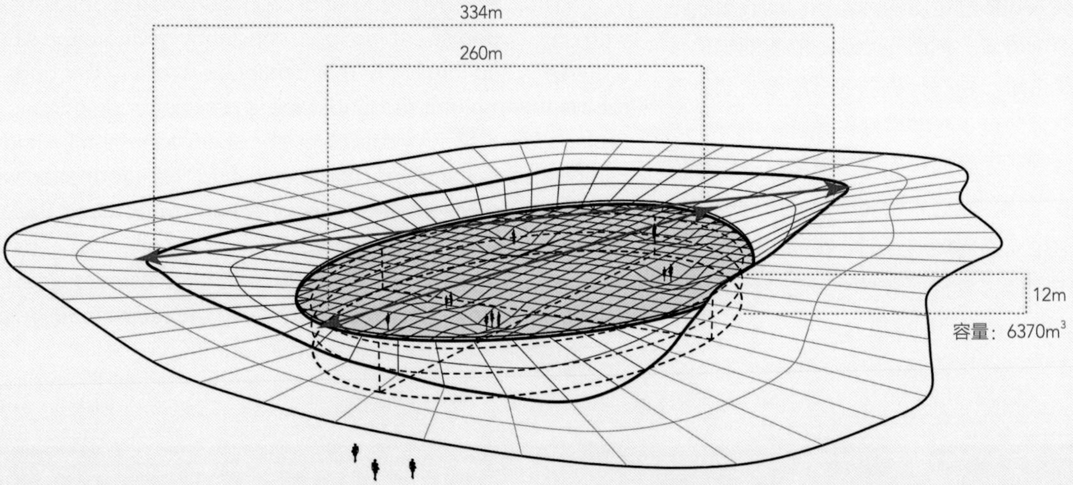
334m
260m
12m
容量：6370m³

图 1.5-3

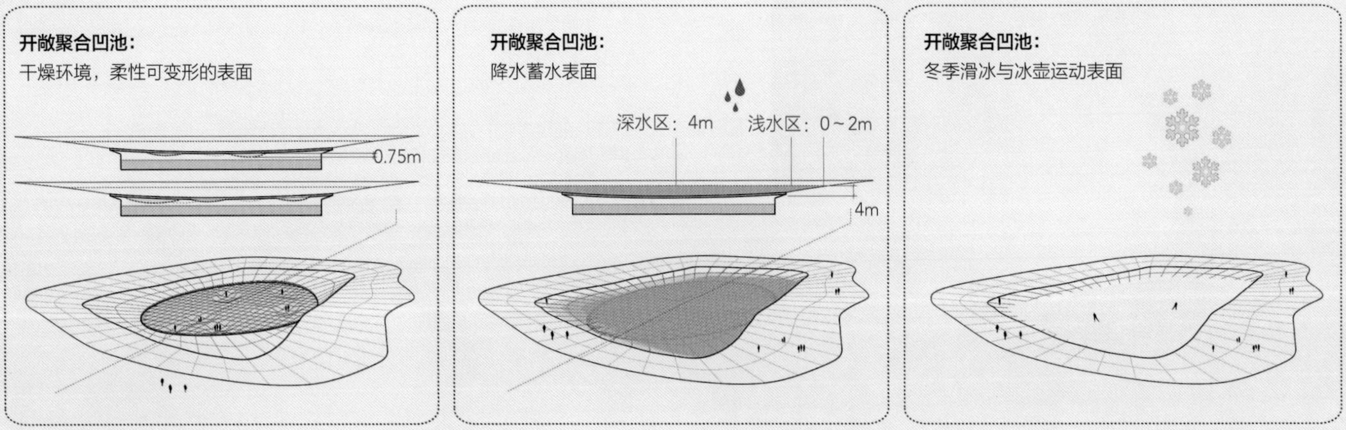
开敞聚合凹池：
干燥环境，柔性可变形的表面
0.75m
开敞聚合凹池：
降水蓄水表面
深水区：4m
浅水区：0～2m
4m
开敞聚合凹池：
冬季滑冰与冰壶运动表面

图 1.5-4

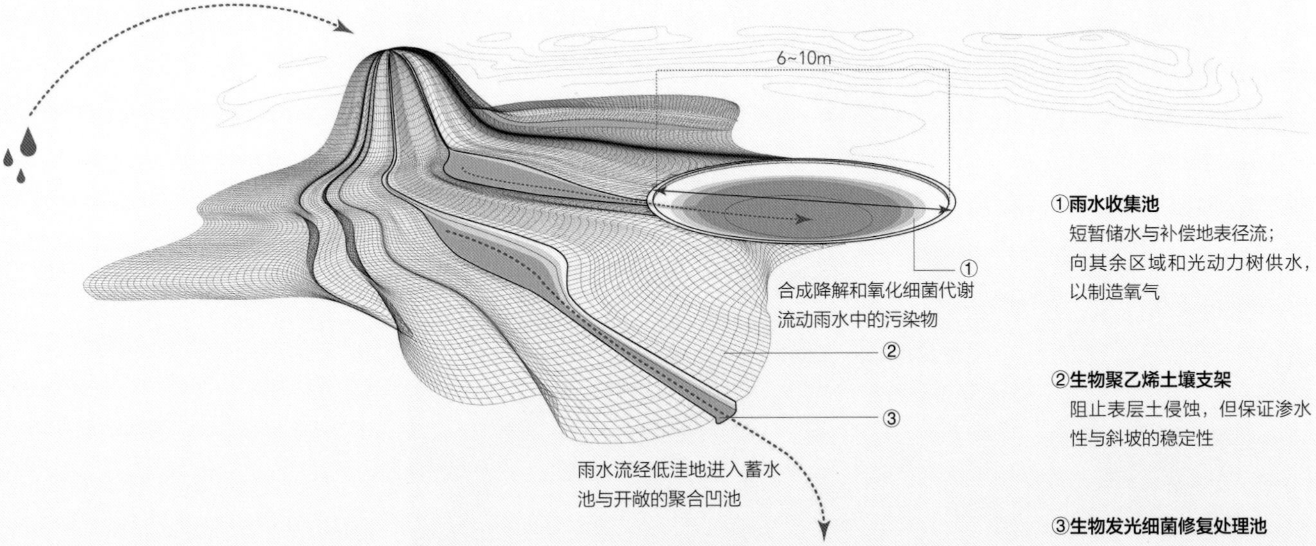

图 1.5-2 密闭的水处理蓄水池
Sealed water treatment reservoir

图 1.5-3 不同状态下的聚合物蓄水池
Polymer seal of reservoir in different states

图 1.5-4 合成降解和氧化细菌代谢流动雨水中的污染物；雨水经过低洼地流向蓄水池与开敞的聚合凹池
Synthetic degrading and oxidating bacteria metabolise contaminants in travelling rainwater; rainwater flows through micro-swales towards catchment areas and the open pit polymer pond

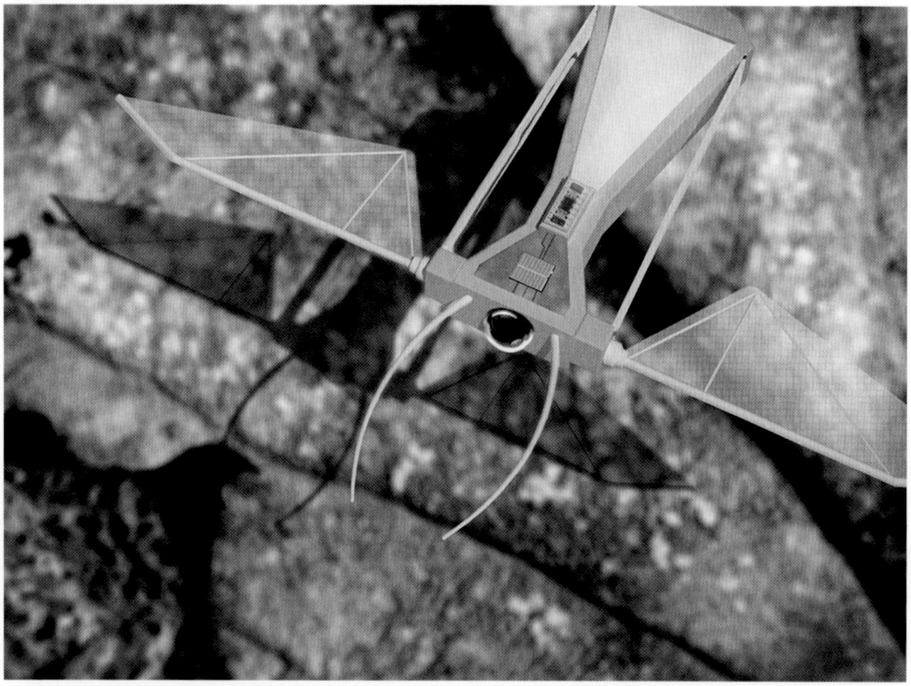

图 1.5-5

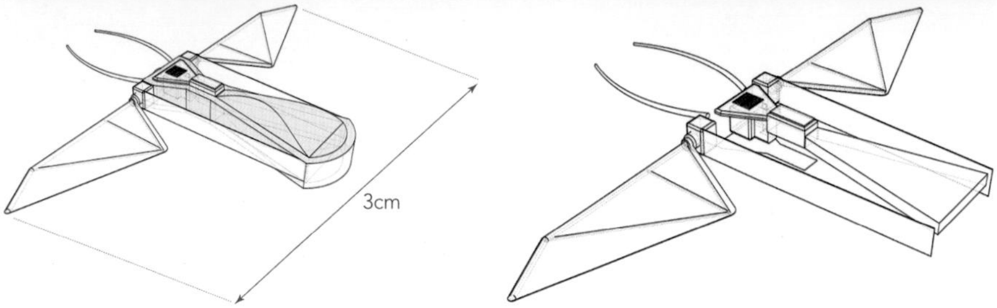

图 1.5-5 人工化学嗅觉昆虫
Artificial chemical-sniffing insects

图 1.5-6 人造景观景象
Views of the artificial landscape

图 1.5-6

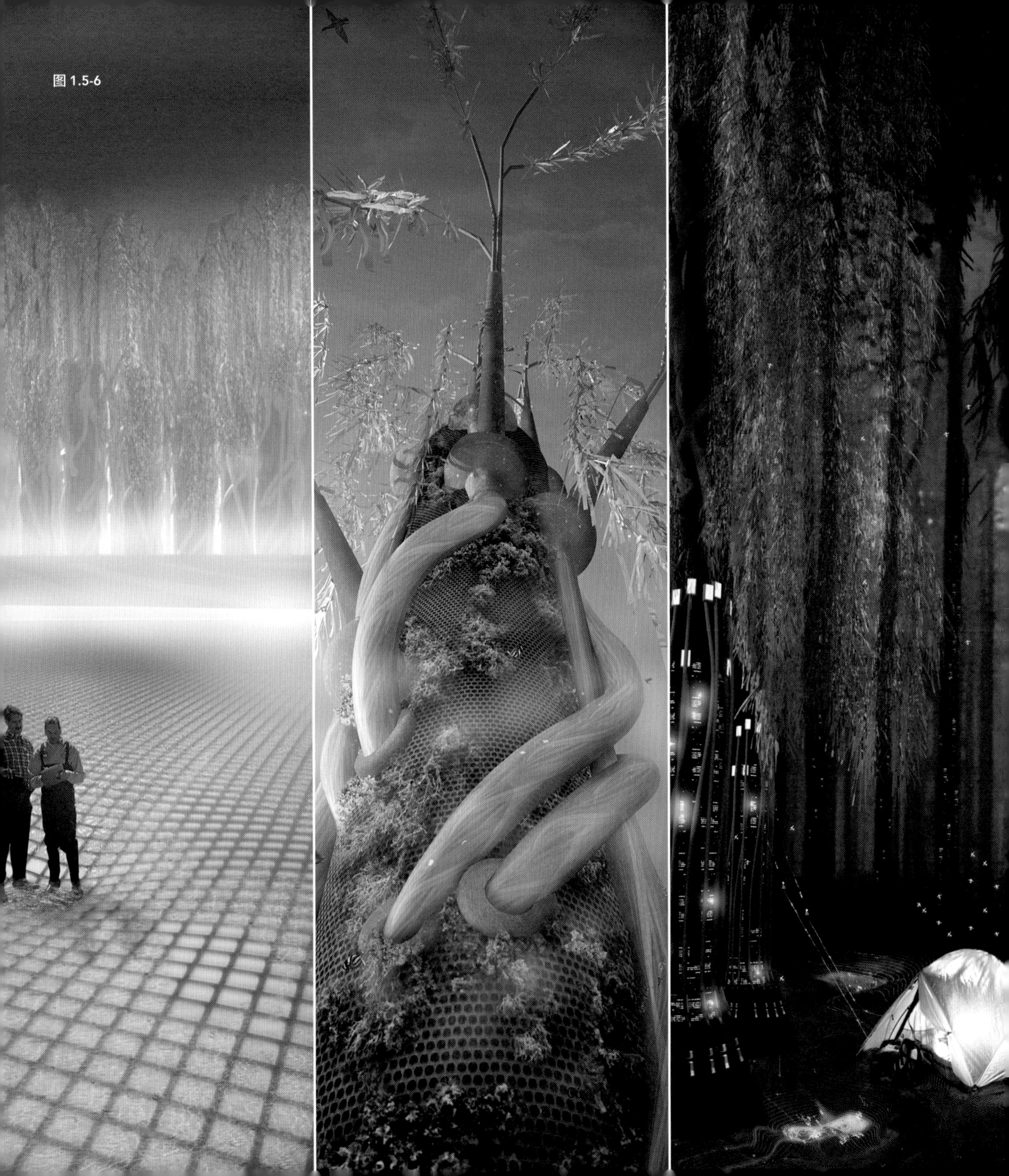

图 1.5-7

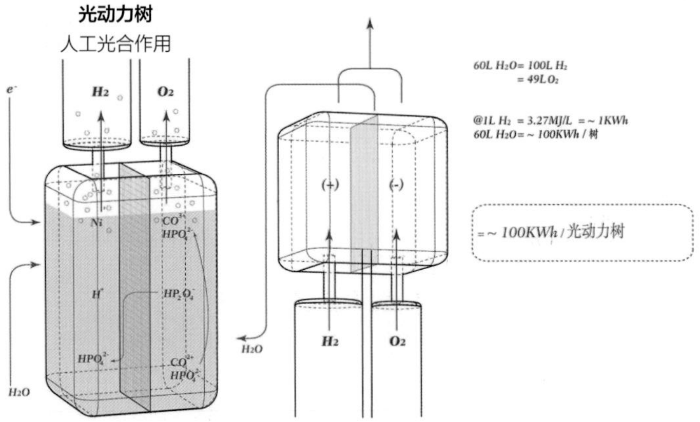

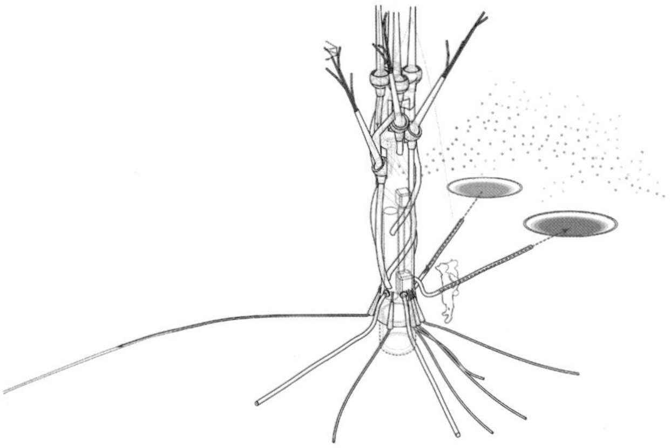

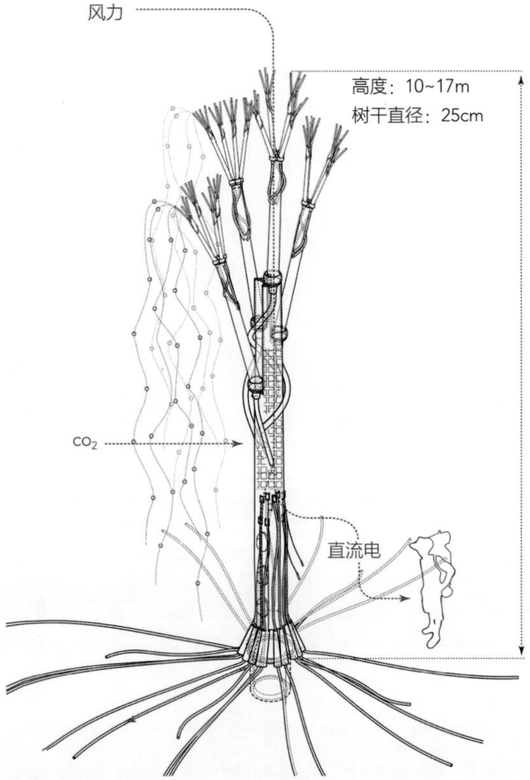

图 1.5-7 能产生能量的叶状压电纤维合成树
Diagram of a synthetic tree showing energy-generating leaf-like piezoelectric filaments

1.6 室外的房间 The Out-House

庭院成为一个能够生产与过滤水体，并在一个半封闭的系统中永久循环利用的有机体。

NMinusOne 工作室，唐 · 施力博格

The yard becomes an organic machine for the harvesting and filtration of a single body of water that is perpetually recycled in a semi-closed system.

Studio NMinusOne with Don Shillingburg

该项目提出了在干旱气候下建造一个倒置的豪宅，在其最后的过度耗能行为中将后院纳入了建筑内部，这种改变恰巧使建筑获得生机。现在纳入的庭院成了一个能够收集与过滤水体，并在一个半封闭的系统中循环利用这些水体的有机体。

原来用于排水的坡屋顶现在被巧妙地改造成集水和收集能量的屋顶。新纳入的庭院与草坪成为一个能够净化中水与再利用废水的郁郁葱葱的生活过滤系统，并为居住者提供了一个伊甸园般的花园。花园由一个厕所维持，一个人类废物处置的场所变为了一个能为动植物生命提供必要营养物质的生产场所，并净化了废水。

曾经住宅的外部现在变成了最私密的空间，最私密的时刻是在过去的户外体验到的。

The project proposes an inverted McMansion in an arid climate which, in its final act of over-consumption, ingests its own rear yard. However this act leads to its own survival. The ingested yard now becomes an activated organic machine for the harvesting and filtration of a single body of water that is perpetually recycled in a semiclosed system.

The original pitched roofs made for shedding water are now opportunistically reconfigured to collect water, and harvest energy. The newly ingested yard / lawn becomes a lush living filtration system cleaning grey and black water for reuse, while offering an Eden-like garden for its inhabitants. Eden is sustained by the toilet; a site for human waste disposal becomes a site for production, providing the necessary nutrients for the maintenance of plant and animal life, which in turn clean the water.

What was once the exterior of the house is now the most private area. The most private moments are experienced in what used to be outside.

In 2004, after five years of severe drought, Las Vegas banned all lawns in any new housing development, allowing 'desert landscaping' only. Residents are paid $1 for every square foot of lawn they dig up.

'Long term, we have to keep growing, but it is going to be a different growth. To put it bluntly, in this town we are going to drink what we flush.'—Oscar Goodman, Mayor of Las Vegas

在经历了五年大旱后，2004 年拉斯维加斯禁止在任何新的住宅开发项目中使用草坪，只允许使用旱景美化，居民每挖掘 1 平方英尺的草坪可获得 1 美元的报酬。

“长期而言，我们必须保持增长，但这将是一种非同寻常的增长，坦率地讲，在这个城镇里我们将会饮用我们的废水。”——奥斯卡 · 古德曼，拉斯维加斯市长

图 1.6-1

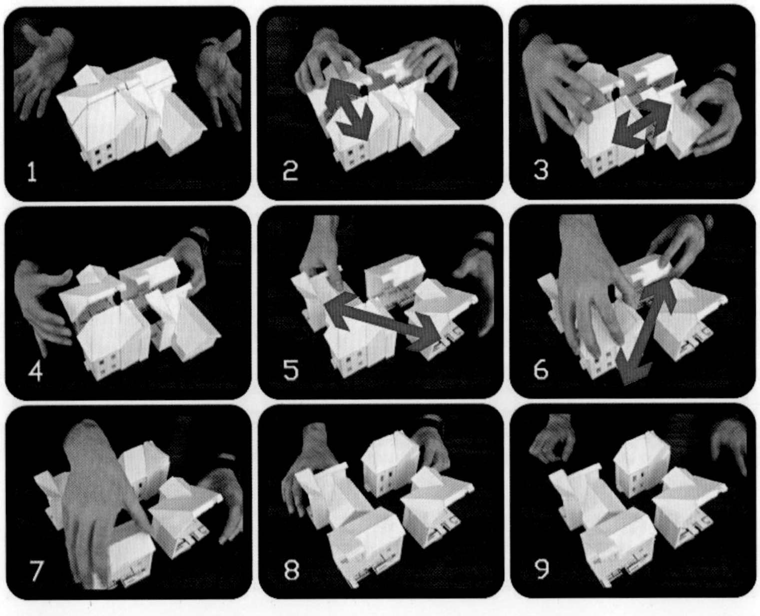

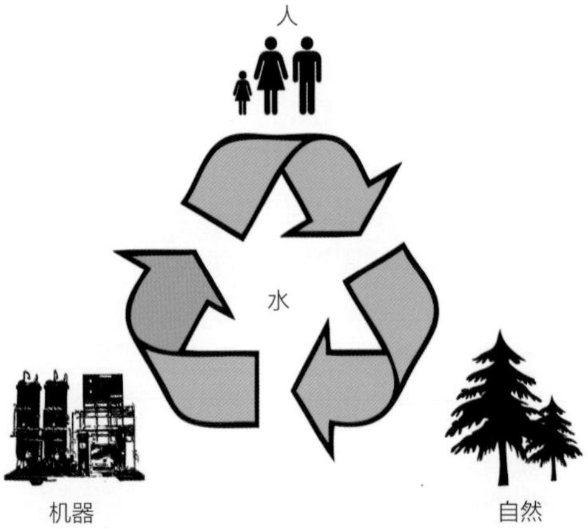

图 1.6-1 豪宅被四等分后就地反转，让庭院居于中心
The McMansion is quartered and inverted on its site. The yard now occupies the centre

图 1.6-2

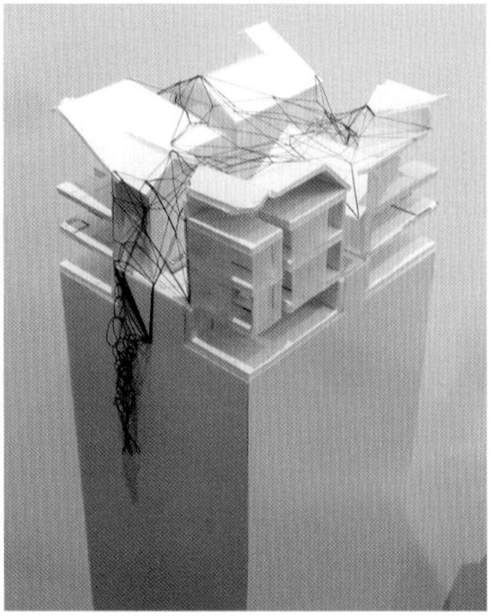

图 1.6-3

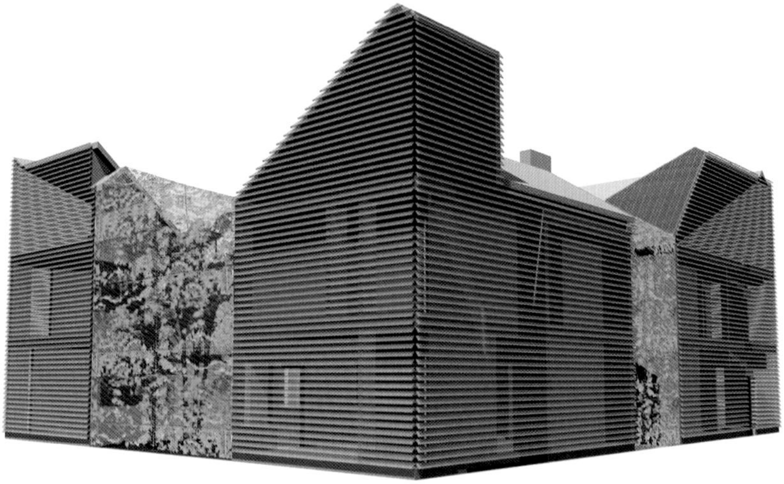

图 1.6-4

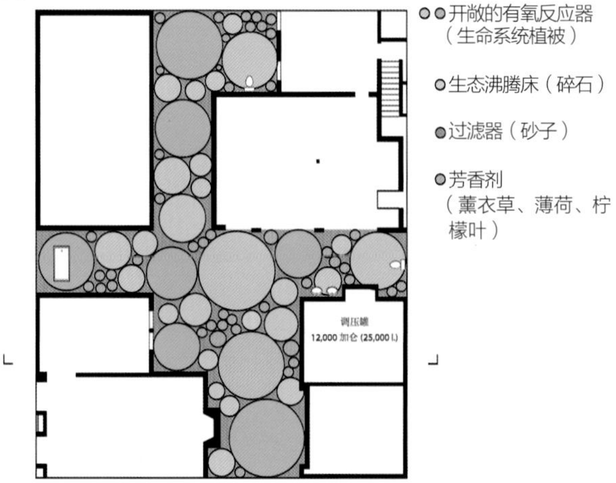

图 1.6-5

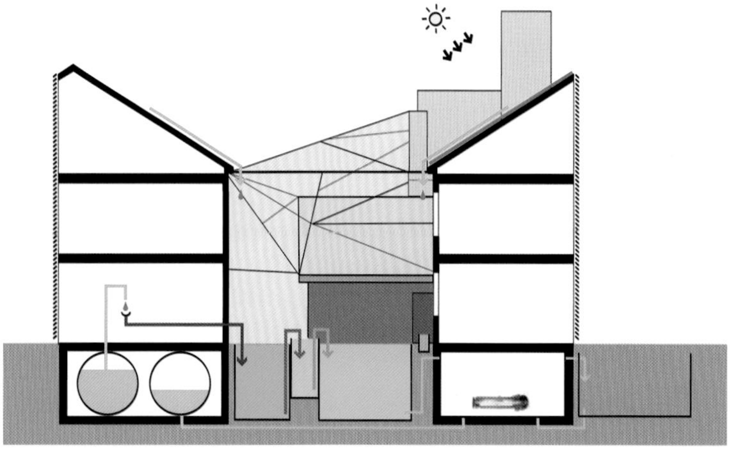

图 1.6-2 室内花园被一个能够阻止水分蒸发的不透水膜构成的网所覆盖，水在室内表面凝结，滴落回花园
The interior garden is covered by the net, an impermeable membrane that prevents water evaporation. Water condenses on the interior surface, dripping back into the garden

图 1.6-3 通过对住宅的重新配置与反转，豪宅形态变得更现代
Through its reconfiguration/ inversion, the McMansion's form evolves towards a more contemporary state

图 1.6-4 庭院由一系列的水箱组成，这些水箱构成了生命系统机器
The yard is composed of a series of tanks that make up the living system machine

图 1.6-5 剖面展示了水的循环利用过程：从地下室储水箱到建筑潮湿区域（厨房与浴室），进入生活过滤系统，并通过紫外线过滤器再回到地下室水箱中
The section shows the cycle of water usage: from the basement storage tank to the wet zones (kitchen and bathrooms), into the living filtration system, through a UV filter and back into the tank

图 1.6-6

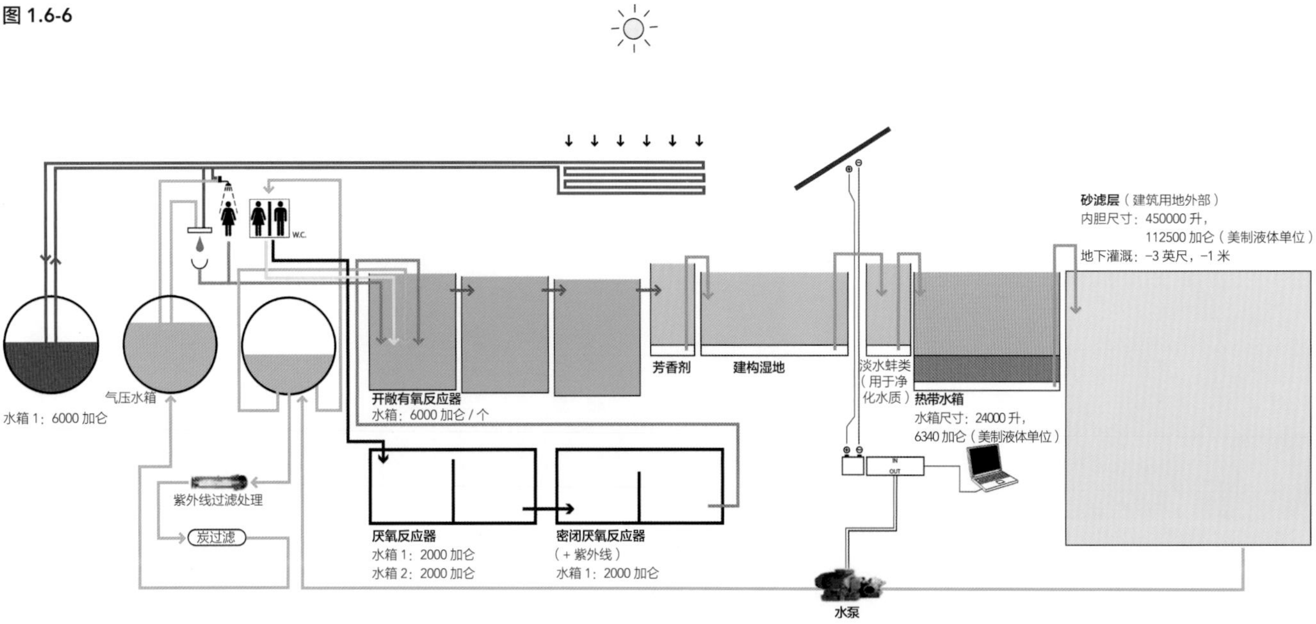

图 1.6-7

图 1.6-8

图 1.6-6 水过滤循环图
Diagram of water filtration loop

图 1.6-7 新纳入的庭院与草坪成为一个能够净化中水与废水再利用的郁郁葱葱的生活过滤系统，并为居住者提供一个伊甸园般的花园，花园由一个厕所维持
The newly ingested yard / lawn becomes a lush，living filtration system，cleaning grey and black water for reuse, while offering an Eden-like garden for its inhabitants. Eden is sustained by the toilet

图 1.6-8 从地下室进入内部庭院下的处理箱看到的景观，水箱顶部是可持续设备的副产品
View from basement into treatment tanks below interior yard. The sublime is a byproduct of the sustainable machinery

1.7 Rgb 花园 RGB garden

花园是由 850 个 6 英寸 ×6 英寸 ×6 英寸的藻类光生物反应器组成，它们是一些半机械半活体的植物机械人。

NMinusOne 工作室，纳希德 · 纳比安

The garden consists of 850, 6″ x 6″ x 6″ algae photo bioreactors. Half-machine, half-alive, these are plant cyborgs.

Studio NMinusOne with Nashid Nabian

反应器收集二氧化碳的效率远远超过一棵树的效率，每个反应装置恰好能用地热、风能或太阳能收集技术来收集能源。

白天，能量视天气状况从不同的渠道采集与储存；夜晚，能量以人工光源的形式释放，将藻类的生长周期延长到超过一天的日照时长，藻类生长在人工加速与增强下完全是超出自然生长速率的状态。光照模式反映出能量来源：绿色 = 地球（湿度）、蓝色 = 风、红色 = 太阳（火），花园成为一个记录天气情况的可视化地图。在法国梅蒂斯国际花园展期间，花园将不断地变化，藻类每天以不同的速度生长，藻类密度与色彩深度的变化给予人们每天全新的体验。

The reactors sequester carbon dioxide with an efficiency far exceeding that of a tree. Each reactor opportunistically gathers energy using ground, wind or solar harvesting technologies.

During the day, energy is captured and stored from these various sources depending on weather conditions. During the night, it is released as artificial light, extending the algae's growth cycle past the day's sunlight hours. The algae's growth is artificially accelerated / augmented beyond its natural rate. The pattern of illumination reflects the energy source it was powered by, Green = Earth (moisture), Blue = Wind and Red = Sun (Fire). The garden becomes a map, a visual recording of the day's weather conditions. During the course of the International Garden Festival 'Jardins De Métis', the garden will be in constant flux; the algae will visibly grow at different rates daily, changing in density and depth of colour to offer a new experience every day.

图 1.7-1

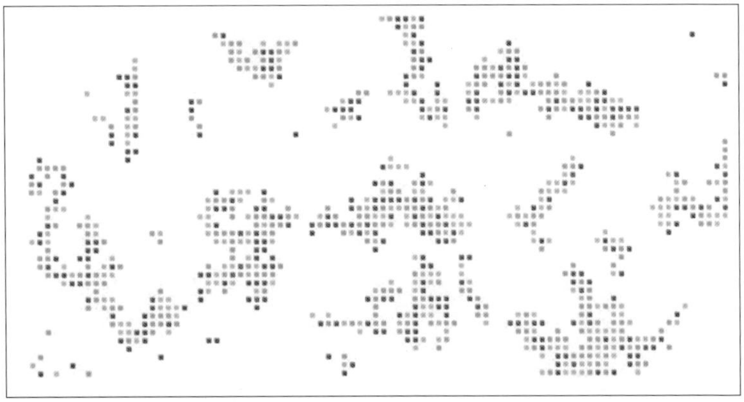

图 1.7-2

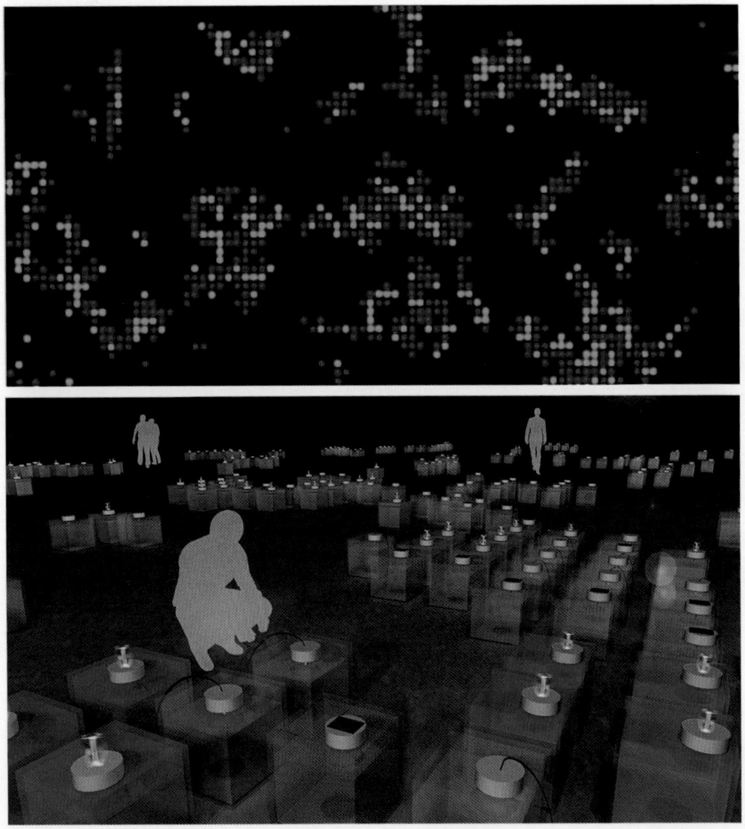

图 1.7-1 白天设计图——藻类生长取决于环境因素，如风、阳光和雨，该设计成为一个长期记录以往天气变化情况的记录

Plan Diagram Day Condition-Algae growth depends on environmental factors: wind, sun and rain. The plan becomes a record of past weather-long term

图 1.7-2 夜晚设计图——夜晚将 LED 灯打开，并用电荷收集来自太阳、风与雨的能量，光照配置能够短期显示日间天气模式

Plan Diagram Night Condition-At night the LEDs turn on and use the electrical charge gleaned from solar, wind and rain energy. The light configuration becomes a display of daytime weather patterns-short term

图 1.7-3

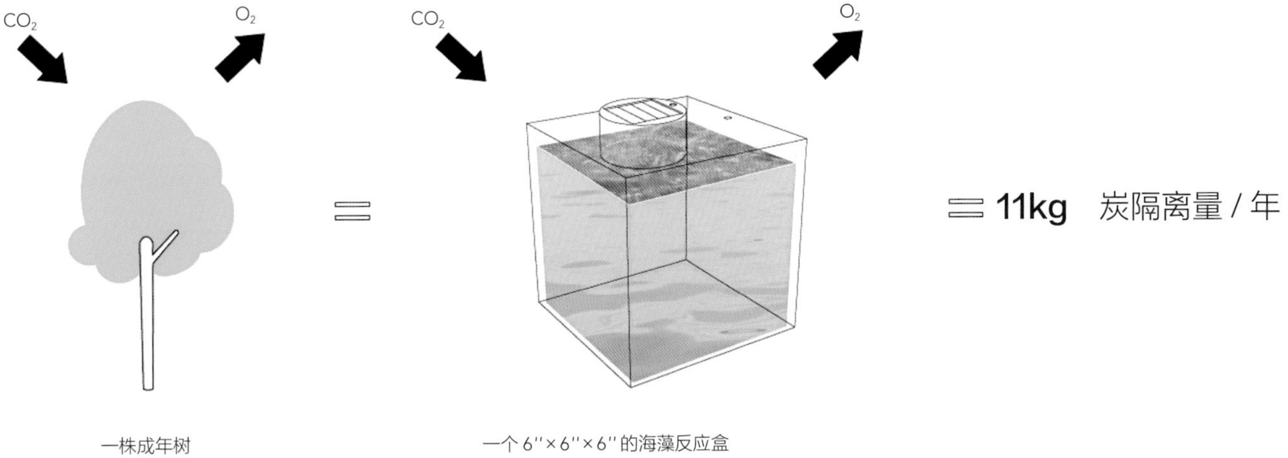

图 1.7-3 一箱处于最佳生长状态下的藻类收集二氧化碳的能力超过一株成年树的能力
One box of algae growing at optimal levels can sequester as much CO_2 as one mature tree

图 1.7-4

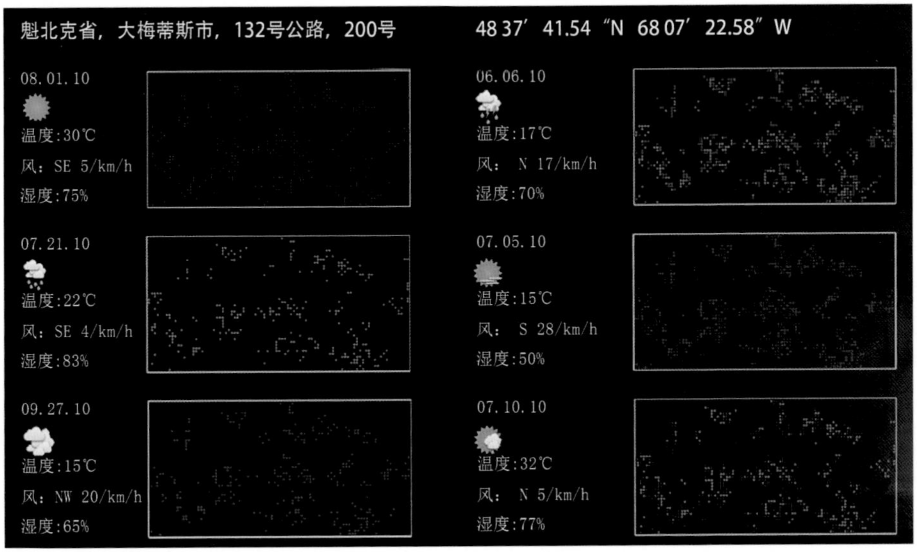

图 1.7-4 花园成为一个记录天气情况的可视地图
The garden becomes a map, a visual recording of the day's weather

图 1.7-5

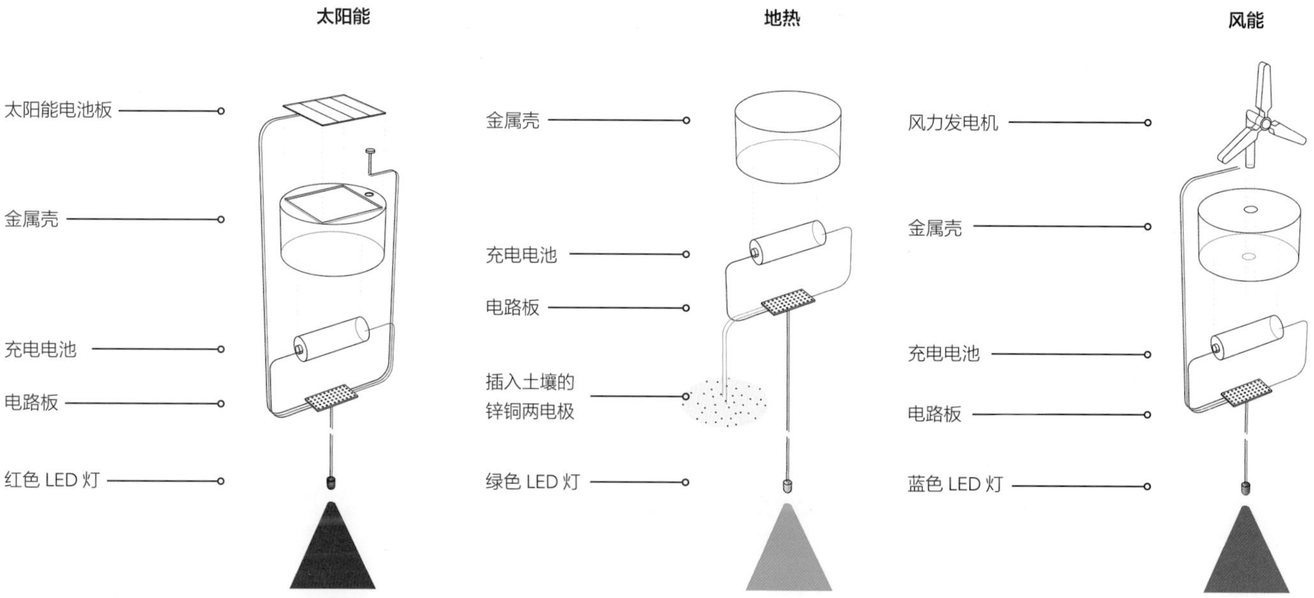

图 1.7-6

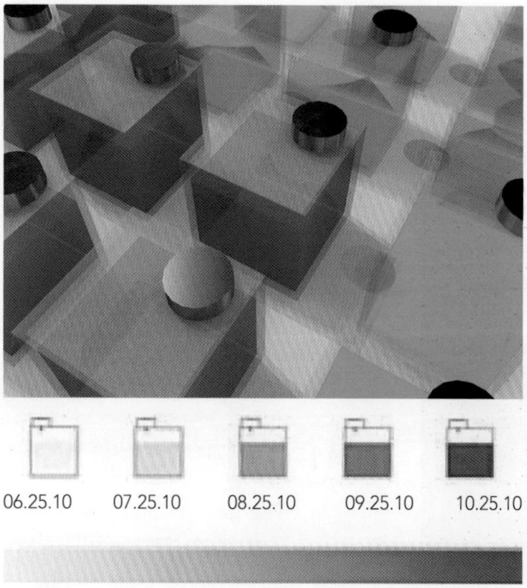

图 1.7-5 三种不同的能源：地热、风能与太阳能，分别来自地球、风与火
Three different energy sources are harvested. Ground, wind and solar: earth, wind and fire

图 1.7-6 藻类的生长速度会随着天气的变化而变化。在很长一段时间内，优势的能量来源将通过最密集的藻类生长显现出来
Algae growth rates will vary depending on the weather. Over an extended period of time the predominant source of energy will become visible though the densest algae growth

传统意义上建筑的形态与情境空间容量是通过技术上提升物理组件而拓展的，空间不仅由墙体和屋顶等材料元素创造，还利用非物质效能来实现。对气候、空气和氛围的精确技术控制创造了感觉上全封闭的环境，重新定义了我们对世界的体验。这些项目探索用分子、生理与心理的因素来创造具有高度调节性的环境，它们被精准地调节来响应使用者的需求，并与环境的动态变化相协调，这些都为更多地从社会学与生态学的角度研究建筑学做出了有益的探索。

Architecture's traditional capacity to shape and condition space is expanded through technologically enhanced physical components. Spaces are created not only with material elements such as walls and roofs, but also by means of non-substantial effects. The precise technological manipulation of climate, atmosphere, and ambience generates sensorial, all-enveloping environments that redefine our experience of the world. These projects explore molecular, physical and psychological elements to generate highly modulated atmospheres. They are finely tuned to respond to the user and resonate with the dynamic fluctuations of the environment, paving the way for a more socially and ecologically engaged architecture.

2

沉浸式空间

IMMERSIVE SPACES

2.1 天空之宅 Sky House

天空之宅实现了最古老的建筑理想，即使身体飘浮在空中。

MinusOne 工作室

The Sky House achieves one of the oldest ambitions of architecture... To suspend the body in space.

Studio MinusOne

The architectural components used to achieve this are also the essential ingredients for a low-carbon footprint house in arid regions.

Sustainability in the service of pleasure. The house is half house, half pool. The interior temperature of the house is regulated by the pool water. The exterior wall of the pool forms the interior walls of the house. The wall works as a heat exchange device. To maximise exchange it is shaped as a series of stacked, continuous, cooling ribs, assuming a form for maximum surface to air exchange. The filtration system of the pool runs the water through a tank of Phase Change Materials (PCM) encapsulated in 3' diameter plastic capsules. During the day the PCM absorb the solar energy embedded in the water from the sun. At night the PCM release this energy back into the water and subsequently back into the house.

用于实现这一目标的建筑构件也是干旱气候区的低碳足迹住宅的基本构成部分。

为了造福于人类的可持续性，住宅一半是房子，一半是游泳池，住宅的室内温度由池水进行调节，游泳池的外墙就是住宅的内墙，墙体起到热交换器的作用。为了实现墙体最大限度的热交换，墙体形状被设计成一系列层叠的、连续的、具有制冷效能的肋板构件，以实现空气对流的最大表面积。泳池的过滤系统让水流经一个装有相变材料（PCM）的水箱，相变材料被封装在直径为 3 英寸（1 英寸为 2.54 厘米）的塑料胶囊中。白天，相变材料吸收水中的太阳热能；夜晚，相变材料将热量释放回水中，再释放到住宅室内。

图 2.1-1

图 2.1-1 住宅屋顶被 1 英寸的水所覆盖，使得游泳池、热水浴缸与屋顶形成一个单向反射面，天空与它的倒影在水中融合得天衣无缝，在屋顶上行走的人仿佛飘浮在空中
The roof is covered by 1″ of water, making the pool, hot tub and roof into a single reflective surface. The sky and its reflection in the water merge seamlessly. People walking on the roof deck appear to be floating in space

图 2.1-2

图 2.1-3

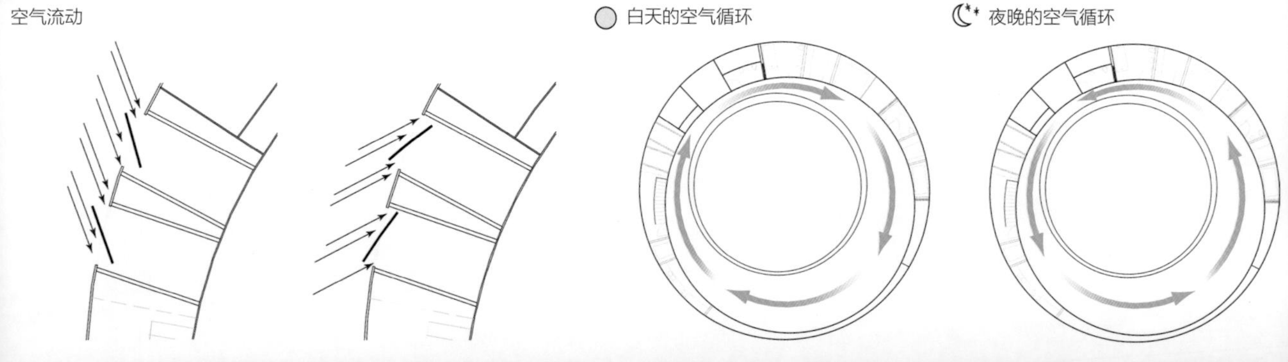

图 2.1-2 住宅顶面被刷上黑色以提高水面的反射率，雾化器和超声波雾化器的组合消除了地平线
The roof surface is painted black to increase the reflectivity of the water surface. A combination of mist makers and ultrasonic atomisers eliminate the horizon line

图 2.1-3 白天，室外炎热的环境与室内凉爽的游泳池之间形成温差，从而在起居空间中产生顺时针方向的热动力空气循环；夜晚，室外寒冷的环境与室内温暖的游泳池之间形成温差，产生逆时针方向的空气循环模式。空气循环保证了空间夜间温暖，白天凉爽
During the day, the temperature difference between the hot exterior climate and the cool pool water generates clockwise thermo-dynamic air circulation within the living space. At night the temperature difference between the cold exterior and the warm pool water generates a counter-clockwise air circulation pattern. The air circulation keeps the space warm at night and cool during the day

图 2.1-4 泳池水与相变材料的循环图
Circulation diagram of pool water and phase change materials

图 2.1-5 游泳池周围的肋板充当热交换器，以调节住宅室内温度
The ribs around the central pool act as heat exchangers, regulating the temperature of the house

图 2.1-4

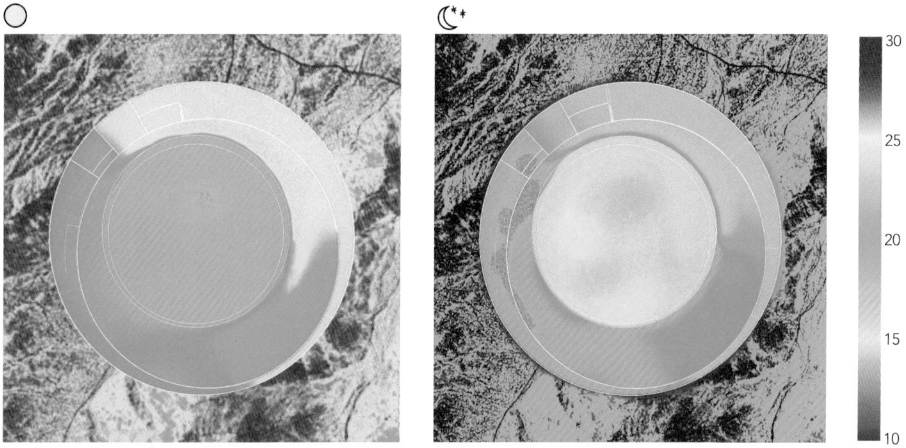

图 2.1-5

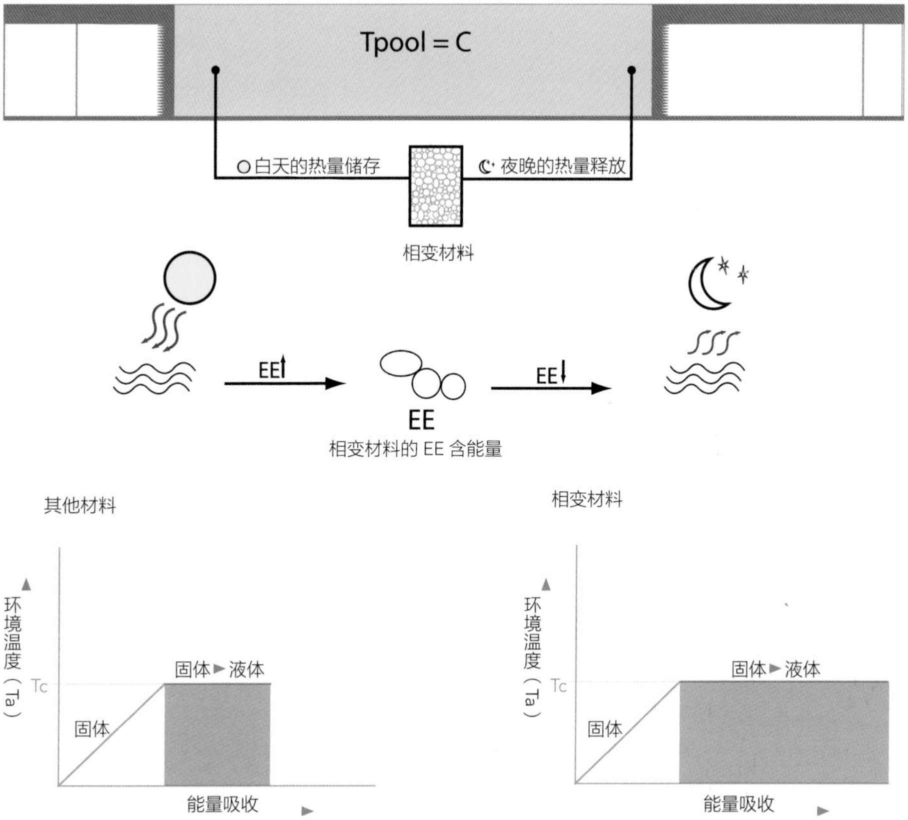

图 2.1-6

图 2.1-6 从起居厅向泳池底部看去
Bottom of the hot tub as seen from the living room

2.2 全视之屋 All Seeing House

建筑砖块模仿昆虫的复眼结构，并嵌入传感器技术来创造一种具有设计感的透明性。[1]

贾斯汀 · 陈

Building bricks that mimic the compound eye, embedded with sensor technology to create an engineered transparency.[1]

Justin Cheng

在“全视之屋”中，独立的砌体单元都装备了一系列最新的嵌入式技术，能够检测 γ 射线、紫外线辐射、热流量、湿度和二氧化碳水平。

全视域砖是模仿昆虫的复眼结构，每块砖都面向略微不同的方向，从而将圆顶的视觉范围投射到一个盒子上，砖块将获取的信息传输到连接RGB LED灯[2]的内表面，创造了一个对外部世界而言具有设计感的透明性。

建筑方案从概念上被分为三部分空间：外部

In the ‘All Seeing House’, individual brick units are equipped with an array of the latest sensor technology, capable of detecting gamma radiation, ultraviolet radiation, heat flux, moisture levels, and carbon dioxide levels.

The All Seeing Bricks mimic the compound eyes of insects, each facing in a slightly different direction, thereby projecting the visual range of a dome onto a box. The bricks transmit the information to an interior surface of correlated RGB LEDs, producing an engineered transparency to the outside world.

The house is conceptualised as a scheme divided into three spaces: the outside world, an interior void that produces the immersive,

❶ 由克里斯托 · 马尔科普洛斯与卡罗尔 · 穆海贝尔指导。
❷ 译者注：RGB LED 灯是以红、黄、蓝三原色共同交集成像。正是由于 RGB 的特性，在光的混色上，具备更多元的特性，LED 背光板就像画家的调色盘一样，完美呈现最真实的彩色世界。

❶ Guided by Christos Marcopoulos and Carol Moukheiber .

heightened experience of the outside world, and the space between the two, where the residential programmatic requirements are packed tightly alongside - if not relegated to the status of - the mechanical infrastructure of the house. Void tubes that connect the sensor skin to the cosmic display not only allow circulation and entry of natural light, but also express the omnidirectional vision of the house.

世界；一个能够产生身临其境的感觉，提升对外部世界体验的室内空间；介于以上两者的空间。如果不降低现状需求的话，该项目的居住需求被压缩并与住宅的基础设施并置。连接传感器表面与外部世界的中空管不但让自然光线流通和进入，而且也展现了建筑全方位的视觉景象。

图 2.2-1

图 2.2-2

图 2.2-3

图 2.2-4

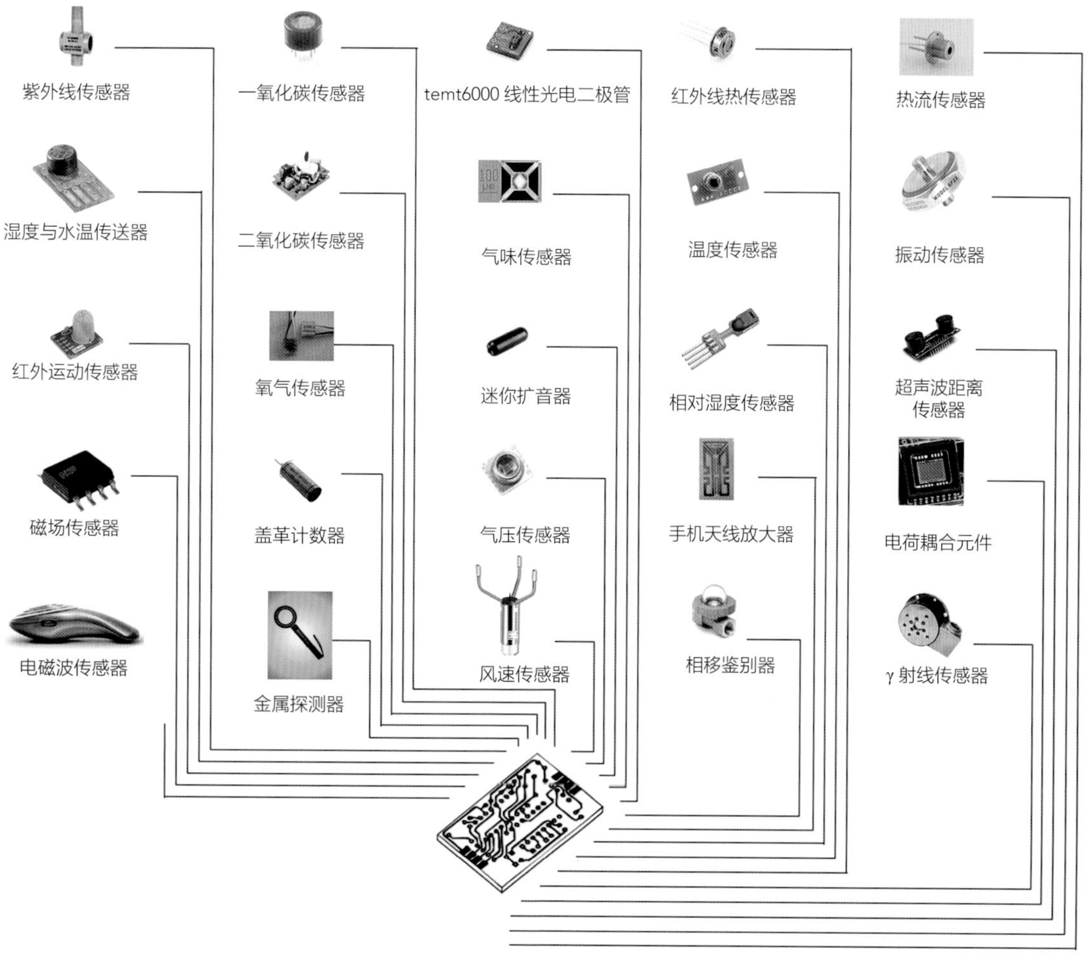

图 2.2-1 图中显示的建筑的主要起居空间就像沐浴在一个不断变化的色彩丰富的万花筒中，这些色彩能够根据外部世界发生的自然现象进行编码
The main living space is bathed in a continuously changing kaleidoscope of colour which is coded according to the natural phenomena that occur outside

图 2.2-2 剖面模型展示室内起居厅
The sectional model reveals the interior living chamber

图 2.2-3 剖面展示了内部大厅与外部墙体之间的关系
Sections exposing the relationship between the inner chamber and the outer walls

图 2.2-4 大量传感器被整合到“砖”内
A multiplicity of sensors are incorporated into the ‘brick’

图 2.2-5

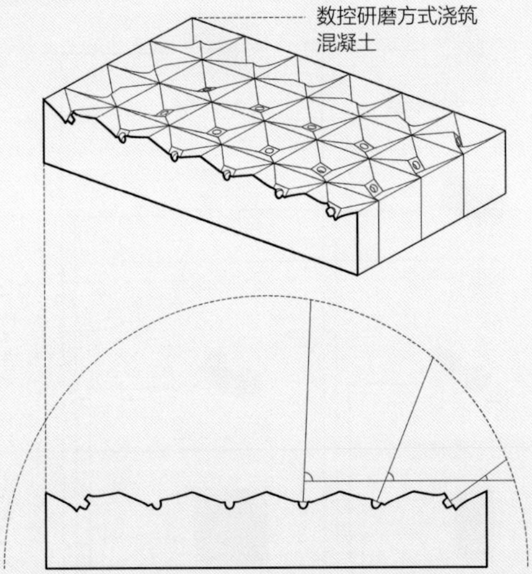

图 2.2-6

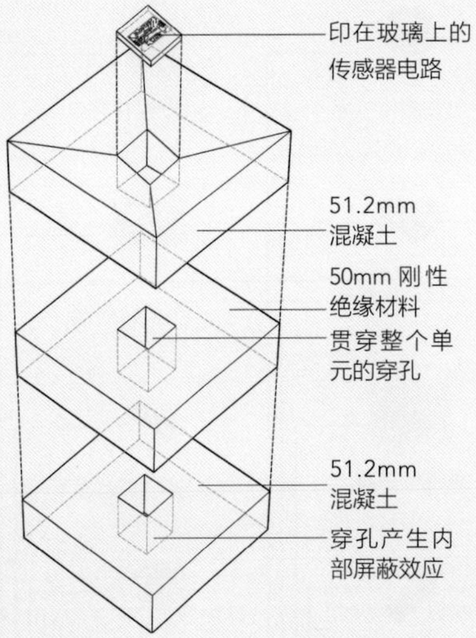

图 2.2-7

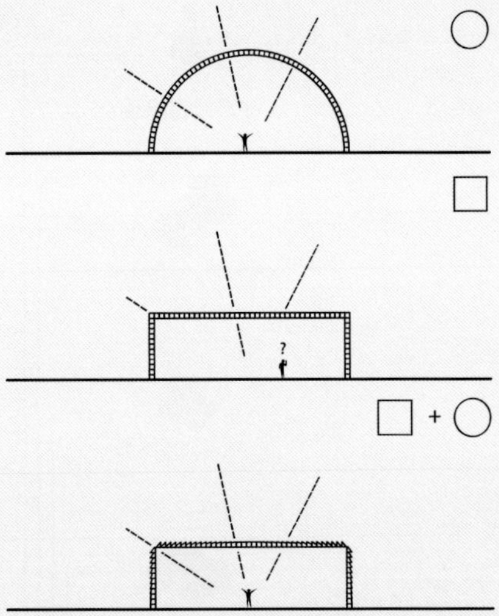

图 2.2-8

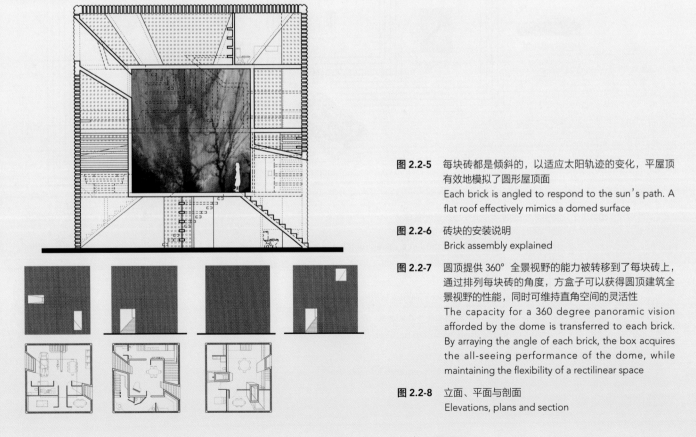

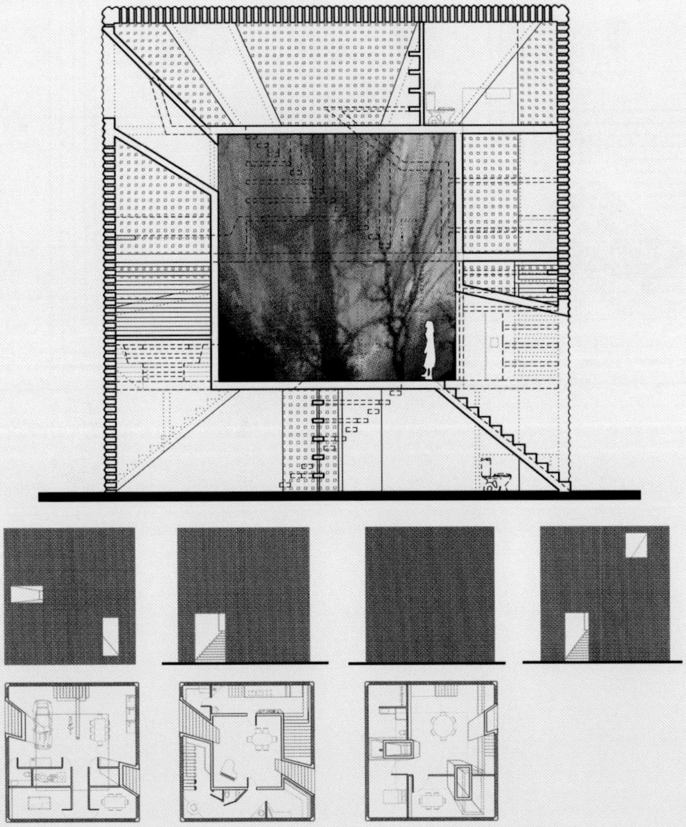

图 2.2-5 每块砖都是倾斜的，以适应太阳轨迹的变化，平屋顶有效地模拟了圆形屋顶面
Each brick is angled to respond to the sun's path. A flat roof effectively mimics a domed surface

图 2.2-6 砖块的安装说明
Brick assembly explained

图 2.2-7 圆顶提供 360° 全景视野的能力被转移到了每块砖上，通过排列每块砖的角度，方盒子可以获得圆顶建筑全景视野的性能，同时可维持直角空间的灵活性
The capacity for a 360 degree panoramic vision afforded by the dome is transferred to each brick. By arraying the angle of each brick, the box acquires the all-seeing performance of the dome, while maintaining the flexibility of a rectilinear space

图 2.2-8 立面、平面与剖面
Elevations, plans and section

图 2.2-9

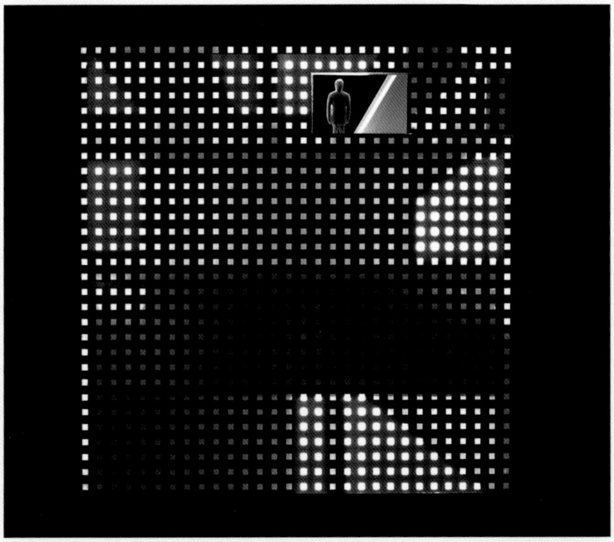

图 2.2-10

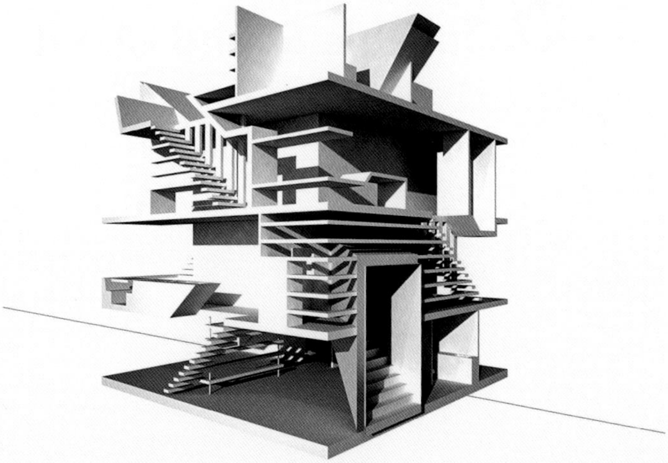

图 2.2-11

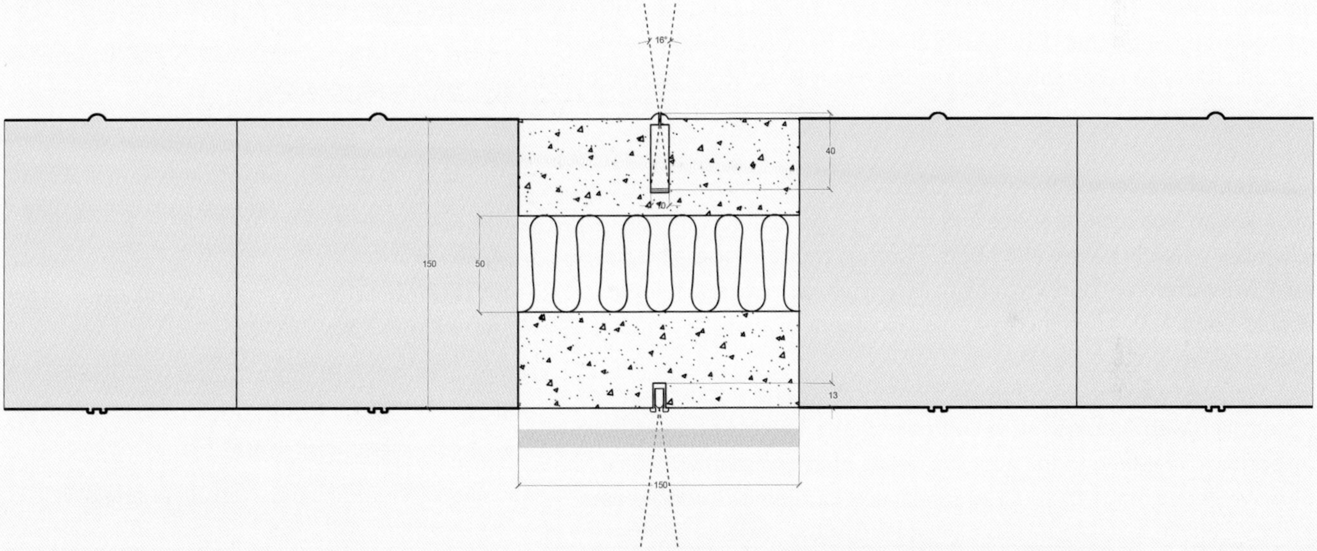

图 2.2-9 立面
Elevation

图 2.2-10 立面、平面与剖面
Elevations, plans and section

图 2.2-11 墙体剖面
Wall section

2.3 可调声云 Tunable Sound Cloud

增强听觉体验的声学衬垫。[1]

曼妮 · 曼妮

An acoustic liner designed to enhance one's auditory experience.[1]

Mani Mani

The 'Tunable Sound Cloud' (TSC) is an acoustic liner designed to enhance auditory experience within a generic space. Comprised of triangulated panels that dynamically change shape to adjust to various acoustic needs, it is capable of simulating highly specific auditory spatial configurations.

Up until the early 20th century, there was an intimate relationship between an architectural space and the music generated within it (chamber music in palace chambers, Baroque music in opera houses, etc.). Contemporary performance halls typically achieve flexibility in acoustic effects through the strategic placement of acoustic panels.

Actuated with a matrix of shape memory alloys and stepper motors - controlled with Arduino micro-controllers - the Tunable Sound Cloud responds in real-time to the acoustic needs of its environment. It can thus simulate a range of auditory spatial configurations. Auditory data

“可调声云”是一种声学衬垫，旨在增强通用空间中人的听觉体验，它由能够动态改变形状的三角形面板构成，以适应各种声学需求，还能够模拟特定的听觉空间配置。

直到 20 世纪初，建筑空间与其所产生的音乐（宫殿式的室内乐、歌剧院的巴洛克音乐等）之间有着密切的关系，当代表演大厅通常是通过设置吸声板的对策实现声学效果的灵活性。

可调声云是由形状记忆合金与步进式电动机矩阵排布来驱动，它们由 Arduino[2] 微控制器控制，以实时回应环境的声学需要。因此，它能够模拟一系列的听觉空间配置。通过扩音器传输听觉数

---

❶ Supervised by Christos Marcopoulos and Carol Moukheiber.

❶ 由克里斯托 · 马尔科普洛斯与卡罗尔 · 穆海贝尔监制。

❷ 译者注：Arduino 是一款便捷灵活、方便上手的开源电子原型平台。

据，并通过 MAX / MSP 软件进行分析，MAX / MSP 软件可以根据不同的场景调整建筑表皮，以进行反射或者吸收声音，建筑表皮又成为控制建筑声学效果的另一手段。

is streamed through microphones and analysed through MAX / MSP software which adjusts the skin surface for sound reflection/ absorption in different scenarios. The architectural surface becomes yet another instrument for the deployment of an acoustic effect.

图 2.3-1

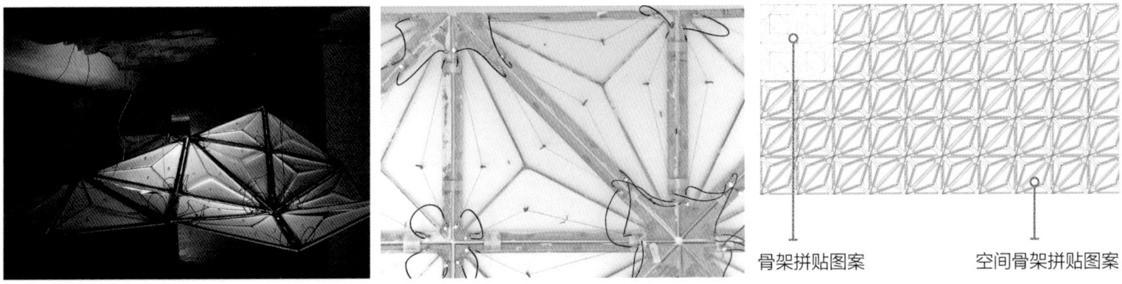

图 2.3-2

图 2.3-1 原材料由椴木、聚酯薄膜、电缆、强力线缆或形状记忆合金组成
The prototype materials consist of basswood, mylar, cables, and muscle wire or shape memory alloy

图 2.3-2 云应用程序能手动调控定制空间矩，或通过编程让软件通过一套反馈机制来自动校准系统，可调声云的模块化设计允许大规模地推广与实现现场定点安装
The Cloud application can be manually adjusted for customised spatial moments, or programmed to allow the software to autonomously calibrate the system through a series of feedback loops. TSC's modular design allows for large scale expansions and site specific installations

2.4 微风穿过的墙 Breeze-Thru Wall

可渗透墙体模块是一种密闭与保温建筑的替代品。❶

亚瑟 · 曾

These permeable wall modules are an alternative to the sealed and insulated house.❶

Arthur Tseng

The 'Breeze-Thru Wall' explores the potential for ventilation and heat recovery to be integrated within a performative building module.

The wall consists of modules acting as air intake units which lock together for structural integrity. The modules are part of an overall envelope system of exchanging and recovering indoor heat. Each module is a funnel designed to capture incoming air efficiently. The air is then held within copper tubing for heat exchange before entering the interior wall. The cycle of hot and cool air from exhaust and intake provides constant ventilation and recovery.

The heat recovery ventilator unit, designed at the scale of the house, is run in response to exterior wind conditions. The module, along with the heat recovery cycle, can be reversed as the prevailing wind changes; the two exterior funnel-shaped faces of the module act as both intake and exhaust openings. The heat recovery cycle uses forced air, which runs on a loop around the envelope of the house, carrying the necessary heat and cold to be exchanged with each module.

❶ Guided by Christos Marcopoulos and Carol Moukheiber.

"微风穿过的墙"在努力探寻将通风与热回收整合在一个性能良好的建筑模块中的潜力。

墙体由作为进气单元的模块组成，这些模块咬合在一起以保证结构的完整性。该模块是整体围护结构室内热交换与回收系统的一部分，每个模块都被设计成能够提高进气效率的漏斗形状，空气在进入室内墙体前，先在铜管内进行热交换，来自排气口与进气口的冷热空气循环提供了持续不断的通风与热交换。

运行以建筑尺度设计的热通风模块以应对外部风环境，随着热交换循环，模块能够随着主导风向的改变而改变；模块的两个外部漏斗形的表面既可以作为进气口，也可以是排气口。热回收循环采用强制风，它在建筑的围护结构周围形成风环流，携带必要的热量和冷空气与每个模块进行热交换。

❶ 由克里斯托 · 马尔科普洛斯与卡罗尔 · 穆海贝尔指导。

图 2.4-1

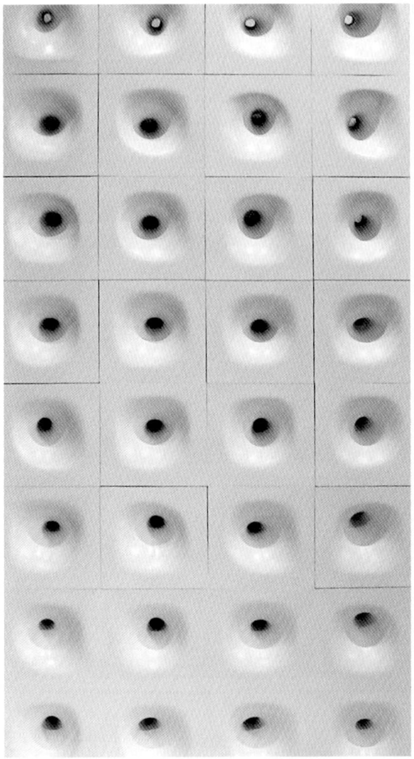

图 2.4-2

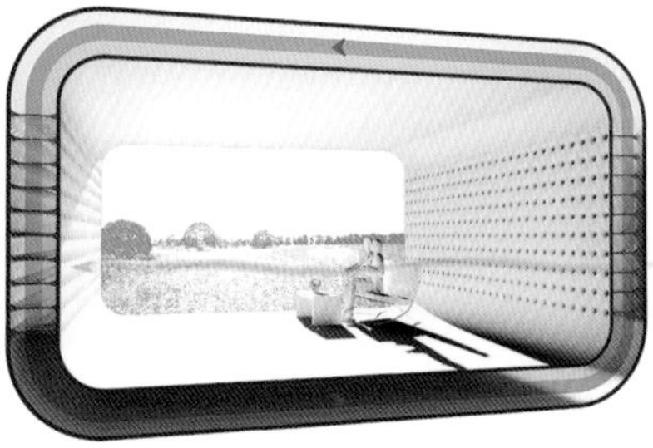

图 2.4-3

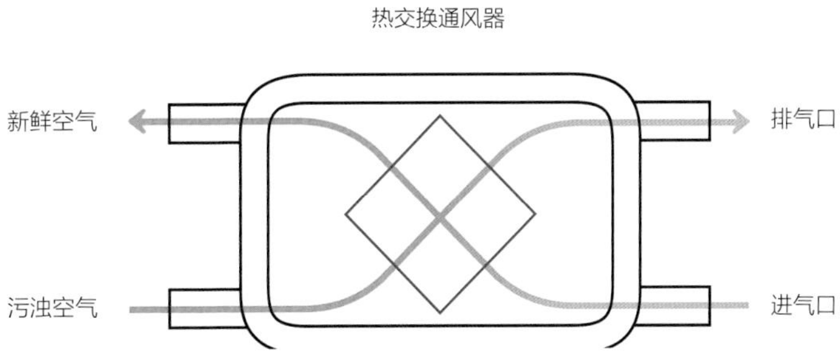

图 2.4-4

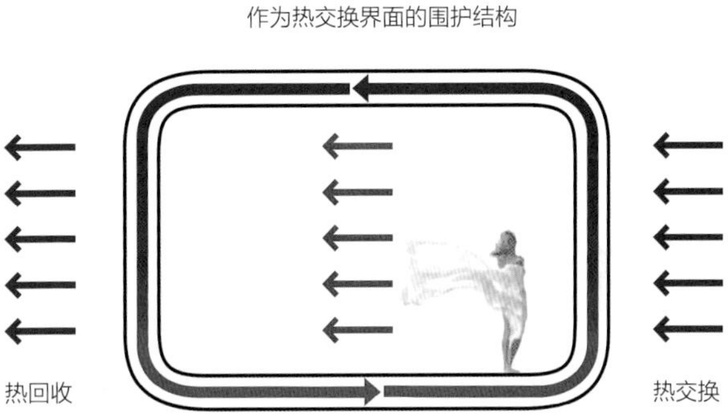

图 2.4-1 墙体立面，咬合的模块
Wall elevation, interlocking modules

图 2.4-2 进入的冷空气经过建筑围护结构变成了温暖的微风
Incoming cool air turns into a warm breeze as it passes through the building envelope

图 2.4-3 传统的热回收通风装置。湿度较高的污浊空气将进入的低温新鲜空气加热，从而提高新鲜空气的通风效率
The traditional set-up of the Heat Recovery Ventilation Unit. Warm, stale exhaust air heats up incoming cooler fresh air, thus making fresh air ventilation more efficient

图 2.4-4 热交换循环：进入的冷空气穿过墙体，在此过程中被墙体内的热空气加热；热回收循环：室内暖空气通过墙体排出，在此过程中加热了墙体内部的空气
Exchange Cycle. Incoming cool air blows through the wall. In the process, it is heated by warm air housed inside the wall // Recovery Cycle: Interior warm air exits through the wall. In the process, it heats up the air housed inside the wall

图 2.4-5

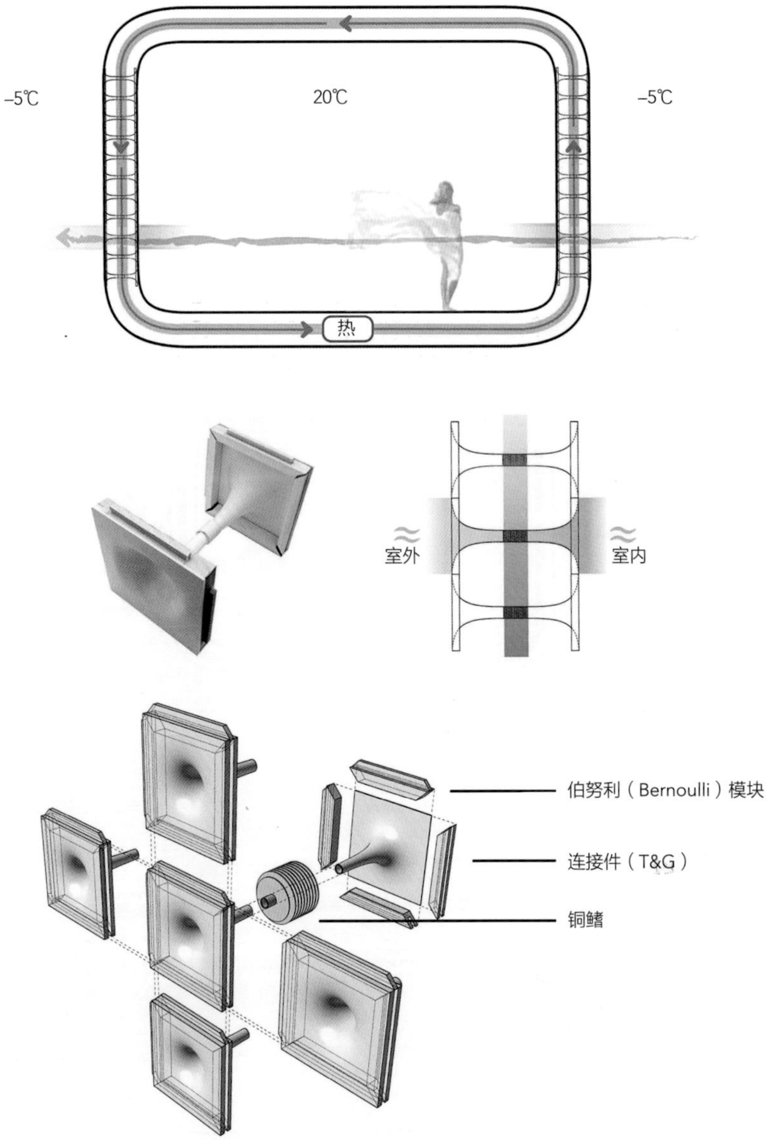

图 2.4-5 模块分解
Anatomy of a module

图 2.4-6

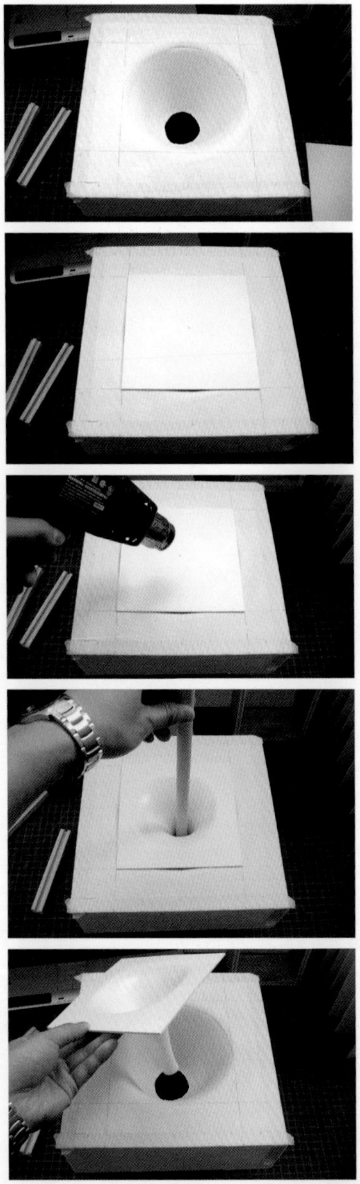

图 2.4-6 采用热塑性塑料制作原型的过程
Prototype fabrication process using thermoplastics

2.5 忙着游戏 Play on the Fly

将身体从日常的平静中释放出来，在公共领域中创造日常“身体游戏”的机会。❶

莱利·萨利曼尼

Releasing the body from its routine composure, daily ‘Body Play’ opportunities are created in the public realm.❶

Lailee Soleimani

“忙着游戏”是一个交互的自动控制环境系统，能够为人们提供在公共领域内“游戏”的机会，基于计算机控制的姿态识别技术，系统可以识别各种有趣的身体姿态，并通过激活经验性的序列来响应身体姿态的变化。

受皮亚杰游戏理论的影响，“忙着游戏”研究了数字增强体系为成年人提供日常身体游戏机会的作用，让“游戏”成为人们日常忙碌生活的一部分，让“无目的的目的”被添加到游戏行为中。该项目地点选择为多伦多的道路系统，即一个连接着多伦多市中心的地下迷宫。该设计在道路廊道上引入了一个交互的环境游乐场网络，游乐场包含了隐藏在城市网络现有项目中的探险，它们集合了一系列的互动游戏。

‘Play on the Fly’ is an interactive, cybernetic, ambient system that offers opportunities for ‘play’ in the public realm. Building on technological advances in computationally administered gesture recognition, the system recognises various playful body gestures and responds to them by activating an experiential sequence.

Influenced by Piaget’s play theory, ‘Play on the Fly’ investigates the role of digitally augmented architectures in creating daily body play opportunities for adults where ‘play’ becomes part of their busy routine life and a ‘purposeless purpose’ is added to the act of playing. The selected site is Toronto’s PATH system, a subterranean labyrinth that connects downtown Toronto. The design introduces a network of interactive ambient playgrounds in the corridors of the PATH. The playgrounds contain adventures hidden within the existing programmes of the urban network. They incorporate a series of interactive games.

---

❶ 由鲁道夫·埃尔-库利监制。

❶ Supervised by Rodolphe el-Khoury.

The proposed ambient game system relies heavily on innovations in computer vision, eliminating the need for interacting with a gaming console by incorporating algorithms that interpret the 'body's gestures as user input. Once a particular gesture is recognised by the system, it activates a certain play scenario that is conveyed to the user via the embedded digital displays.

该设计的环境游戏系统很大程度上依赖于计算机视觉方面的创新，无需与游戏机互动，而是通过整合算法将身体姿态解读为用户输入。一旦特定的姿态被系统识别，它就会激活一个特定的游戏场景，并通过嵌入式的数字显示器传达给用户。

The proposal is a series of activities that are inspired by conventional playground equipment. Each activity is an interactive game where a body movement in a certain direction activates a screen on which the player can navigate through the 3D model of the site or get contextual information by moving his body. The interactive screens are hidden inside the public corridors of PATH system as part of their structure: walls, floors, and ceilings, and only become activated when a person starts playing.

该方案是一系列受传统游戏设备启发而得到的活动，每个活动都是一个互动游戏，在这个游戏中，身体朝某个方向移动就会激活一个屏幕，玩家可以在这个屏幕上浏览网站的 3D 模型，或者通过移动身体获取相关信息。互动屏幕作为结构的一部分，隐藏在道路系统公共廊道的内部，如墙壁、地板、屋顶，只有当有人开始游戏时互动屏幕才会被激活。

The playgrounds are hidden adventures within the existing program of the urban network, and the play activities of each individual become departures from the everyday routine of navigating the corridors to and from work, as anyone can choose to be a player, or an audience member along the way.

游乐场是在城市网络现有程序下隐藏的探险，每个人的游戏活动都脱离了每天穿梭于上下班通道的日常行为，任何人都可以选择成为一名玩家或沿途的观众。

图 2.5-1

图 2.5-1 从地下道路网络看到的景象。当玩家在原地旋转时，一个虫眼大的孔洞浮现在顶棚上，当玩家旋转时洞越变越大。顶棚上浮现的场景可以让人看到室外，通过数字技术将人工照明的室内与室外连接起来

A scene from the underground PATH network. As the user twirls in place, a worm hole appears on the ceiling becoming larger as the player twirls. The scene on the ceiling allows one a glimpse into the exterior, digitally reconnecting the artificially lit interior with the outside

图 2.5-2

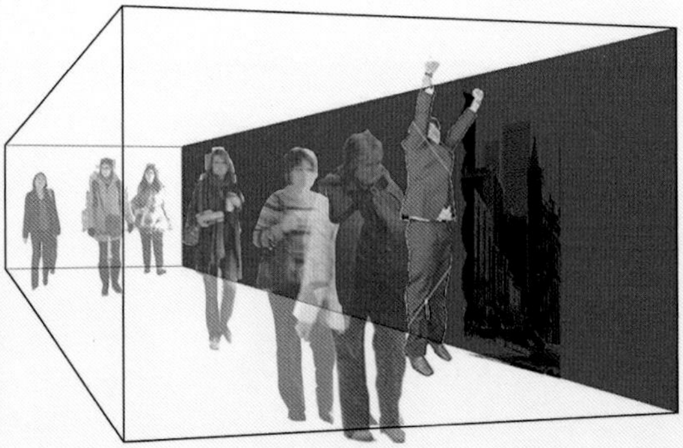

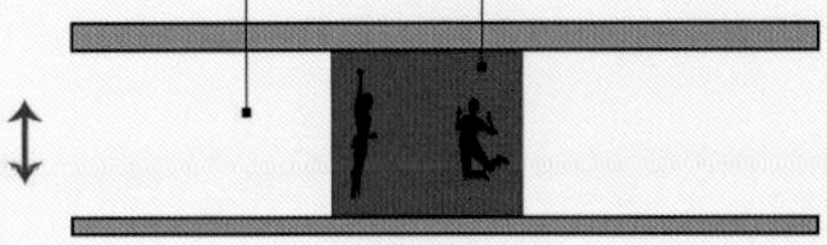

描述：
当玩家开始在墙前面跳跃时，屏幕就会被激活，显示出玩家。随着玩家不断跳跃，屏幕会显示一个更高的视图，就像玩家每次跳跃都增加了自己的高度。增加的高度可以被更高的跳跃和更有力的尝试所操控。玩家可以达到他所想要的高度，但是一旦他停止游戏，视图会落到它的原始位置，并且墙会关闭启动状态。

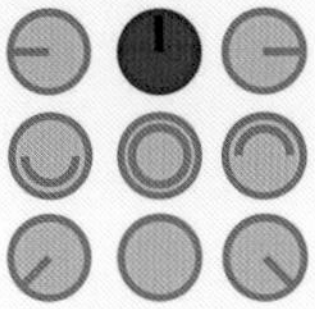

运动 : 跳跃
活性表面 : 墙

图 2.5-2 场景一
Scenario one

图 2.5-3

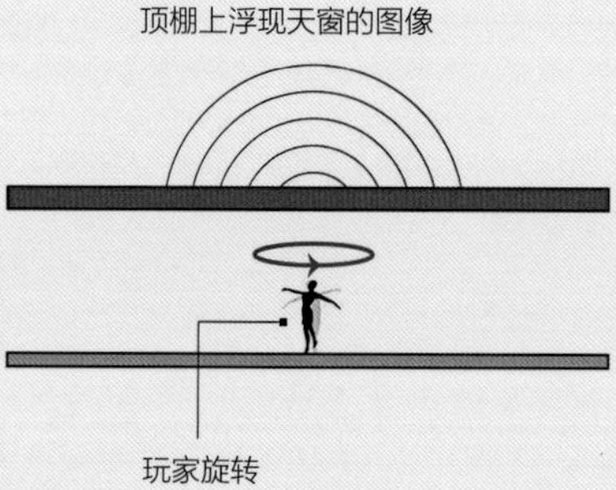

描述：
玩家开始旋转，一个天窗图像出现在上方的顶棚上，形成一个小圆圈。当玩家继续旋转时，圆圈会变大，就像天窗被打开，玩家能够看到上方地面的景象。

运动：旋转
活性表面：顶棚

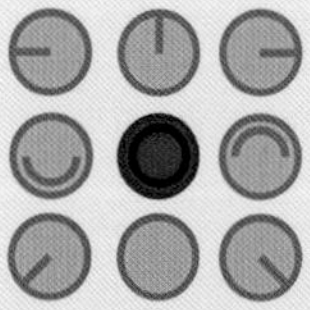

图 2.5-3 场景二
Scenario two

图 2.5-4

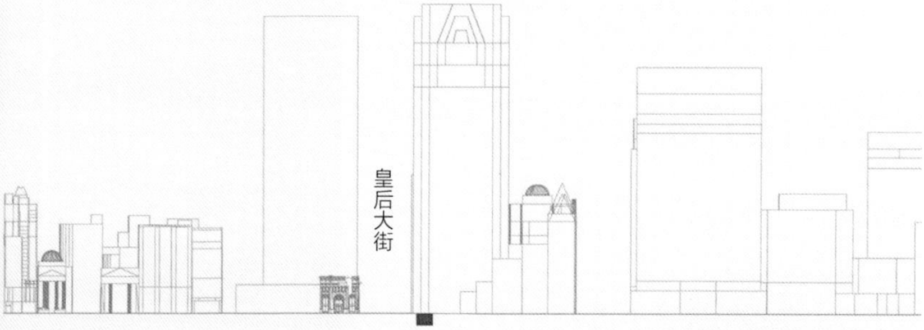

街景：Yonge 街

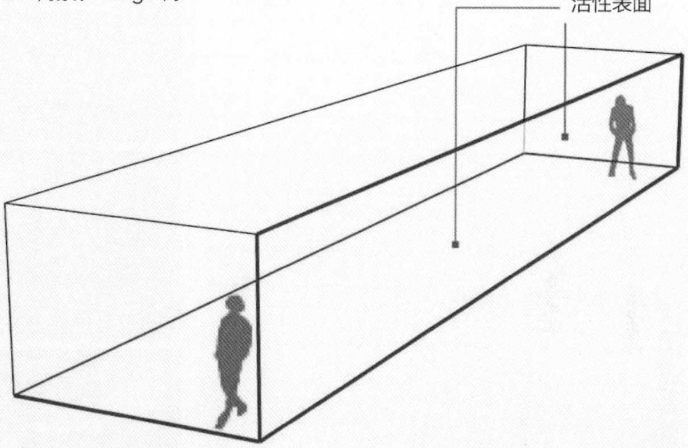

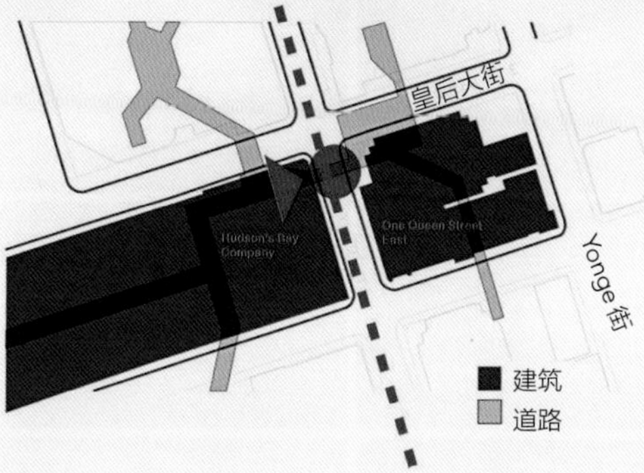

外形尺寸（宽 × 长 × 高）：3.60m×19.80m×2.30m

空间分析：每个方向上的等距移动。低顶棚

直接通道：皇后大街东一、海湾公司

附近主要目的地：伊顿中心、皇后地铁、海湾公司

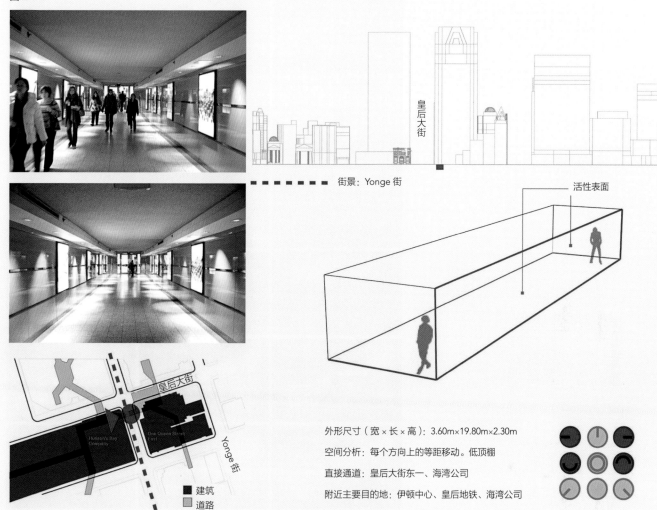

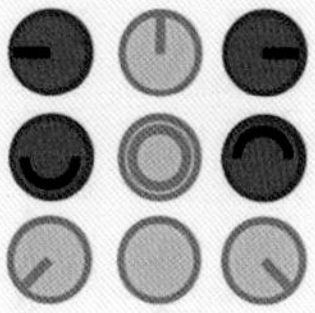

图 2.5-4 运动的显示器位置与用途

Diagram showing location and purpose of active display surfaces

图 2.5-5

图 2.5-5 在道路系统中探索三种不同的场景，这是一个连接多伦多市中心的地下迷宫
Three different scenarios are explored within the PATH system, a subterranean labyrinth which connects downtown Toronto

2.6 看不见的住宅 Invisible House

在看不见的住宅里，照明是一个人空间体验的主要塑造者。

NMinusOne 工作室

In the Invisible House, lighting is the primary shaper of one's spatial experience.

Studio NMinusOne

无形的概念可以用两种不同的方式来探索：一种是极端熟悉度和同质性使事物变得无形，另一种是字面意义上的消失了。

在室内，建筑被系统地从人们的视觉感知中移除，只有那些逃离任何永久性形式的物体才会显现。晚上，客厅是一个黑色的空间，而白天，它变成了一个白色的空间。到了晚上，只有必要的东西才会被照亮。物体被红外光束探测到，并根据物体的存在和位置进行照明。地板和顶棚由 12 英寸 ×12 英寸对应瓷砖组成。顶棚上的瓷砖由 16 个装有红外传感器的 LED 灯组成。

地板的瓷砖上有 16 个红外接收器，与顶棚上的瓷砖传感器相对应，在空间内创造了一个不可见的垂直红外光束场。当一个物体打断这些光束时，顶棚上相应的窄长 LED 射灯就会亮起来。

The notion of the invisible is explored in a couple of different ways: on the one hand the extreme familiarity and homogeneity that renders things invisible and on the other, the literal white out.

On the interior, architecture is systematically removed from one's visual perception, only the objects which escape any form of permanence will appear. During the night, the living room is a black void, and during the day it becomes a white void. At night, only the essential is illuminated. Objects are detected by infrared beams, and illuminated according to their presence and location. The floor and ceiling are composed of 12″ x 12″ corresponding tiles. The ceiling tile contains a grid of 16 LED lights with infrared sensors.

The floor tile has 16 infrared receivers which correspond to the ceiling tile sensors, creating a field of invisible vertical infrared beams within the space. When an object interrupts those beams, the corresponding ceiling LED narrow spotlights turn on.

图 2.6-1

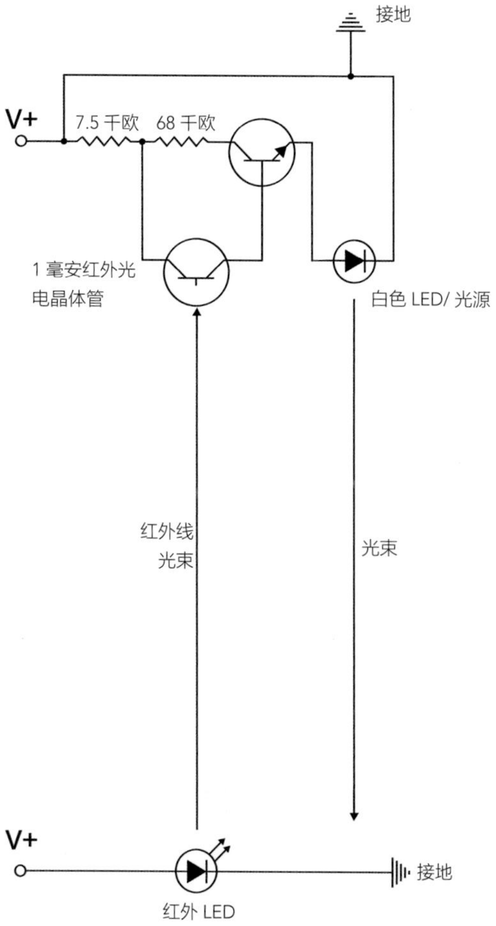

图 2.6-2

图 2.6-3

图 2.6-1 电路原理图
Circuit schematic

图 2.6-2 城市中看不见的住宅，罗迪欧大道，洛杉矶，2007 年
Urban Invisible House. Rodeo Drive, L.A. 2007

图 2.6-3 顶棚和地面的瓷砖动态地照亮一个行走的人
Ceiling and floor tiles dynamically illuminate a person walking

图 2.6-4

图 2.6-5

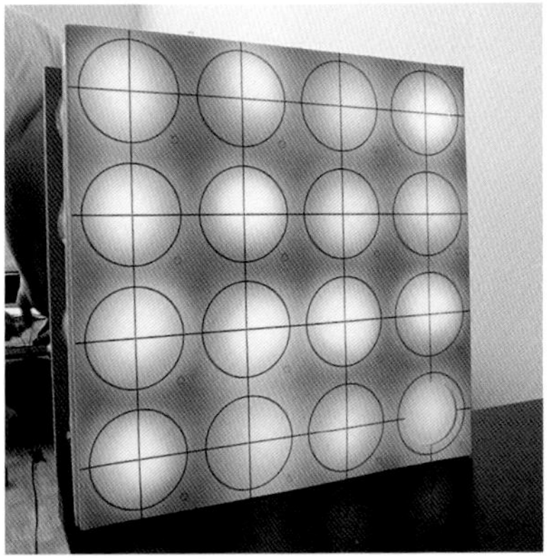

图 2.6-6

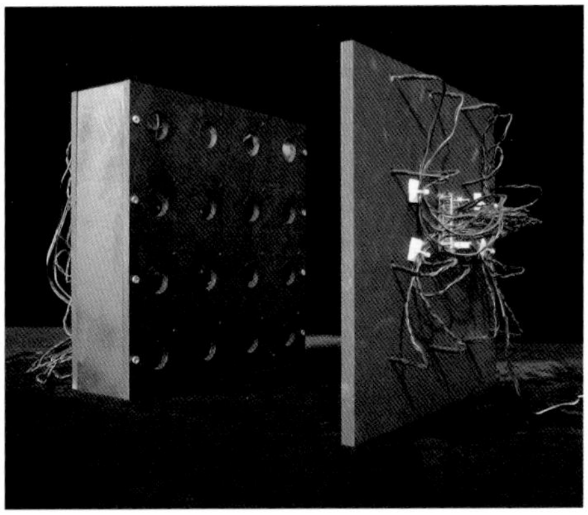

图 2.6-7

图 2.6-4 顶棚瓷砖的第一个原型。使用的材料包括泡沫芯、硬纸板管和一组镜头
The first prototype of the ceiling tile. The materials used included foamcore, cardboard tubing, and an array of lenses

图 2.6-5 地砖的第一个原型。红外传感器与顶棚瓷砖对齐（所有的灯都亮着）
The first prototype of the floor tile. Infrared sensors are aligned with the ceiling tile (all lights are on)

图 2.6-6 地板和顶棚瓷砖的第二个原型。一个坚固的木框架代替了泡沫芯
The second prototype of the floor and ceiling tiles. A rigid wood frame replaced the foamcore

图 2.6-7 通过一种郊区的伪装形式，房子的外观看起来也是隐形的
Through a form of suburban camouflage, the exterior of the house appears as being invisible too

图 2.6-8

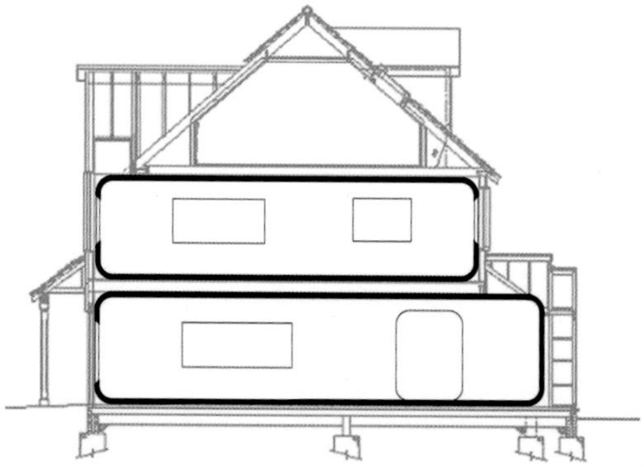

剖面 A–A

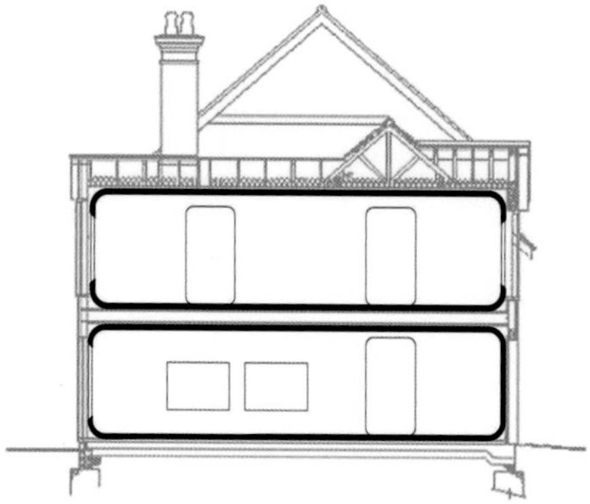

剖面 B–B

图 2.6-9

图 2.6-10

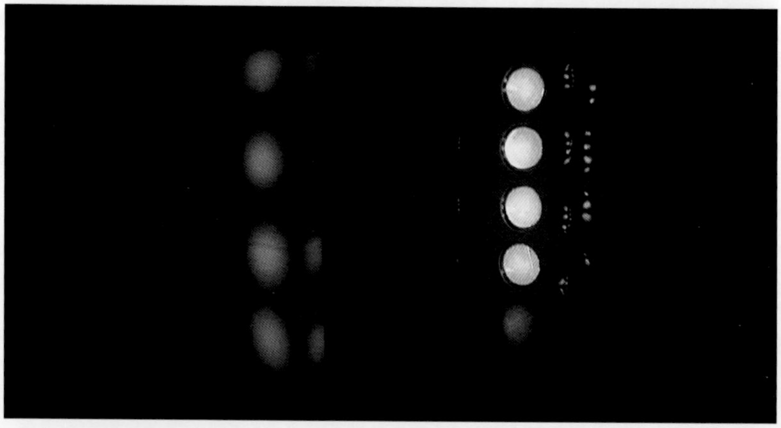

图 2.6-8 一个典型的郊区示范住宅的室内空间被改造成圆角。墙壁、地板和顶棚之间柔和的边缘将室内变成了虚拟的空间
The interior spaces of a typical suburban spec house are reshaped to create rounded corners. The soft edges between wall, floor and ceilings turn the interior into a virtual void

图 2.6-9 顶棚的瓷砖动态地照亮一个人
Ceiling tiles dynamically illuminate a person

图 2.6-10 瓷砖
The tiles

图 2.6-11

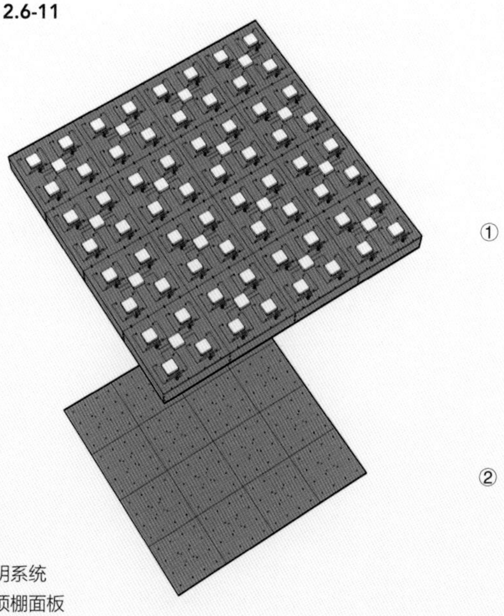

照明系统
①顶棚面板
②地板面板

图 2.6-12

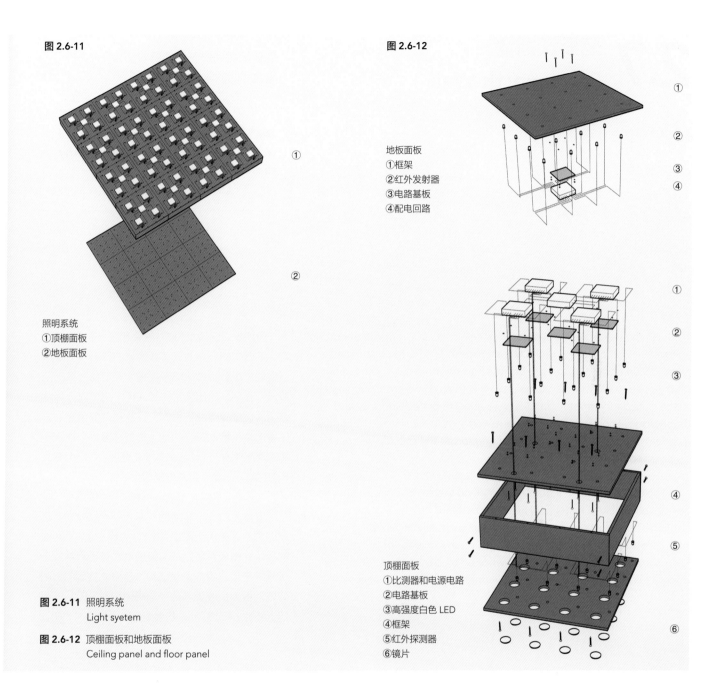

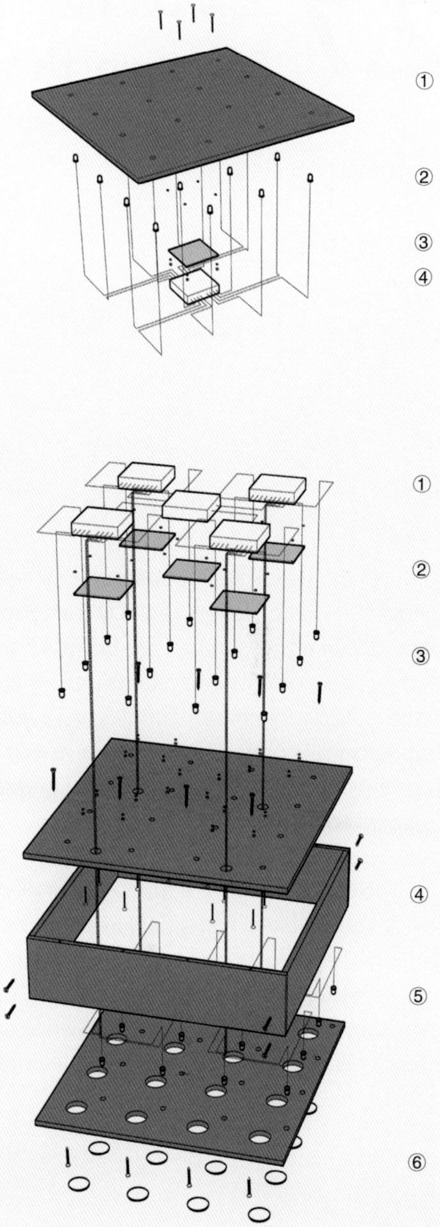

地板面板
①框架
②红外发射器
③电路基板
④配电回路

顶棚面板
①比测器和电源电路
②电路基板
③高强度白色 LED
④框架
⑤红外探测器
⑥镜片

图 2.6-11 照明系统
Light syetem

图 2.6-12 顶棚面板和地板面板
Ceiling panel and floor panel

2.7 范斯沃斯墙 Farnsworth Wall

范斯沃斯墙是一种能够收集太阳能进行室内照明的建筑墙体模块。

NMinusOne 工作室，曼妮 · 曼妮

The Farnsworth Wall is an architectural wall module that harvests solar energy to illuminate the interior.

Studio NMinusOne with Mani Mani

Through the use of solar panels on one side and LED lights on the other, an opaque wall becomes digitally transparent. The Farnsworth Wall takes its cue from Mies van der Rohe's Farnsworth House. The famous glass house made its inhabitants uneasy with its complete transparency to the outside. Perimeter curtains were hung to provide opacity and psychological comfort. With the Farnsworth Wall, the outside becomes a dynamic digital wallpaper allowing one to see the outside without the outside seeing in.

The wall consists of a modular panelling system built as a Structural Insulated Panel (SIP) in a standard four foot by eight foot size. This module can be cut and applied to virtually any existing or new construction. A layer of solar panels on the exterior of the module harvests and stores radiant energy, while an embedded system of low-energy LED lights on the other side illuminates the interior space.

通过一侧的太阳能电池板和另一侧的 LED 灯，不透明的墙体变成了数字化的透明墙。范斯沃斯墙的设计灵感来源于密斯 · 凡 · 德 · 罗的范斯沃斯住宅。这座著名的玻璃房子由于对外界全透明而使居住者感到不安，只能悬挂窗帘来为居住者提供不透明感与心理上的舒适感。范斯沃斯墙的外部成为一个动态的数字化壁纸，可以允许人在室内看到外部，但是外部却看不到室内。

墙体是由一个标准的 4 英尺 ×8 英尺（1 英尺约为 0.3 米）大小的结构隔热板（SIP）构成的模块化镶板系统组成，模块可以被切割并应用于几乎任何已建或新建结构，在模块外层的太阳能电池板面层收集与储存热辐射能量，而另一侧的低能耗 LED 灯嵌入式系统可以提供室内照明。

图 2.7-1

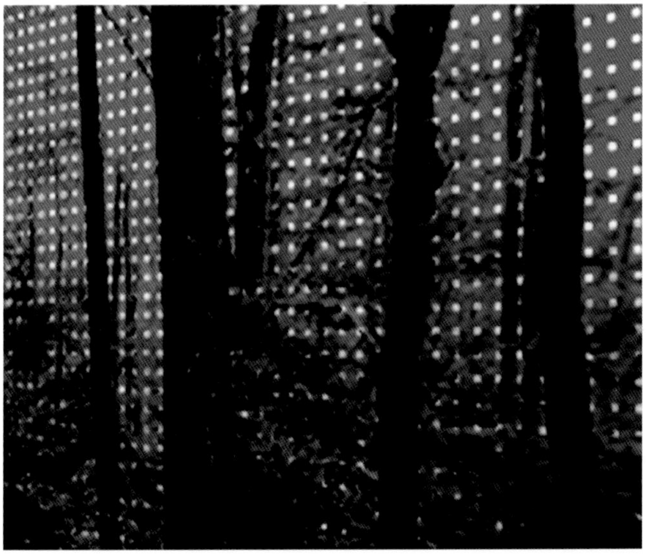

图 2.7-2

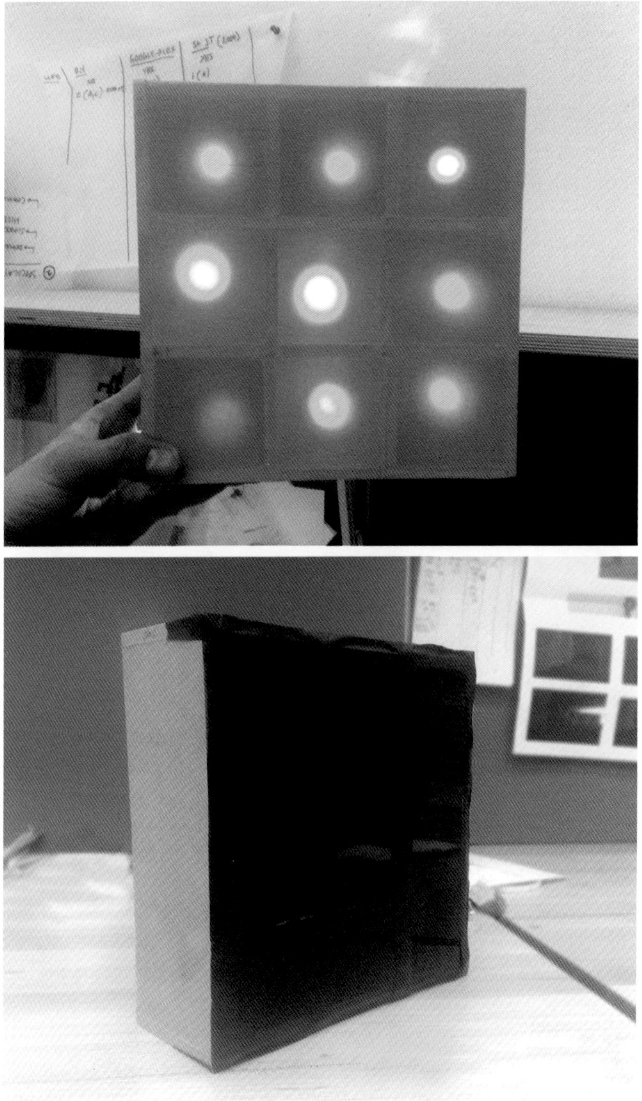

图 2.7-1 人体与阴影遮挡了墙体外表面的光进入通道，导致 LED 灯在室内相应区域内变暗，这就产生了单向透明的效果
Bodies or shadows disrupt the passage of light on the exterior surface, causing the LED lights to dim in the corresponding areas on the interior. This gives the effect of one-way transparency

图 2.7-2 范斯沃斯墙体模块的一部分 1：1 原型的模型：室内和室外
A 1 : 1 prototype of a portion of the Farnsworth Wall module: interior and exterior

图 2.7-3

图 2.7-3 范斯沃斯墙
Farnsworth Wall

2.8 范斯沃斯帘幕 Farnsworth Curtain

帘幕是可以将外部在内表面上留下记录的隔断。

NMinusOne 工作室，鲁道夫 · 埃尔 - 库利

The curtain forms an enclosure yet allows the outside to register itself on its interior surface.

Studio NMinusOne with Rodolphe el-Khoury

“范斯沃斯帘幕”是范斯沃斯墙体的软件版，其创造的曲线形空间产生了开放且更私密的空间。

帘幕形成一层可以将外部在内表面上留下记录的隔断，通过红外线传感器与 LED 组成的网格，帘幕以外的物体与身体会以阴影的形式显现在帘幕内表面——通过电子调节器来创造透明度。通过抑制所有声音，建筑内部空间进一步分化，封闭空间在保持与外界联系的同时，成为一个感官上的庇护所，远离主楼层的喧嚣。

The ‘Farnsworth Curtain’ is a soft version of the Farnsworth Wall project. The curtain carves out a curvilinear space which allows for both open and more intimate spaces.

The curtain forms an enclosure yet allows the outside to register itself on its interior surface. Through a grid of infrared sensors and LEDs, objects and bodies on the exterior of the curtain appear as shadows on the curtain’s interior surface - creating transparency through electronic mediation. The space within is differentiated further by dampening all sound. The enclosed space then becomes a sensory refuge from the bustle of the main floor, while still maintaining contact.

图 2.8-1

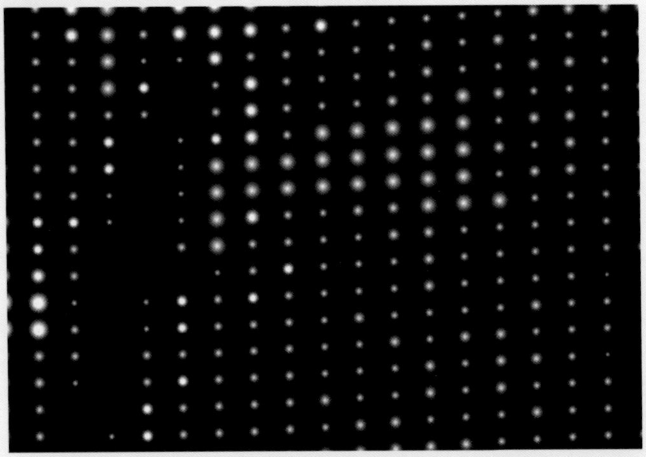

图 2.8-1　阴影似乎漂浮而过，外部越嘈杂，内部体验却越让人昏昏欲睡
Shadows will seemingly float by, the busier the outside the more hypnotic the experience

图 2.8-2　帘幕是由两个完全分离的隔层构成，外层是红外线传感器与 LED 的支撑结构，内层形成可以使 LED 投射光线的屏幕。当物体干扰了红外线，相应的 LED 会变暗，导致相应的黑色形状成像在窗帘内层上
The curtain is composed of two separate layers. The outer layer is the supporting structure for the infrared sensors and the LEDs. The inner layer forms the screen upon which the LEDs project light. When an object interrupts the infrared beams, the corresponding LED lights dim, causing a dark shape to emerge on the inner layer of the curtain

图 2.8-2

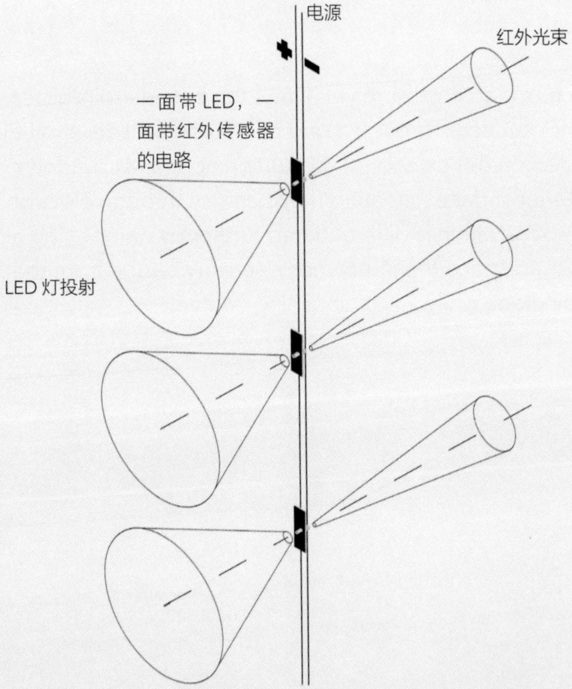

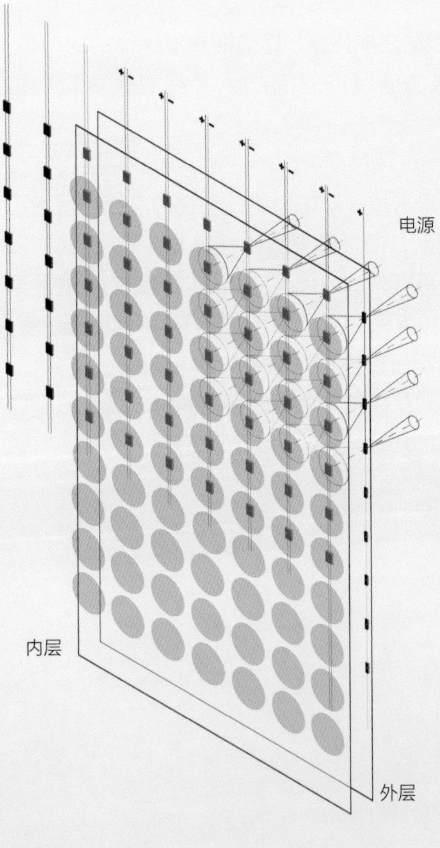

图 2.8-3

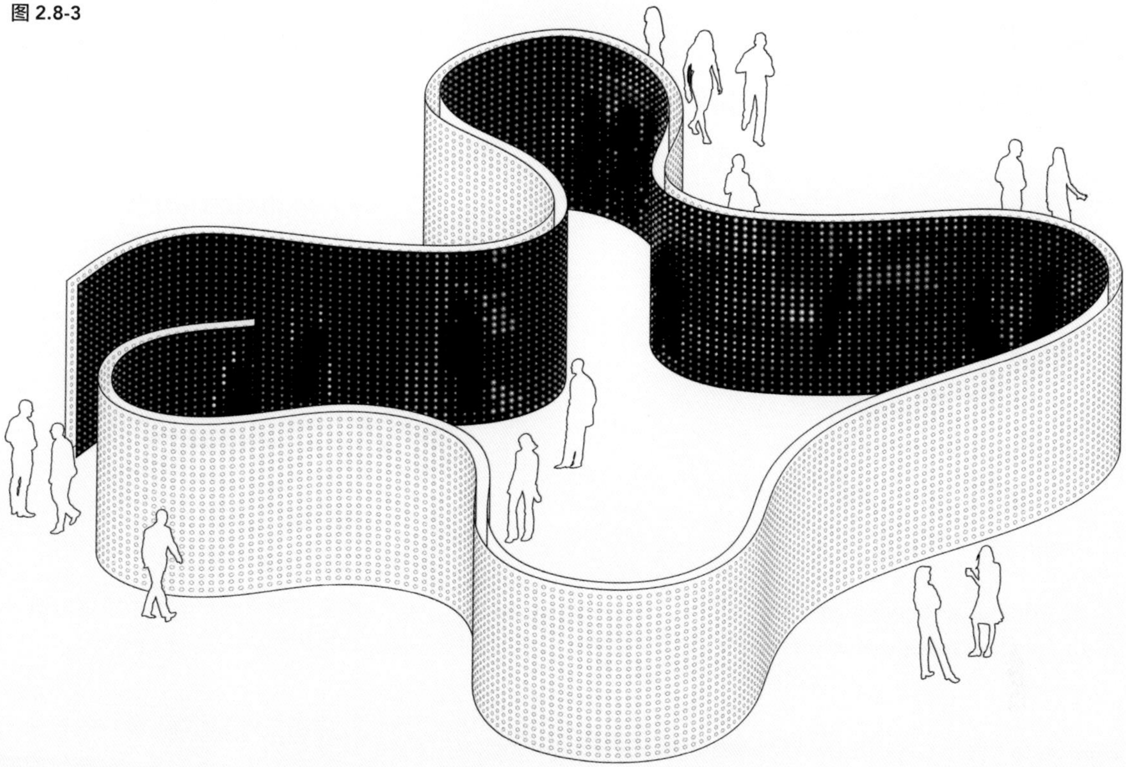

图 2.8-3 通过利用电子降噪技术和吸声材料，封闭空间几乎是寂静无声的
Through the use of electronic noise cancellation technology and sound absorbing materials, the enclosed space is almost silent

2.9 双重状态房间 Dual State Room

该空间通过不断调节其表面来响应亮度波动，以提升亮度与环境氛围。

NMinusOne 工作室

A space that responds to fluctuating light levels by continuously modulating its surface to enhance light and mood.

Studio NMinusOne

The Dual State Room is a systematically constructed deep infrastructure for controlling the most superficial of tectonic conditions: the wall surface.

The Dual State Room can change its appearance. One moment it is a warm, sound-absorbent walnut veneer, and the next, a cool, reflective fluorescent yellow enamel. The space flip flops between these states, responding to light level fluctuations over the course of the day. When the space is too bright, the wall adopts the wooden side, reducing glare, and offering a feeling of warmth. Conversely, when dark, the wall flips to a highly reflective surface maximising light levels in dark zones. The Dual State system takes on the critical role of light modulation.

Its cellular construction - the result of the re-appropriation of discarded technology - is capable of responding locally by changing its material qualities depending on the weather, altering in turn the perception of the inhabitant.

双重状态房间是一个系统建构的深层基础设施，用于控制最表层的构造状况的是墙体表面。

双重状态房间能够改变它的外观，有时是温暖的、吸声的胡桃木饰面，有时是凉爽的、反射荧光的黄色珐琅。这些状态间的空间翻转是对一天中的亮度波动做出的反应。当空间太亮时，墙体就采用木面，减少眩光，给人温暖的感觉；相反，当空间太暗时，墙体就翻转到具有高反射率的表面以增大黑暗区的亮度级别。双重状态系统对光线的调节起到关键作用。

其组织结构是废弃技术再利用的成果，能够根据天气状况及时响应改变材料性能，进而又改变了居住者的感受。

图 2.9-1

图 2.9-1 在物理层面上，木饰面既吸收光也吸声；在心理层面上，采用了类似玛莎·斯图尔特的室内设计风格，给人一种温暖和平静的感觉
On a physical level, the wood veneer absorbs both light and sound. On a psychological level it offers a feeling of warmth and calm adopting a Martha Stewart-like interior design rhetoric

图 2.9-2

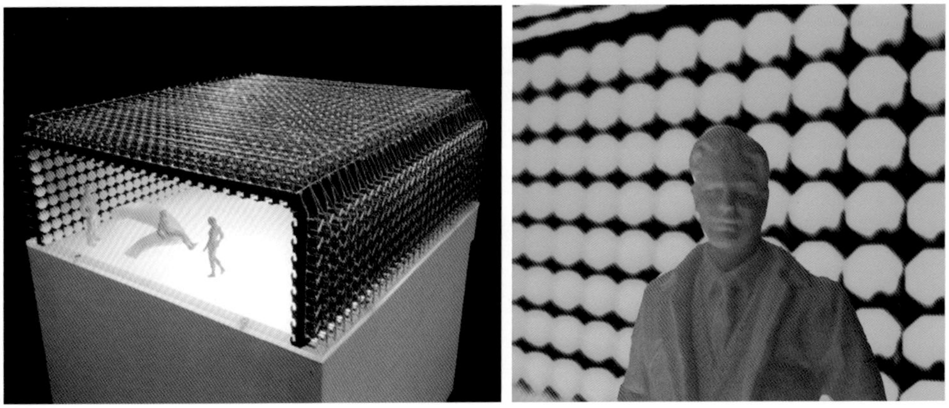

图 2.9-2 身处黄色空间中，对象的情绪也改变了
Bathed in yellow, the subject's mood is altered

图 2.9-3

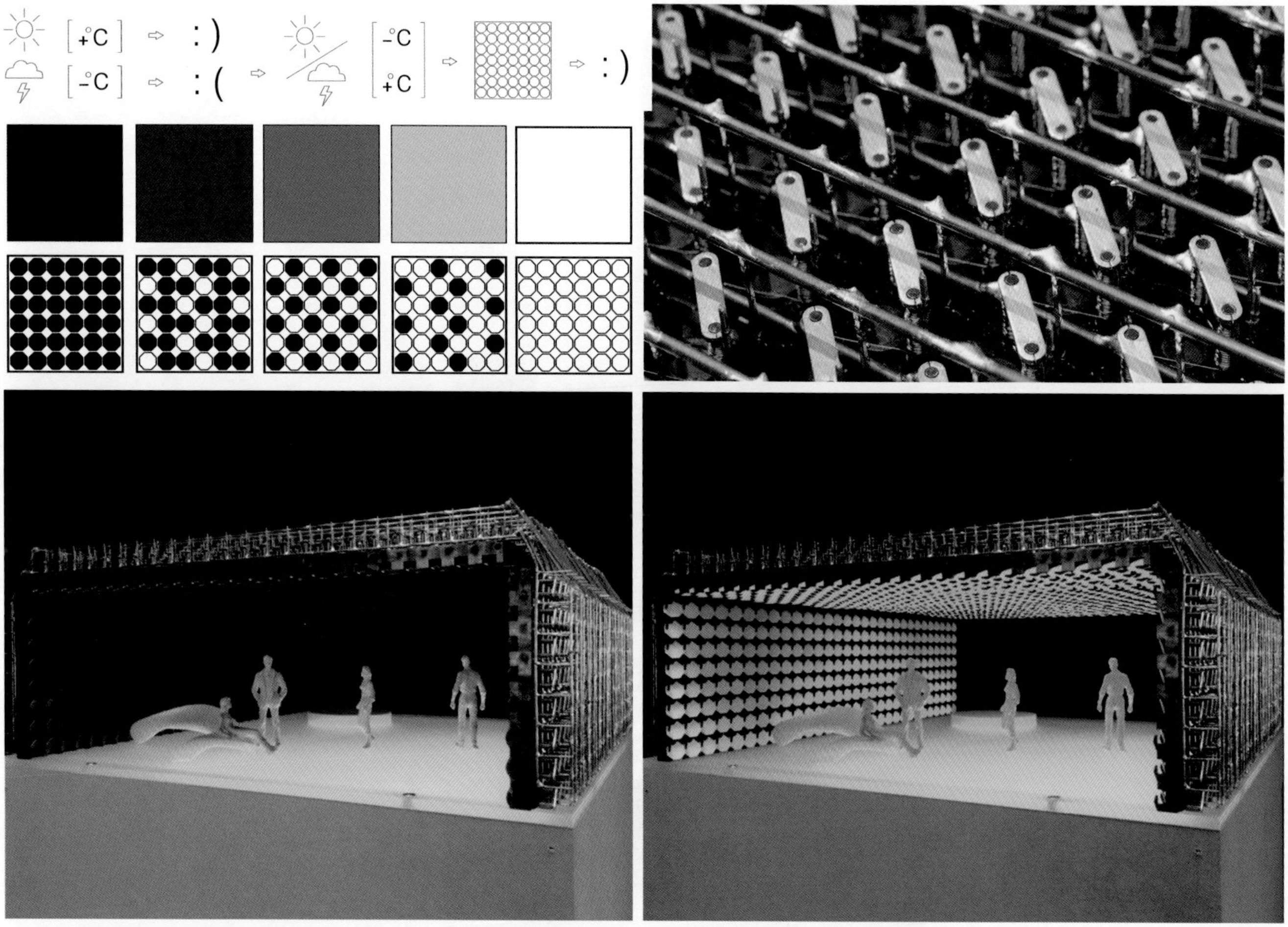

图 2.9-3 双重状态房间探索了天气、室内亮度与情绪之间的关系
The Dual State Room exploits the relationship between weather, interior light levels, and mood

图 2.9-4

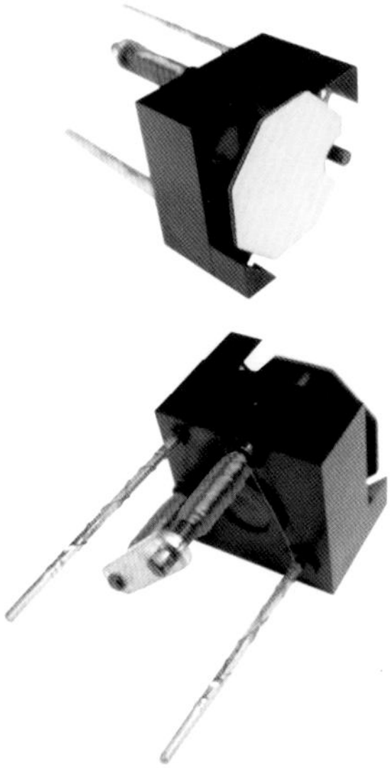

图 2.9-5

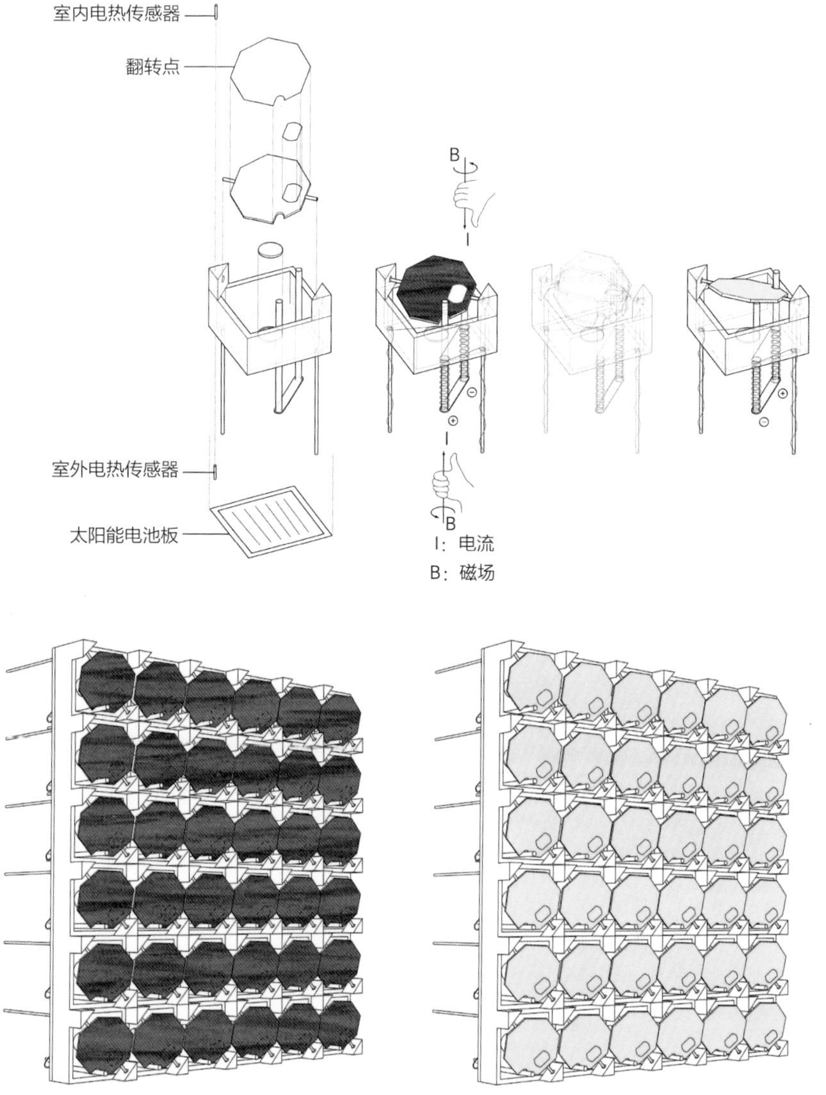

//A 面 //
材料：木
状态：最大吸收率

//B 面 //
材料：亮荧光漆
状态：最大反射率

图 2.9-4 工作原型细部
Detail of a working prototype

图 2.9-5 每个单元都是自主运行的，容纳了室内与室外温度及光传感器，单元对亮度波动水平做出反应并相应地翻面
Each cell operates autonomously, containing exterior and interior temperature and light sensors. The cells respond to fluctuating light levels, and change sides accordingly

图 2.9-6

图 2.9-6 当空间太暗时，墙体就翻转到具有高反射率的表面，以增大黑暗区的亮度级别
When dark, the wall flips to a highly reflective surface maximising light levels in dark zones

2.10 西安气味花园 Xi'an Scent Garden

西安气味花园是通过将芳香的当地植物与靠运动传感器激活的气味柱相结合而实现的自然与人工的整合性作品。

KLF 事务所，朱 · 亚当斯、詹姆斯 · 狄克逊和法迪 · 马苏德

The Xi'an Scent Garden marries nature with artifice through a combination of fragrant local plant species alongside motion sensor-activated scent poles.

Khoury Levit Fong with Drew Adams, James Dixon and Fadi Massoud

西安气味花园是受 2011 年西安世界园艺博览会委托建设，其散发芳香气味的同时又可以愉悦感观。

排列整齐的松树林、迷迭香与百里香组成的花床建立了一个独特的嗅觉世界，当地的针叶树通过嗅觉、视觉、触觉与西安相连接。运动传感器所激活的气味柱扩散了植物提取物，将这些气味再配上可以让人想起不同地方的方言构成了一个中国嗅觉地图。一系列发光的气味柱创造了气味花园的标志形象，花园作为公共空间，其微妙的感观输出提供了庇护性的个人体验。

气味柱可以兼作照明装置，它们由不锈钢管、集成的气味盒和 LED 灯组成，气味可以随风自然扩散，更实时地由进场的参观者扩散，或者被传感器制动的风扇扩散。灯光的强度随风速而改变，它是通过柱顶端的风感应器来调节的，灯的动态响应将气味扩散模式转变为一种气味视觉现象。这些人造香味被装瓶包装，在一个玻璃展馆里出售给游客。

The Xi'an Scent Garden, commissioned for the 2011 Xi'an World Horticultural Exposition, delights all the senses while favouring smell.

Its terraced pine groves and beds of rosemary and thyme build and sustain a unique olfactory world. The local conifers create an olfactory, visual, and tactile connection to Xi'an. Motion sensor-activated scent poles diffusing plant extracts complement these fragrances with exotic accents evoking other locales to create an olfactory map of China. The constellation of glowing scent poles creates an iconic image of the garden as a public space, while its nuanced sensory output provides sheltered and individual experiences.

The scent poles double as light fixtures. They consist in stainless steel tubes with integrated scent cartridges and LED lights. The scent is diffused naturally by the breeze or, more promptly for approaching visitors, by a sensoractivated fan. The light intensity varies with the velocity of the wind. It is modulated by a wind sensor at the top of the pole. The lighting's dynamic response translates the scent propagation pattern into a visual phenomenon. The collection of artificial scents are bottled and packaged for sale to visitors in a glass pavilion.

图 2.10-1

图 2.10-2

图 2.10-3

图 2.10-4

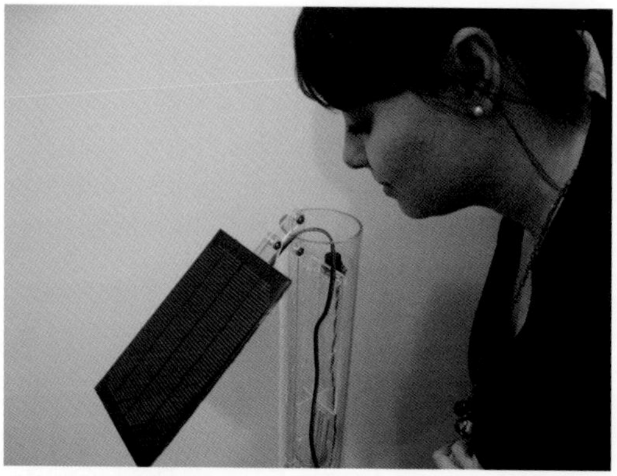

图 2.10-1 夜景透视图
Night time perspective view

图 2.10-2 主通道景观
View of main path

图 2.10-3 玻璃展馆内景
View of glass pavilion

图 2.10-4 气味管原型
Scent tube prototype

图 2.10-5

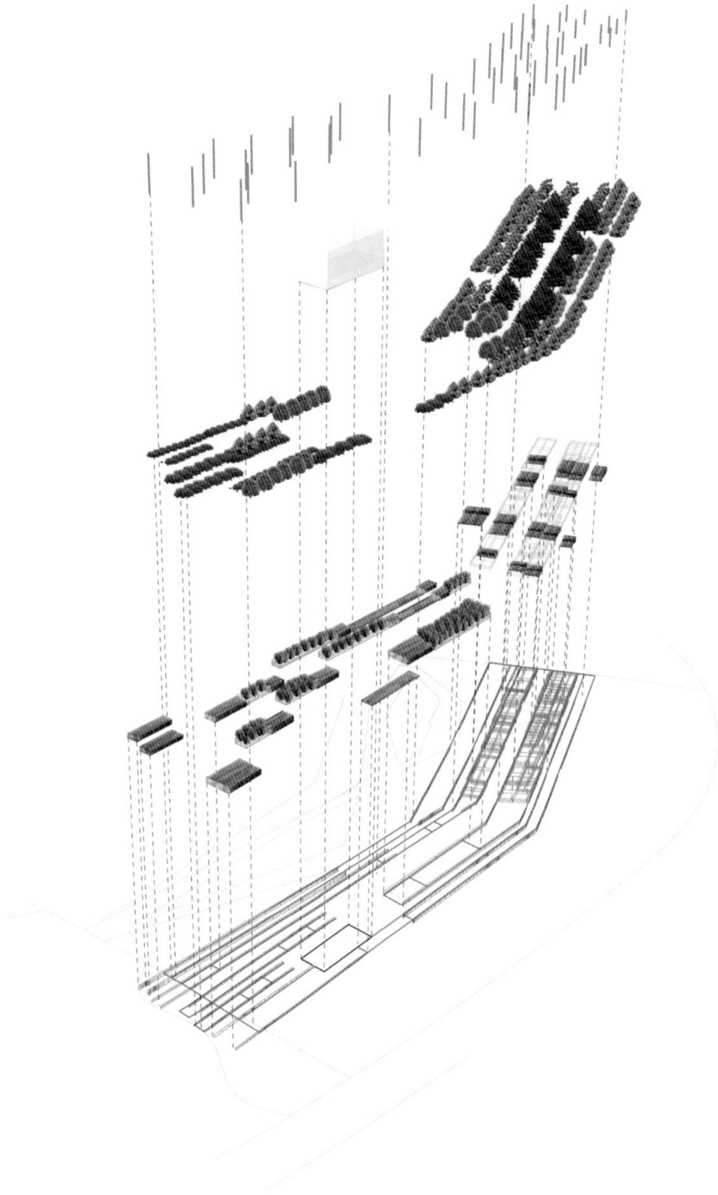

图 2.10-5 显示不同系统的分层爆炸轴测图
Exploded axonometric drawing showing the layering of different systems

图 2.10-6

图 2.10-7

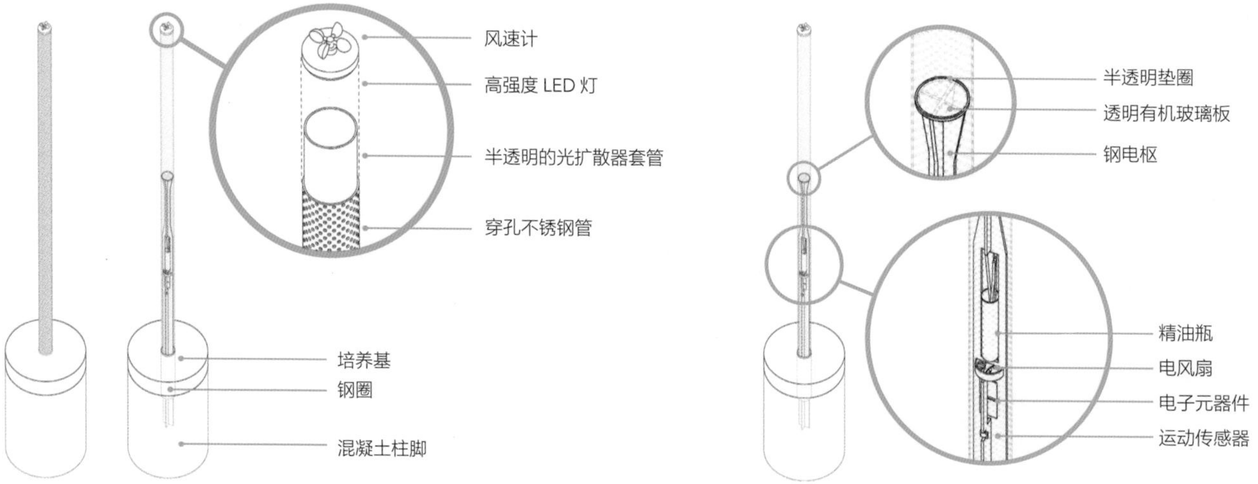

图 2.10-6 山上的透视景观
Perspective view from hill

图 2.10-7 数字气味柱的装配组件
Digital scent pole assembly

2.11 纪念碑 + 比特 Monuments + Bits

这个展览探讨了对物理对象的两种数字增强策略。

KLF 事务所

The exhibit explores two strategies for the digital enhancement of physical objects.

Khoury Levit Fong

“纪念碑与比特”的展览将参数化建模、制造技术与新兴的响应式媒体技术结合为一体，在交互式数字增强环境中呈现建筑与城市设计项目。

The ‘Monuments and Bits’ exhibition combines parametric modelling and fabrication techniques with emerging responsive media to present architecture and urban design projects in an interactive and digitally augmented environment.

展览于 2009 年 9 月在多伦多大学的艾瑞克 · 亚瑟美术馆开幕，并在密歇根大学的艾尔弗雷德 · 陶布曼建筑与城市规划学院的 CMYK 美术馆举办巡展，直到 2009 年 12 月展览才闭幕。

The exhibition opened at the University of Toronto’s Eric Arthur Gallery in September 2009 and travelled to the CMYK Gallery of the University of Michigans A. Alfred Taubman College of Architecture and Urban Planning, running until the end of December 2009.

“纪念碑与比特”的展览将集体纪念碑事件及与个人体验的即时性之间的紧张关系更戏剧化。传统是通过在建筑形式生成过程中新的参数量来重塑的，然而当今互动媒体的发展刷新了对建成环境的固有认知。

‘Monuments and Bits’ dramatises the event of collective monuments and their tense relationship with the immediacy of individual experience. Conventions are recast through new parametric capacities in the production of architectural form, while current developments in interactive media refashion perceptions of the fixity of the built environment.

展览设计利用艾瑞克 · 亚瑟美术馆的两个主要空间的不同特征，来探讨将数字技术与影像整

The exhibition design takes advantage of the different characters of the Eric Arthur Gallery’s two main spaces to explore contrasting strategies for the integration of digital technology and imagery into physical environments: while electronic and AV equipment is prominently displayed as part of a space-defining superstructure in one room (room A), it is seamlessly embedded in the other (room B).

合在物理环境中的对照策略：当电子与 AV 设备作为限定空间的上层建筑的部分，被置于 A 房间的显著位置时，它也被自然地嵌入房间 B 中。

图 2.11-1

图 2.11-1 增强现实的前景与感应式投影的后景
View of AR station in foreground and responsive projection behind

图 2.11-2

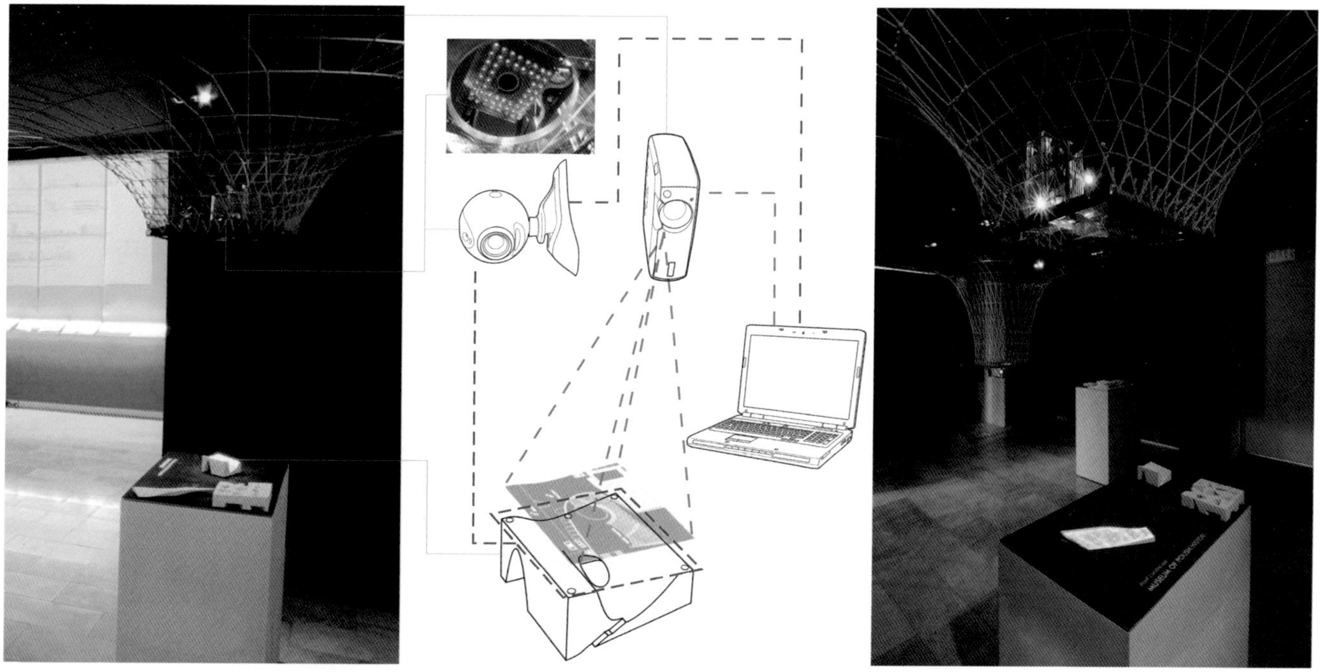

图 2.11-3

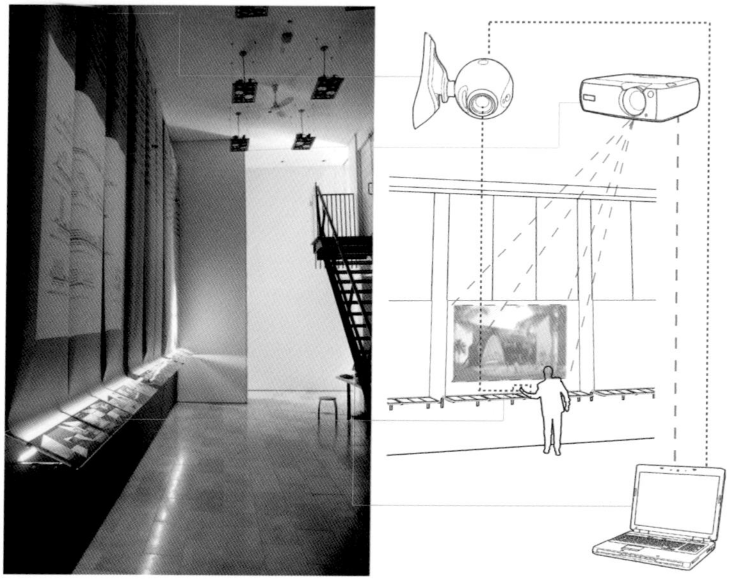

图 2.11-2 增强现实基站。该系统识别物理模型，并为附加信息投射数字动画图像。投影锁定选择对象，并在投影区域内跟踪对象

AR station. The system recognises the physical model and projects a digital animated image for additional information. The projection locks onto the selected objects and follows them within the projection area

图 2.11-3 系统识别与跟踪参观者的手指，并将与项目相关的图像传输给投影仪展示出来

The system recognizes and tracks the visitors' pointing finger to select and project pertinent images complementing the project on display

图 2.11-4

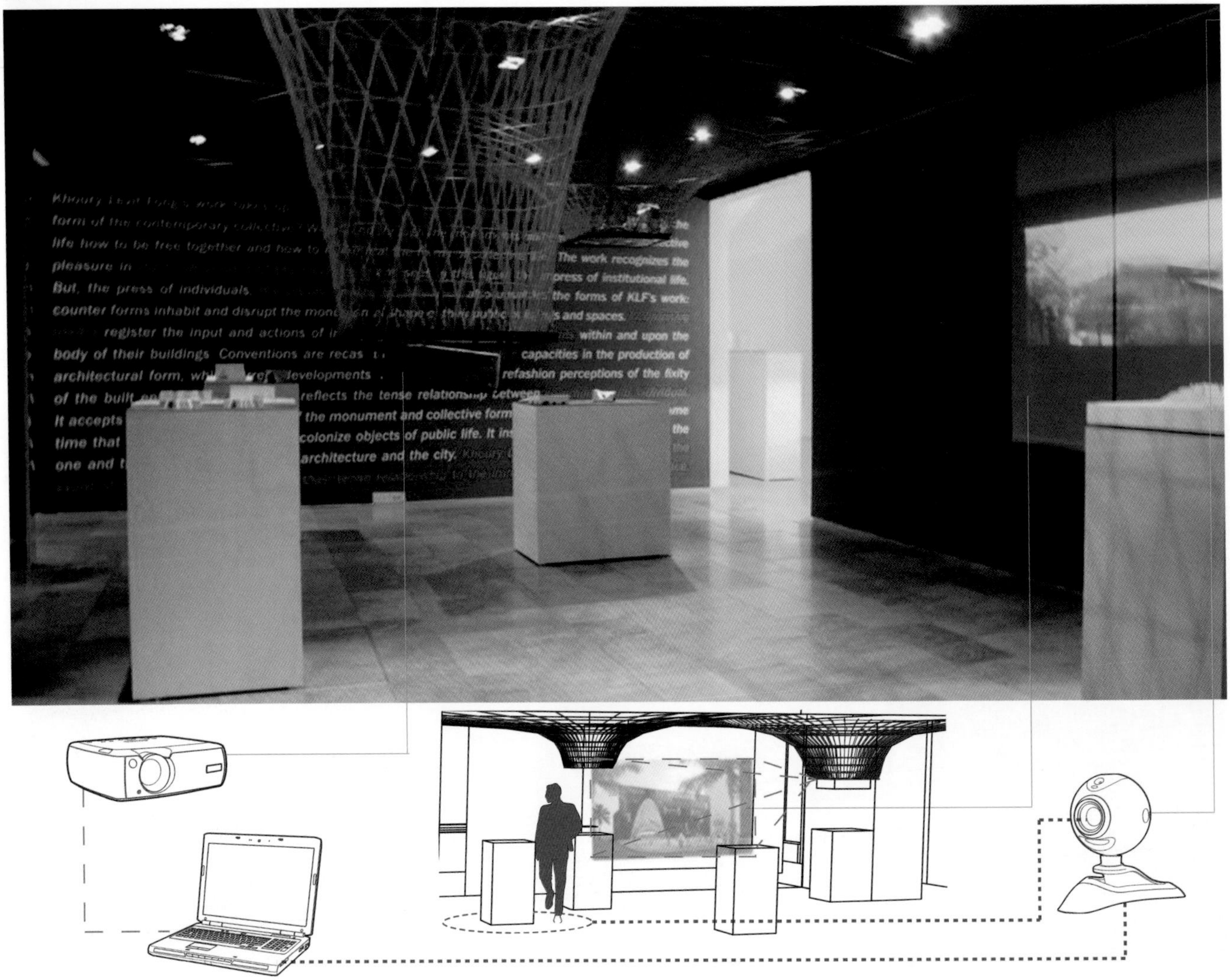

图 2.11-4 系统能追踪参观者的方位，并显示出与参观者当前观看的模型相关的图像
The system tracks the position of the visitor to display the images related to the model he or she is currently viewing

图 2.11-5

图 2.11-5 相关图片
The pertinent images

2.12 闪烁 Blink

凹状眼形表面将光线反射到管道上，加热管道内的物质，并为位于管道顶部下方的电动涡轮机提供动力。

KLF 事务所

Concave eye-shaped surfaces reflect light upon pipelines, heating their contents and powering electric turbines housed beneath their crests.

Khoury Levit Fong

A constructed dunes cape with southern slopes of mirror-bright polished stone adapts the machinery of solar power generation, embedding it within an artificial landscape.

A rich pattern of mirrored eyes is luminous with the sun by day and shifting patterns of LEDs by night. The dunes cape blinks.

‘Blink’ adapts the machinery of solar power generation. The machine does not sit in nature; rather, the dunes look natural but are a work of geometry and artifice. Biomorphic eye-like surfaces blinking in the dunes cape uncannily animate the inanimate.

These dunes are stabilised: an open mat of concrete, mostly below the surface, holds their form and patterns their surface. Concrete ribs and planks form the substrate for the parabolic curvature of the mirrored surfaces. Under several of the larger dunes-over ten metres in height-powerful electrical turbines are housed. The mirrored surfaces reflect intense desert sunlight onto the pipes which carry superheated liquid to drive electricity generating turbines.

一个由镜面抛光的石头建构的南向斜坡沙丘采用了太阳能发电设备，将其嵌入人工景观中。

丰富的镜像眼模式白天是由阳光照射而发光，夜晚则换成 LED 模式，沙丘闪烁着光芒。

“闪烁”采用太阳能发电设备，这台设备在自然环境中并不凸显，而且沙丘看起来更自然，仿佛巧夺天工。在沙丘上，仿生眼表面在闪烁，它赋予了非生命体以神奇的活力。

这些沙丘体非常稳固：一个混凝土开口底板，大部分在表面以下，它支撑整个形体并形成了表面图案。混凝土肋与厚木板构成镜面抛物线曲率的底板。在高度超过 10 米高的几个更大的沙丘体下安置了强大的电力涡轮机。镜面将沙漠强烈的光线反射到管道内，管道承载着过热的液体驱动发电涡轮机的运转。

夜晚，“闪烁”变成了信息化的装饰，阿布扎比与国际飞行目的地之间的空中旅行轨迹被绘制在沙丘镜面上，成为路径化图案。这种新的电子景观表现了可持续的能源生产与环境艺术间的完美结合。

At night, Blink turns into an informational ornament. The trajectory of air travel between Abu Dhabi and international destinations is mapped as trailing light patterns upon the mirrors of the dunes. This new electrical landscape signals the convergence between sustainable energy production and environmental art.

图 2.12-1

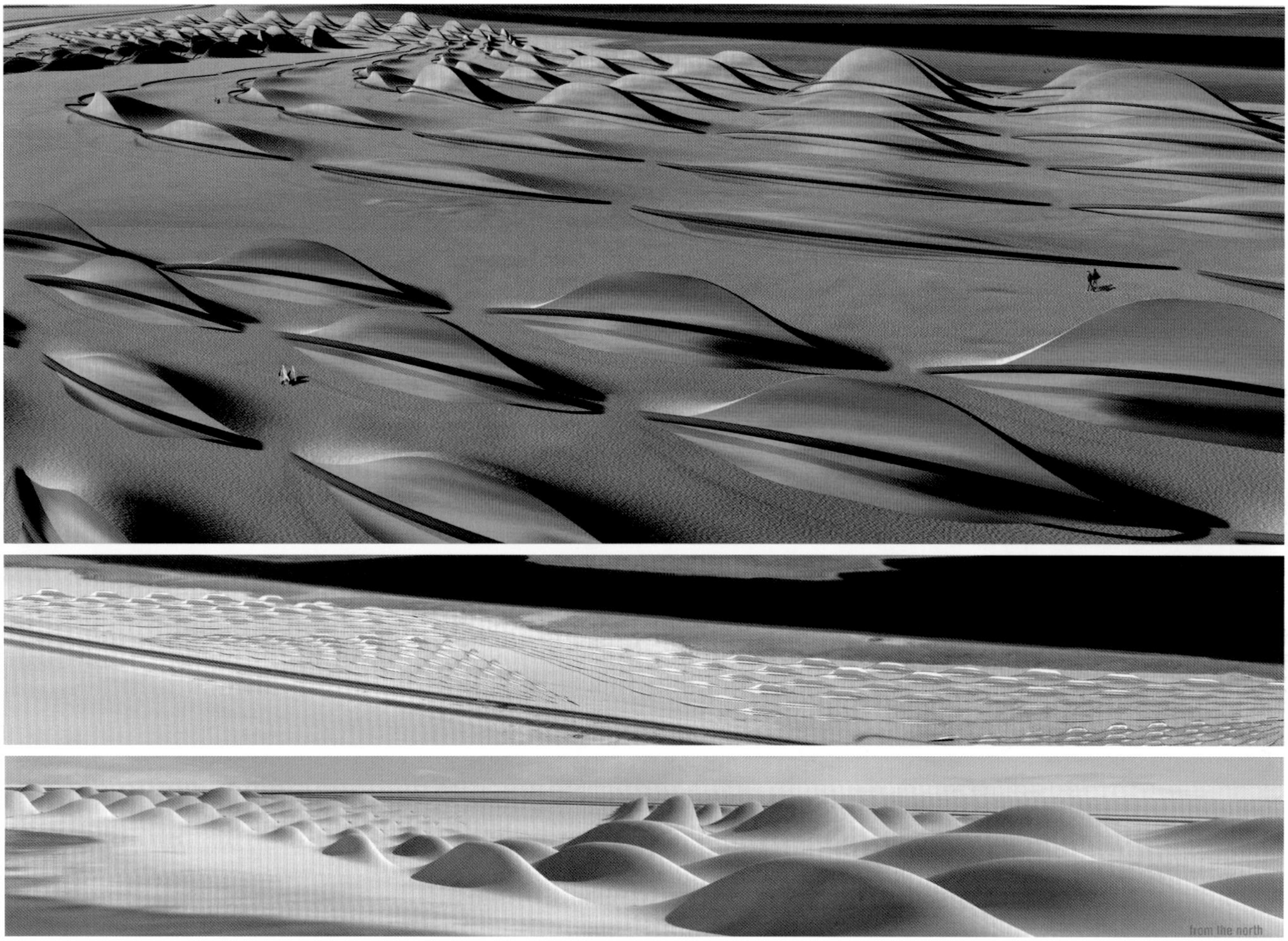

图 2.12-1 日落时的沙丘：沙丘上升与下沉，形成了丰富多变的形态，既可用于观赏也可填充空间
Dunes at sunset: the dunes rise and subside, making for a richly varied array of forms, both to look at and to occupy

图 2.12-2

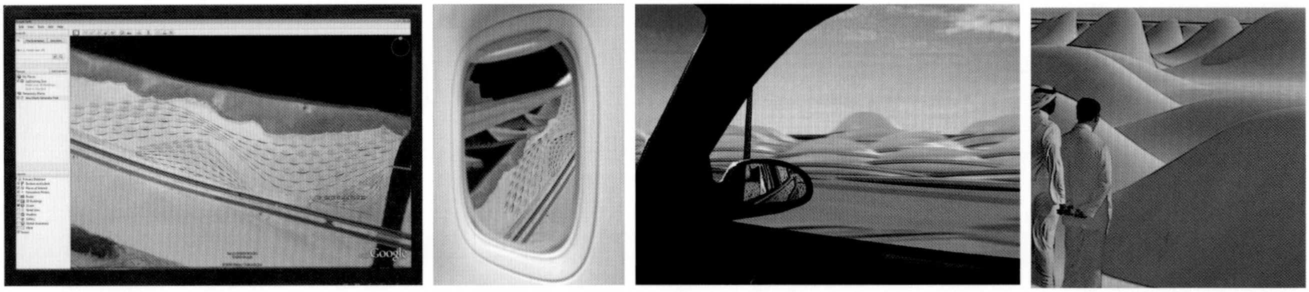

图 2.12-2 太阳能沙丘体的设计是为了让人从不同速度、不同尺度、不同媒介，如从谷歌地图、飞驰的汽车、飞机舷窗或步行等媒介来体验
The solar dunes cape is designed to be experienced at different speeds, at different scales and in different media: from Google Earth, a speeding car, the window of an airplane, or by foot

图 2.12-3

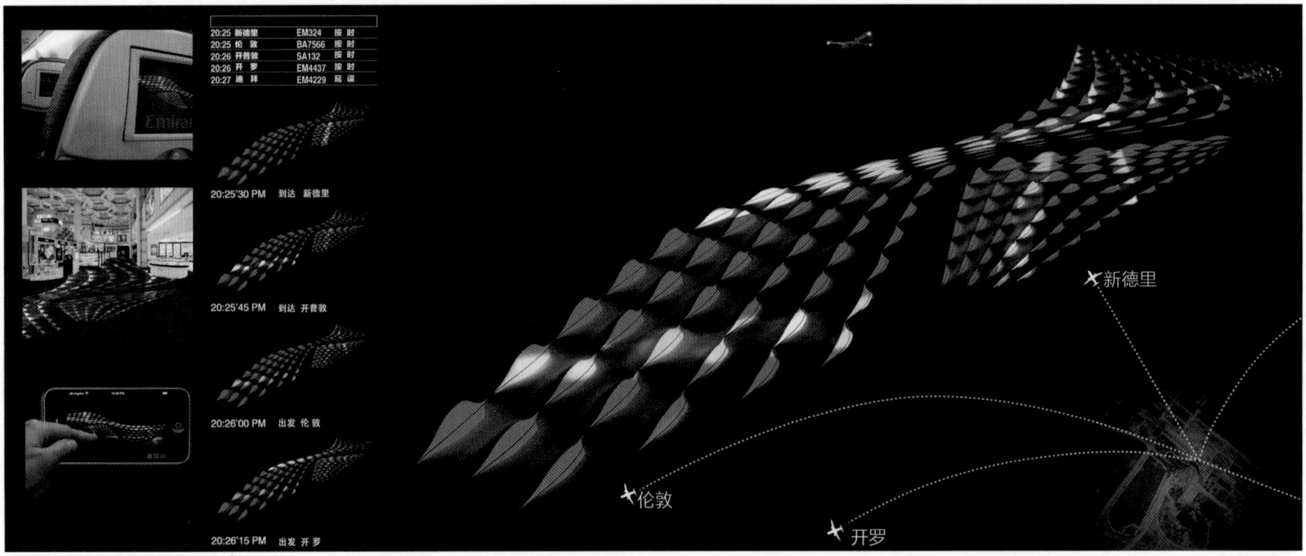

图 2.12-3 每面镜子前的管道里都嵌入了 LED 灯，它们可以根据阿布扎比机场飞机的进出港情况变亮或变暗
LEDs embedded in the pipes in front of each mirror brighten and dim according to the arrivals and departures from Abu Dhabi airport

图 2.12-4

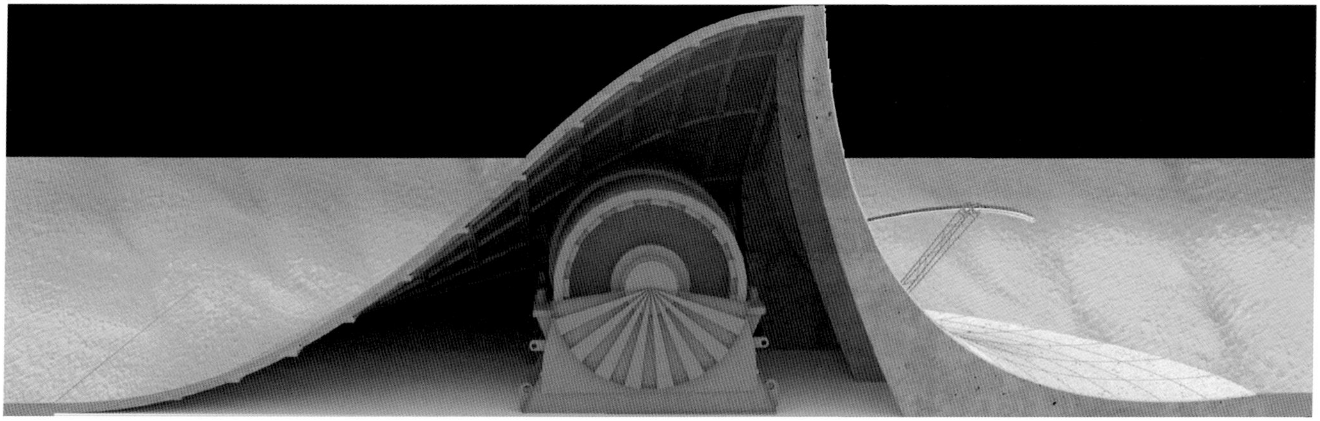

图 2.12-4 设置于沙丘体内的涡轮发电机
Electric generator turbine housed in a dune

图 2.12-5

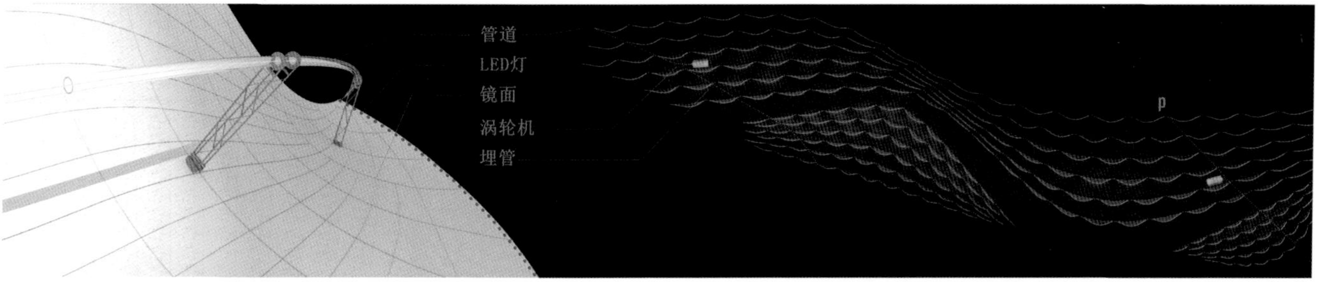

图 2.12-5 沙丘体解剖图
Anatomy of a dunes cape

2.13 光之隧道 Tunnel of Light

在设计展上由一台电脑控制的，被发光外壳包裹着的，并能对参观者的运动做出反应的自动扶梯。

KLF 事务所

A computer-controlled luminous shell envelops escalators at a design show and reacts to visitors' movement.

Khoury Levit Fong

The two-layer tubular shell consists of nested rotational paraboloids whose lines are made from strands of fibre-optic filaments.

The complexity of a curvature that is made through the combination of straight lines makes for a visual and cognitive experience that permits immediate understanding while creating an ineffable effect that resists this easy grasp. This construct is animated with the application of proprietary responsive technology and techniques developed for constructing immersive and interactive environment by means of fiber-optic filaments. They serve to produce intricately shaped spaces and multi-sensory effects with quasi-immaterial means: ultrathin cables and light.

双层管状壳体由嵌套的旋转抛物线构成，这些抛物线是由光纤丝束组成。

装置通过直线组合而成的复杂曲率，产生了视觉与知觉上的体验，这种体验允许实时理解，并创造了一种难以轻易掌握的、不可言喻的效果。该结构采用专有的感应式技术，能够创造一种生动的画面，这种技术是通过光导纤维丝来建构沉浸式空间与交互式环境。它们利用非物质的手段，例如超薄电缆和光导纤维丝来产生形状复杂的空间和多感官的效果。

图 2.13-1

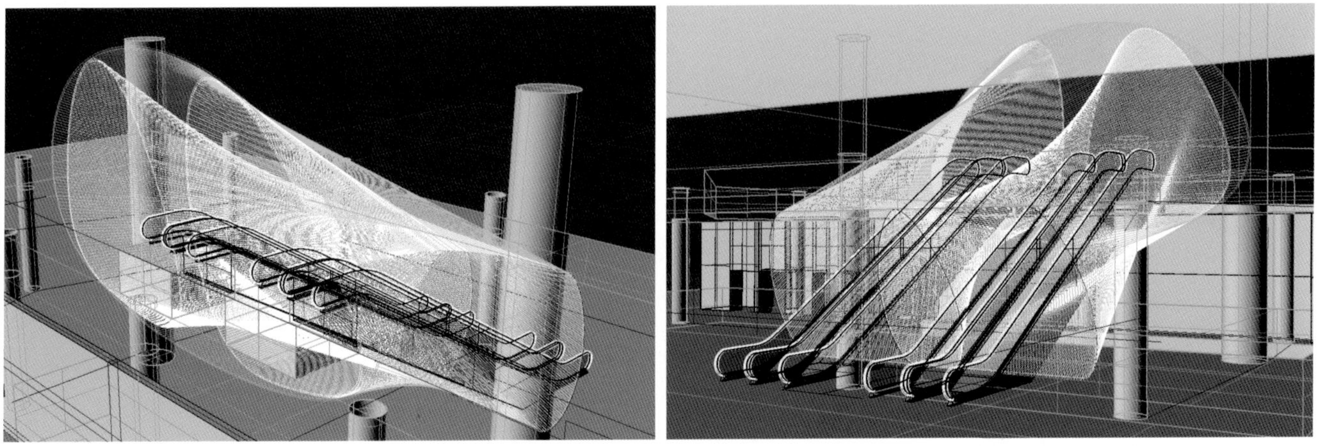

图 2.13-2

图 2.13-1 光导纤维丝的参数化数字模型
Parametric digital modeling of fibre-optic filaments

图 2.13-2 模拟效果图
View of simulated effect

图 2.13-3

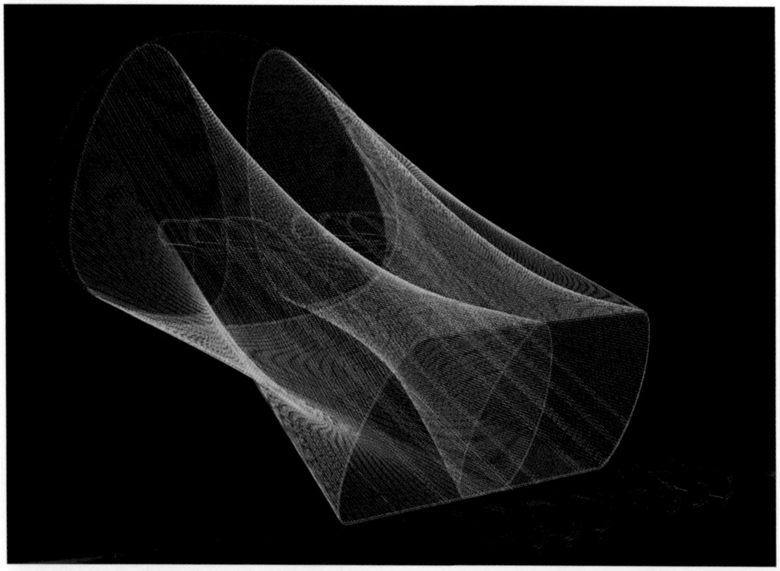

图 2.13-3 双壳的几何学研究
Double-shell geometry study

图 2.13-4

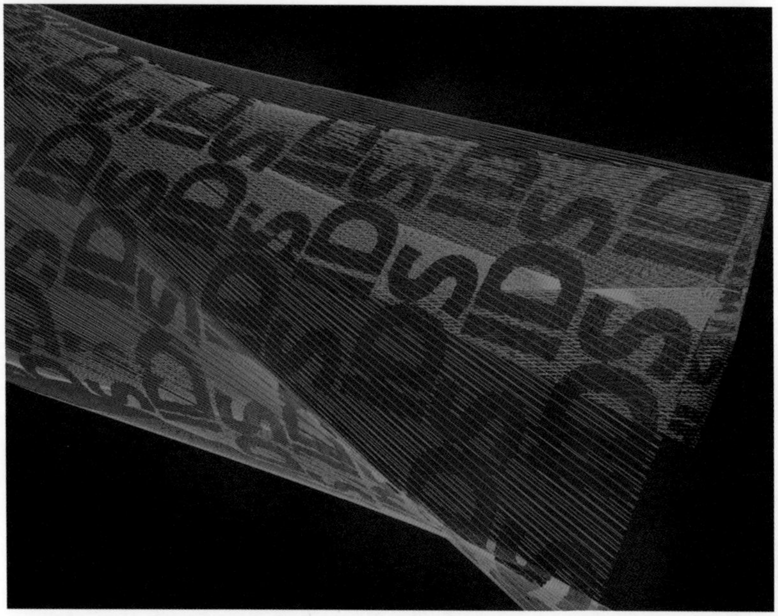

图 2.13-4 品牌灯光图案。光纤电缆是在指定位置进行侧向照明的，这个图形以重复的模式展示了 IDS❶ 缩写
Branding light pattern. The fiber-optic cables are treated for side-lighting in a designated pattern. This rendering shows the IDS acronym in a repeating pattern

❶ 译者注：IDS 是英文“Intrusion Detection Systems”的缩写，意思是“入侵检测系统”。就是依照一定的安全策略，通过软、硬件，对网络、系统的运行状况进行监视，尽可能发现各种攻击企图、攻击行为或者攻击结果，以保证网络系统资源的机密性、完整性和可用性。

图 2.13-5

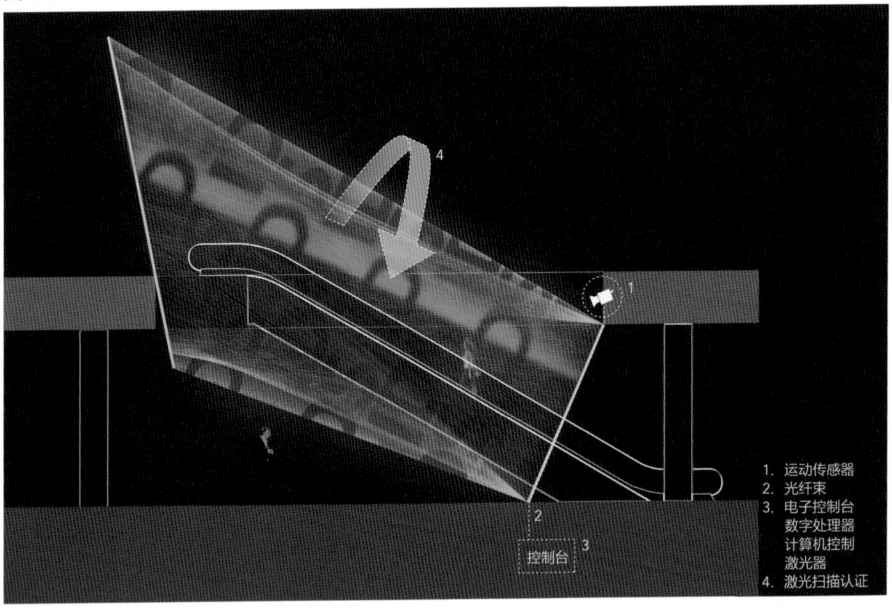

图 2.13-5 光纤网络是由微型激光仪控制，与参观者的在场和移动相互作用。感应是通过视频跟踪设备来校准的，由光缆输送的光在运转扫描中得以强化。参观者数量越多，扫描速度越快

The fiber-optic web is powered by micro-controlled lasers that interact with visitor's presence and movement. Responsiveness is calibrated by means of video-tracking devices. The light carried by the cables intensifies in a rotational sweeping motion. The greater the number of visitors the faster the sweep

图 2.13-6

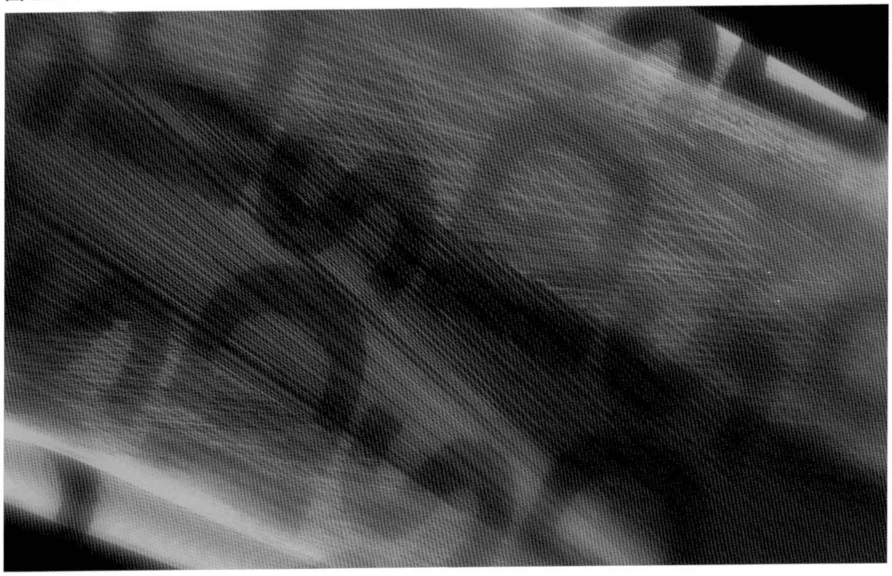

图 2.13-6 光之隧道

Tunnel of Light

2.14 规则 Rule

> **KLF 事务所在麻省理工学院的作品展览中展示了在最近建筑项目中使用的直线曲面的超凡品质，这些直线曲面用于数控激光驱动的光纤交互式装置中。**
>
> KLF 事务所
>
> **This exhibition of work by KLF at MIT demonstrates the phenomenal qualities of ruled surfaces used in recent building projects in an interactive installation of digitally controlled and laser powered fiber-optics.**
>
> Khoury Levit Fong

The ruled surface exacts a discipline that relates architecture directly to the persistent modularity of building materials.

The ruled surface, which invites visual understanding and engagement, also dramatises the event of collective monuments and their tense relationship to individual experience in KLF's work. These forms, made of discrete parts, gather themselves up into figures that choreograph public experiences: they are both a framework and formal allegory for social event.

While geometry joins the pleasures of the intellect with those of the senses, explorations of responsive media and augmented reality are used to make immersive environments of light, image, and smell. New technologies mobilise the senses through media that are stimulated by the activities and presence of people. Scents are released and patterns of light are initiated in this exhibit in response to the presence of visitors. The immediacy of the olfactory, of the moving image, and its response to the presence of individuals, draws together the scales of collective experience to the singularity of the individuals who are at its origin.

直线曲面体现了建筑与建筑材料的持久模块化直接相关的原则。

直线曲面能引起视觉上的感知与参与，同时又能使 KFL 事务所作品中的整体纪念性与个体之间的紧张关系更为戏剧化。这些由离散的部分组成的形式，聚集在一起，构成能够产生公共体验的图形：它们既是社会事件的框架，也是一个形式隐喻。

当几何学融入智力与感官的乐趣中时，感应性媒体与增强现实的研究常被用于创造光、影像与气味的沉浸式环境。新技术是通过由人的存在与活动刺激而产生的媒体来调动感官。在这个展览中，随着参观者的到来，气味被释放出来，光的图案也显现出来。嗅觉与动态图像的即时性以及它对个体存在的反应，将集体性的体验与个人最初的特殊性体验都汇集在一起。

图 2.14-1

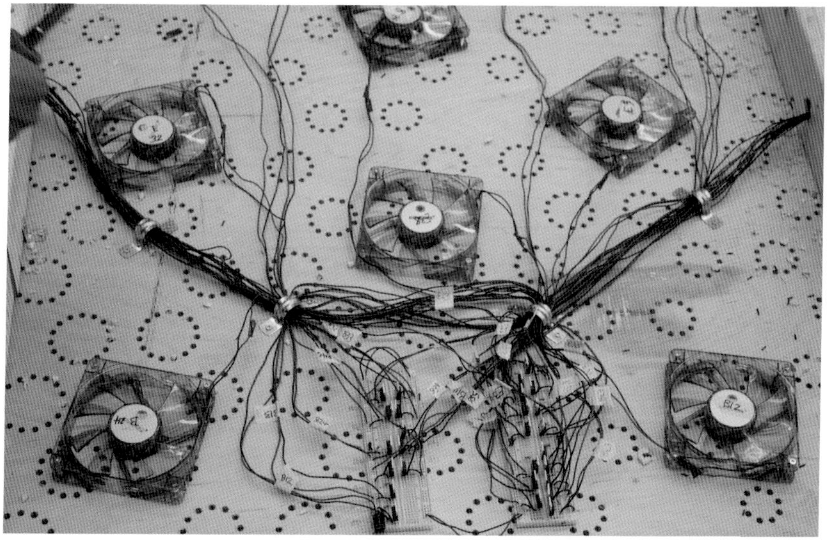

图 2.14-1 嵌进穿孔地板中的微控气味传播器
Micro-controlled scent dispensers embedded in the perforated raided floor

图 2.14-2 装置景象图：由旋转抛物面构成的薄丝拱顶系统
View of the installation: a gossamer vault system made of rotational paraboloids

图 2.14-3 穿孔地板的图案用来校准气味扩散的方向
Pattern of floor perforation for calibration of scent diffusion

图 2.14-2

图 2.14-3

图 2.14-4

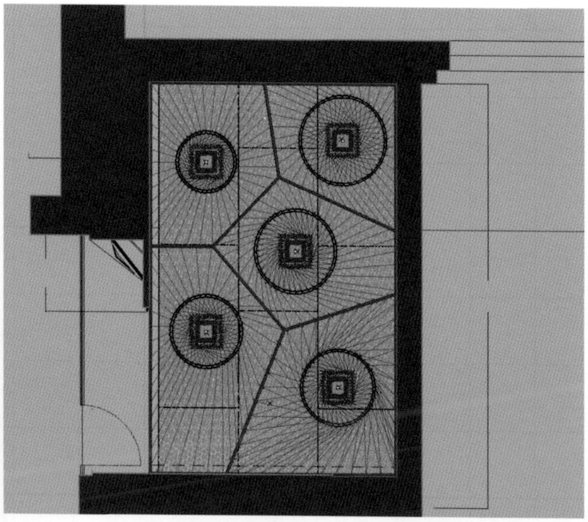

图 2.14-5

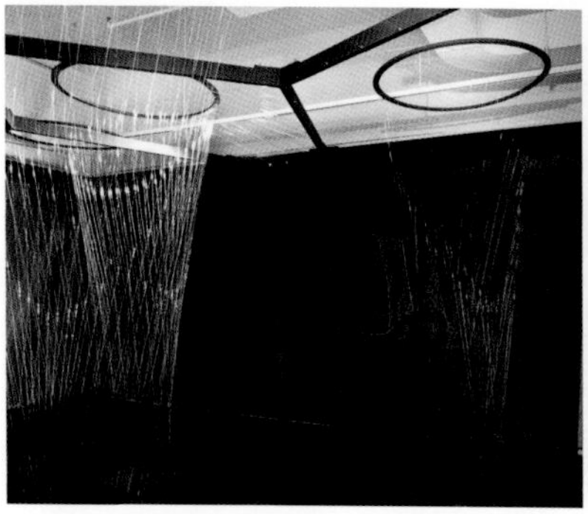

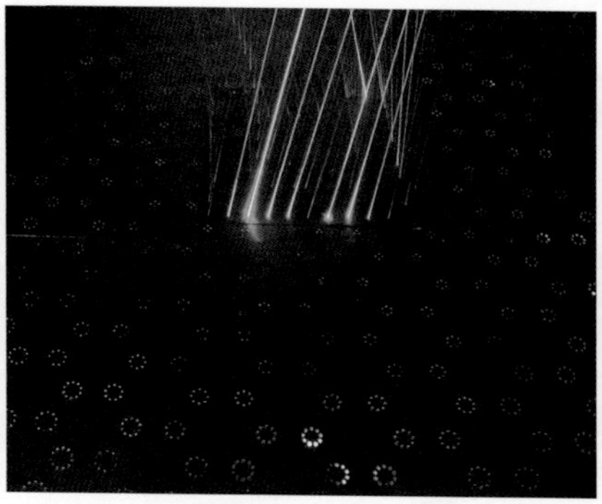

图 2.14-4 参数化的数字模型研究
Parametric digital model studies

图 2.14-5 装置细部
Installation details

2.15 自然 2.0 Nature 2.0

闪烁的光纤电缆与气味填充管能够模仿激光微控的感应式亮灯的自然效果。

KLF 事务所

Flickering fibre-optic cables and scent-filled tubes imitate natural effects with the responsive light-burst of micro-controlled lasers.

Khoury Levit Fong

“自然 2.0”是由人造的光纤丝树构成，它的背景是一个像是被装在树上的蚀刻玻璃板做成的发光体，能够使人产生一种奇妙的发光电子树的感觉。

项目的预期效果是让人产生身临其境的视觉感受，仿佛人们身处森林的边界。这是一个交互式的环境，在此通过数字技术再现了感官刺激的错综复杂性，它将我们与自然的生命力有机融合。悬挂起来的光纤丝网塑造了这个发光的电子树，通过将一组激光传输到悬挂的玻璃管中而发光，激光闪烁的模式是对游客接近时的响应。这些无毒与低过敏性的化合物是由气味艺术家西塞尔 · 图拉斯设计的，人工构建了一个复杂的自然嗅觉图像。在“自然 2.0”中的人造树是一个几何结构——具有菱形底部与椭圆形顶部的旋转抛物面（能够依据展览场地的形状进行调整）。这是一个直纹曲面的人工制品，它的曲面由 400 股光纤丝构成。椭圆形顶部

‘Nature 2.0’ consists of an artificial tree of fibre-optic filaments. Its backdrop is a luminous arboreal moiré of etched Plexiglas panels that gives the impression of a luminous electronic forest behind and beyond.

The intended effect is an immersive vision akin to that of being at the threshold of a forest: an interactive environment where digital technology reproduces the intricacy of sensory stimuli we associate with the teaming vitality of nature. The tree is shaped by strung fibre-optic filaments. They glow with the light they channel from a set of lasers into suspended glass tubes. The lasers flicker in patterns responsive to the proximity of visitors. These contain nontoxic and hypoallergenic chemical compounds designed by smell artist Sissel Tolaas that artificially construct an intricate olfactory image of nature. The artificial tree in ‘Nature 2.0’ is a geometrical construct — a rotational paraboloid with a diamond base and oval top (modified to fit the shape of our exhibition locations). This is a ruled surface artefact whose lines are made from 400 strands of fibre optic filaments. A canopy of filaments hang down from the oval top weighted at each of their ends by test tubes filled with coloured, scented oils. These test tubes hang in a stepped pattern that follows a similar geometry to that of the tree: a modified oval cut-out of a hyperbolic paraboloid.

The effect of this stranded form is to produce a glowing and scented environment in which the underlying energies of nature are echoed by the transmission of light through the fibre optic filaments. The floral scent of the tree is made artificially through the curated combinations of scents in the oils filling the test tubes.

下方悬吊光纤丝，每一根光纤丝的末端都用装有彩色芳香油的试管加重了重量。这些试管呈阶梯状悬挂，其几何形状与树的形状相似，并可以调整椭圆形的双曲面开口。

这种线编织形式的效果创造了一个发光和芳香的环境，在这种环境中自然界的潜在能量通过光纤丝的光传导而得到回应。电子树的花香是人工合成的，是通过装在试管中的芳香油的合成气味来实现的。

图 2.15-1

图 2.15-1 人工森林的景象
View of artificial forest

图 2.15-2

图 2.15-4

图 2.15-3

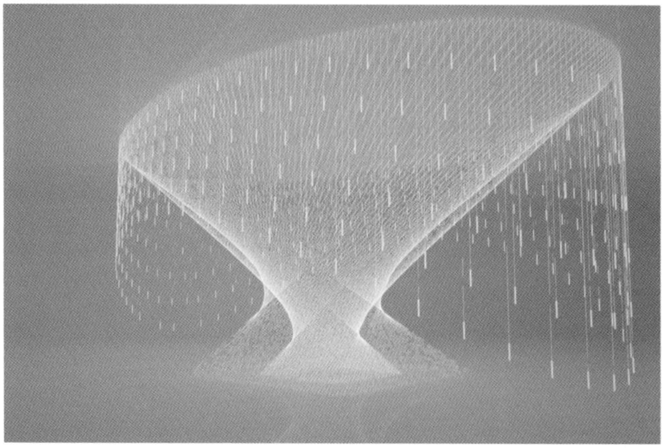

图 2.15-5

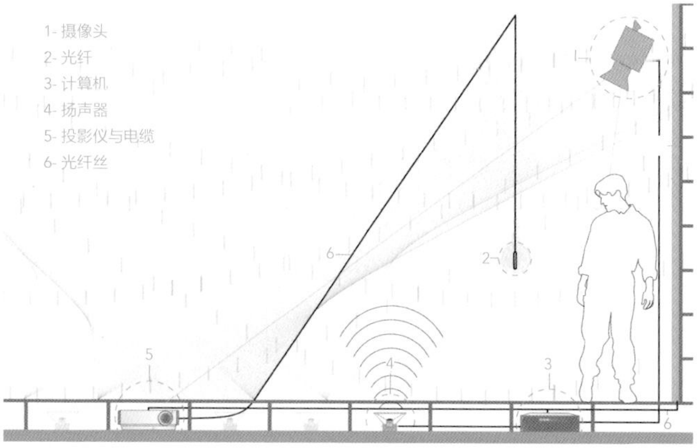

图 2.15-2 嵌入式的 LED 与斜角传感器的气味扩散管细部
Detail of scent diffusing tube with embedded LED and tilt sensor

图 2.15-3 多伦多湖滨中心案例的参数化数字模型研究
Parametric digital model study for Toronto Harbourfront Centre version

图 2.15-4 交互式特征图
Diagram of interactive features

图 2.15-5 2011 年成都双年展案例
Study for 2011 Chengdu Biennale version

图 2.15-6

图 2.15-8

图 2.15-6 旋转抛物面几何体的细部
Detail of rotational paraboloid geometry

图 2.15-7 感应式发光效果
Responsive lighting effects

图 2.15-8 微控激光校准仪
Micro-controlled laser calibration

图 2.15-7

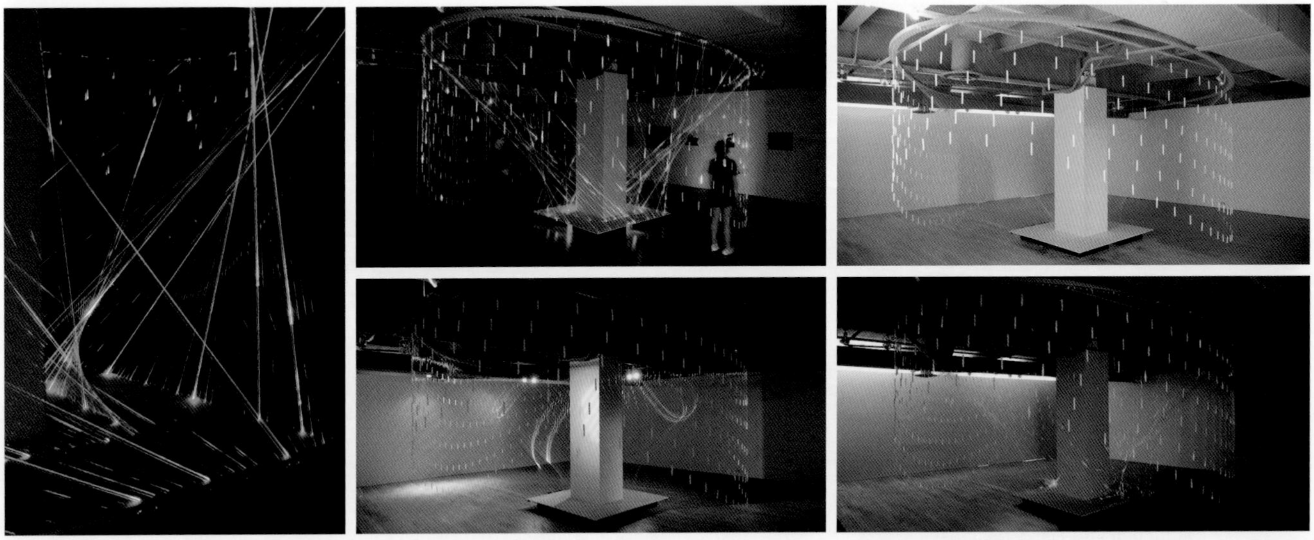

2.16 雾宅 Fog House

喷雾器和空气幕的电子协调作用塑造了自由立面建筑。❶

瑞奇 · 卡 · 冯 · 卢克

The electronically coordinated actions of misters and air curtains shape this wall-free house.❶

Ricky Ka Fung Luk

在无实物的环境中，雾是墙体材料，它是由超声波喷雾器产生的，并由空气幕形成。

“雾宅”主要由地板与顶棚两个水平面组成，两者之间是结构支撑，能够容纳机械设备与固定装置，它们用功能性的设施对自由空间进行空间限定，例如，容纳厨房用具的台墩限定了厨房区域。

住宅的不同房间之间没有实质性的分隔墙体，它们是由计算机控制与机械塑形的雾构成。任何房间的墙体可以根据需要与期望出现或消失，房间处于一种不断“形成”和“分解”的过程中。当它们恰好达到需要的临界点时，房间的边界就会被空气幕明确划定，而当不需要或不可见时，房间边界就会消失在无形的雾中。

Fog is the stuff of walls in this substance-free environment. It is produced by ultrasound misters and shaped by air curtains.

The ‘Fog House’ consists mainly of two horizontal planes: a floor and a ceiling. In between are structural piers that accommodate mechanical equipment, fixtures and appliances. They anchor the loosely defined spaces with functional amenities. For instance, the pier that contains kitchen appliances anchors the kitchen area.

The different rooms of the house have no substantial enclosure. They are constituted by the computer-controlled and mechanically sculpted fog. The walls of any given room thus emerge or disappear according to needs and desires. Rooms are in a constant flux of ‘formation’ and ‘dissolution’. They come into sharp focus on demand, their edges neatly defined by the air curtain, and dissolve into formless fog when not needed or seen.

---

❶ 由鲁道夫 · 埃尔 - 库利指导。

❶ Guided by Rodolphe el-Khoury.

The endless possibilities of responsive space-shaping scenarios include privacy that materialises only in anticipation of an approaching passer-by, rooms that expand or shrink according to need and compartments that materialise and open in response to a gesture.

感应式的空间塑造场景具有无限的可能性：只有在预期有路人经过时才显现私密性的分隔；房间可以根据需要扩大与缩小；一个手势就可以打开或显现墙体分隔。

图 2.16-1

图 2.16-2

图 2.16-1 田园环境中的雾宅景象
View of the Fog House in its pastoral setting

图 2.16-2 空气幕通过造雾机产生的雾为经过的路人开辟一个入口
Air curtains carve out a doorway for an approaching user from the mister-generated fog

图 2.16-3

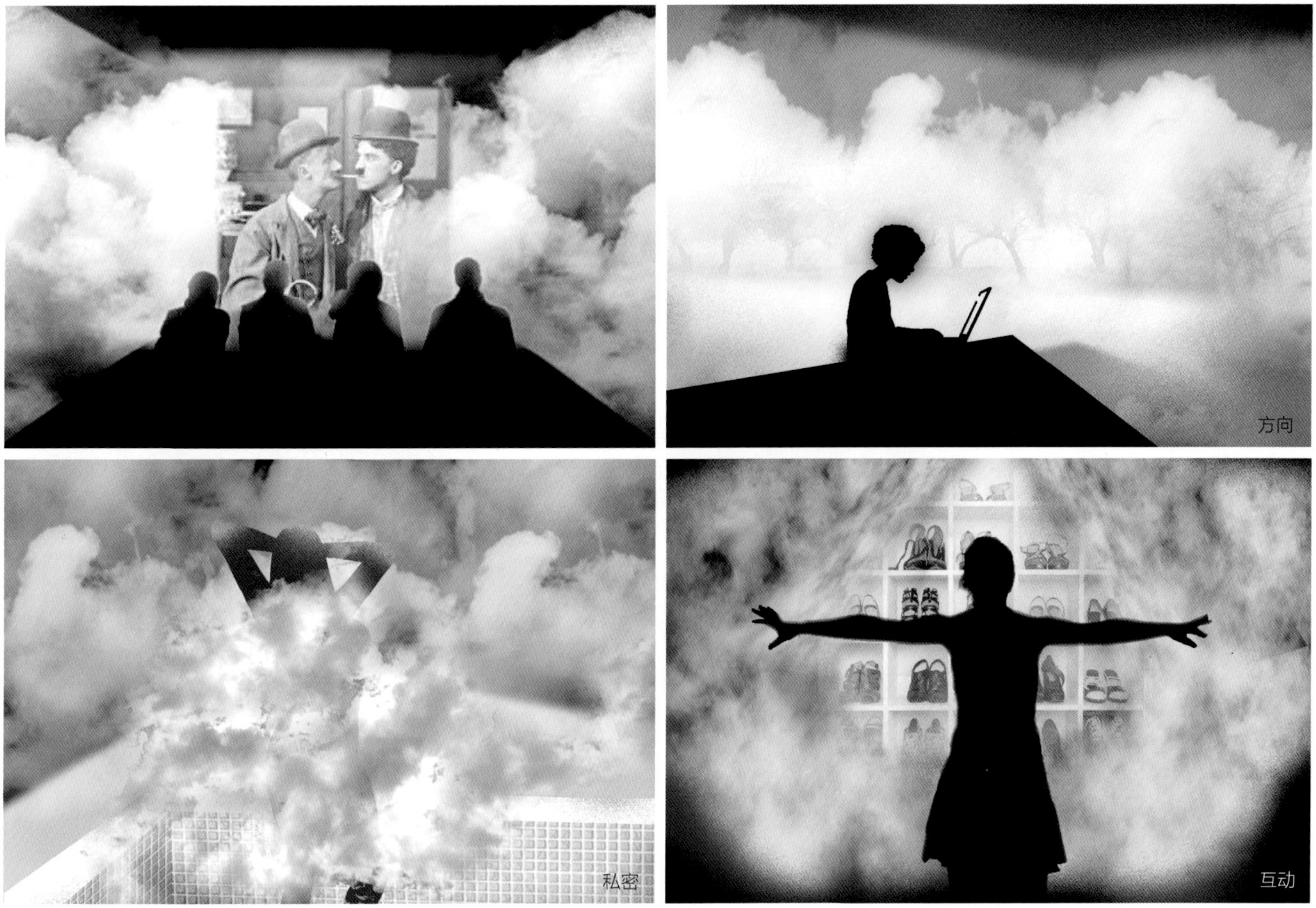

图 2.16-3 限定与控制雾形成墙体的不同场景
Various scenarios for defining and controlling the fog formed walls

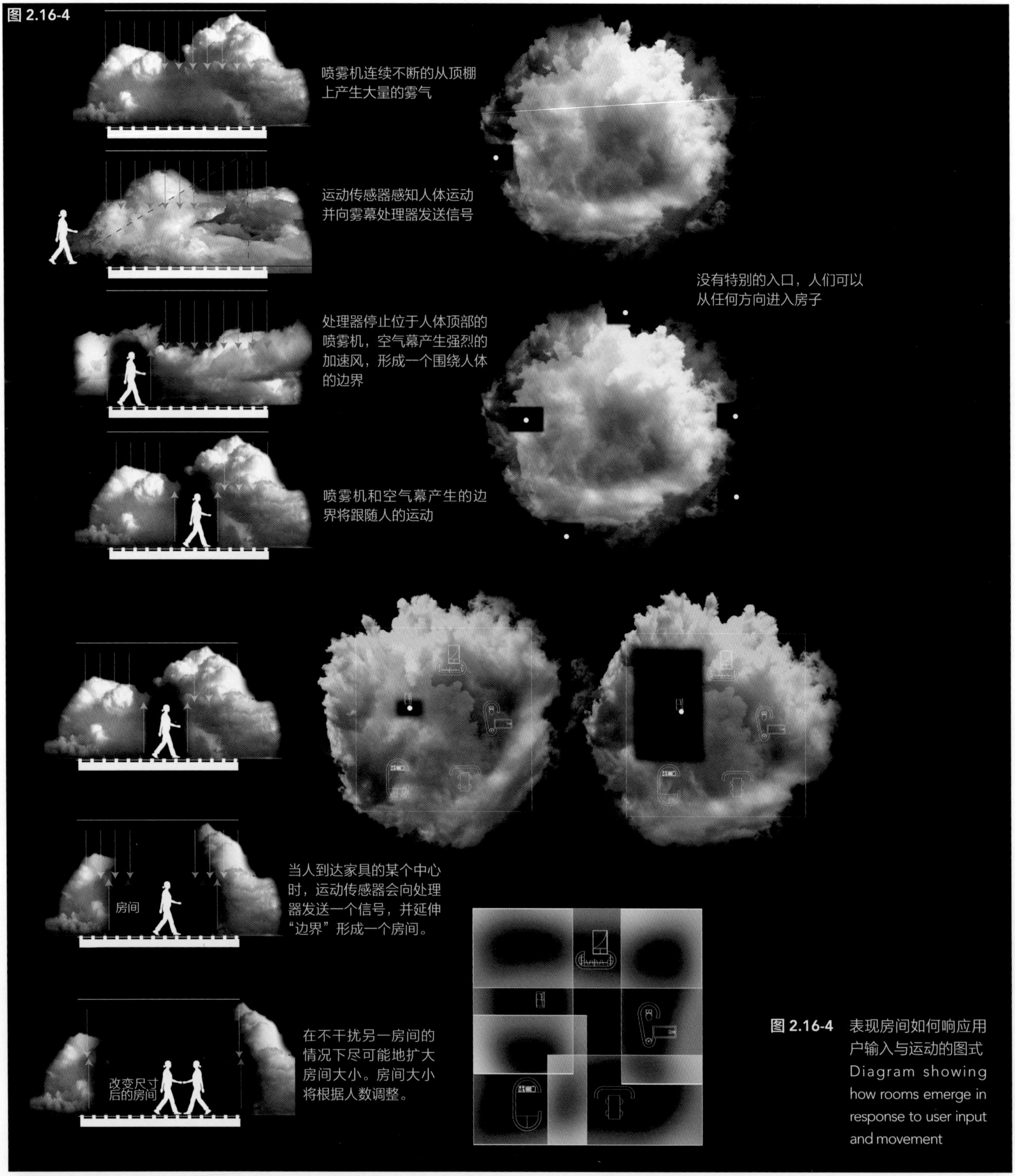

图 2.16-4
喷雾机连续不断的从顶棚
上产生大量的雾气
运动传感器感知人体运动
并向雾幕处理器发送信号
处理器停止位于人体顶部的
喷雾机，空气幕产生强烈的
加速风，形成一个围绕人体
的边界
喷雾机和空气幕产生的边
界将跟随人的运动
没有特别的入口，人们可以
从任何方向进入房子
房间
当人到达家具的某个中心
时，运动传感器会向处理
器发送一个信号，并延伸
“边界”形成一个房间。
改变尺寸
后的房间
在不干扰另一房间的
情况下尽可能地扩大
房间大小。房间大小
将根据人数调整。

图 2.16-4 表现房间如何响应用户输入与运动的图式
Diagram showing how rooms emerge in response to user input and movement

图 2.16-5

图 2.16-6

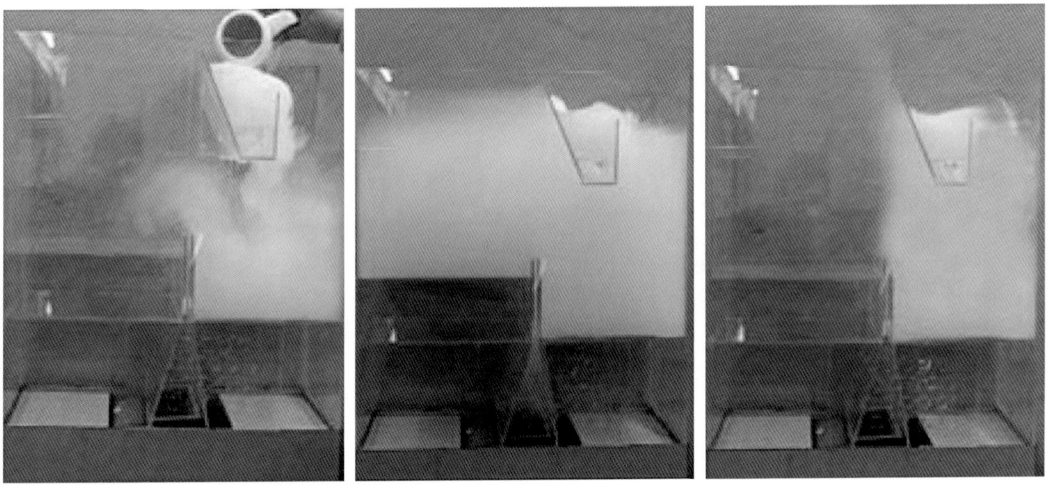

图 2.16-5 雾宅的剖面与平面
Section and plan of Fog House

图 2.16-6 测试与校准塑形雾的仪器
Machine for testing and calibrating the shaping of fog

2.17 同步实例化 Simultaneous Instantiations

"同步实例化"提出了一种空间概念，其中多个空间状态同时存在于一个场所。偏光滤光器用来给特定的光波方向编码，以产生一些独特的视觉感受。

内森·毕晓普

'Simultaneous Instantiations' proposes a spatial idea in which multiple spatial states exist in one place concurrently. Polarizing filters are used to encode specific orientations of light waves with unique visual experiences.

Nathan Bishop

Observers in the space then tune-in to any of these visual channels by rotating the polarized lenses in their uniquely designed glasses. In order for this spatial idea to be totalizing and ultimately effective, maintaining control over the orientation of these light waves is paramount. Maintaining the polarization of light in a space requires careful consideration for the highly specific interactions of light, material and geometry. As such, 'Simultaneous Instantiations' suggests that the manifestation of a desire for a multitude of differentiated visual, and ultimately spatial, experiences need not always be electronic, nor so removed from physical consequence in our material domain.

空间中的观察者通过旋转设计独特的偏光镜可以匹配到任意一个视觉通道。为了使这个空间概念全面并最终有效，保持对这些光波方向的控制是至关重要的。为了在空间中维持这种偏光，我们需要仔细考虑光线、材质、形状之间特定的相互作用。同样的，"同步实例化"表明，对于大量差异化的视觉、完全的空间感及体验愿望的表现，既不总是电子化的，也不是脱离我们的物质领域的物质结果。

图 2.17-1 编码图。建筑空间中的所有光线都将通过一系列不同方向的偏振滤光片进行编码，以形成特定的视觉体验。这些视觉通道通过空间传递给观察者，观察者通过旋转设计独特的眼镜，看到他们理想中的空间景象
Encoding diagram. All light in an architectural space would be encoded with specific visual experiences by passing the light through a series of differently orientated polarizing filters. These visual channels are then passed through space to an observer who can ‘tune-in’ to their desired spatial state by rotating the lenses of their uniquely designed glasses

图 2.17-1

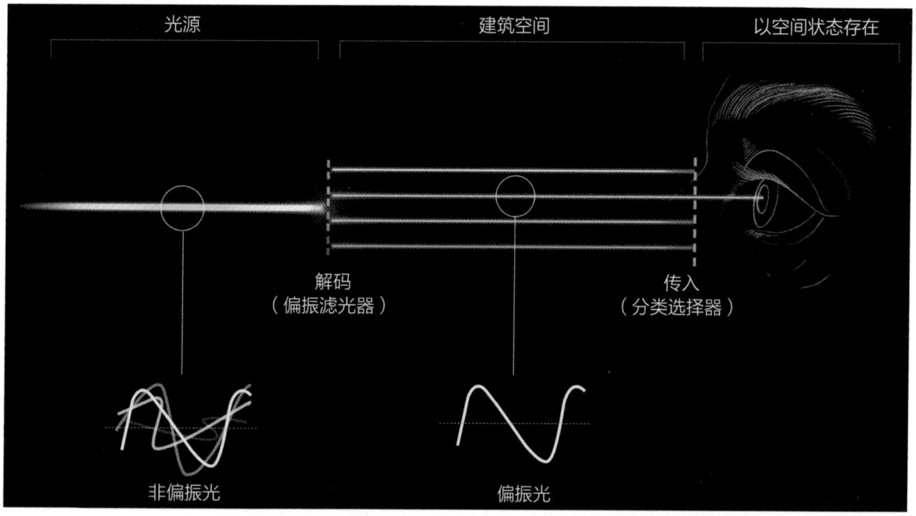

图 2.17-2

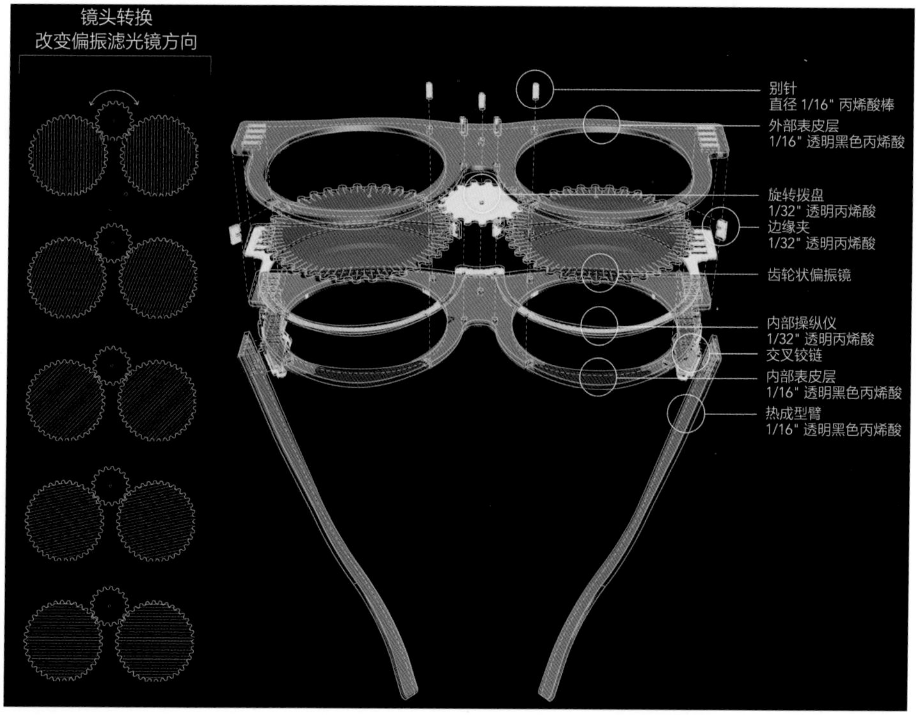

图 2.17-2 选择器。选择器眼镜采用齿轮偏振透镜设计，它通过转动观察者鼻梁上方的调谐旋钮来调整
Selectors. Selectors glasses are designed with geared polarized lenses which can be rotated by turning a tuning dial above the bridge of the observers nose

图 2.17-3

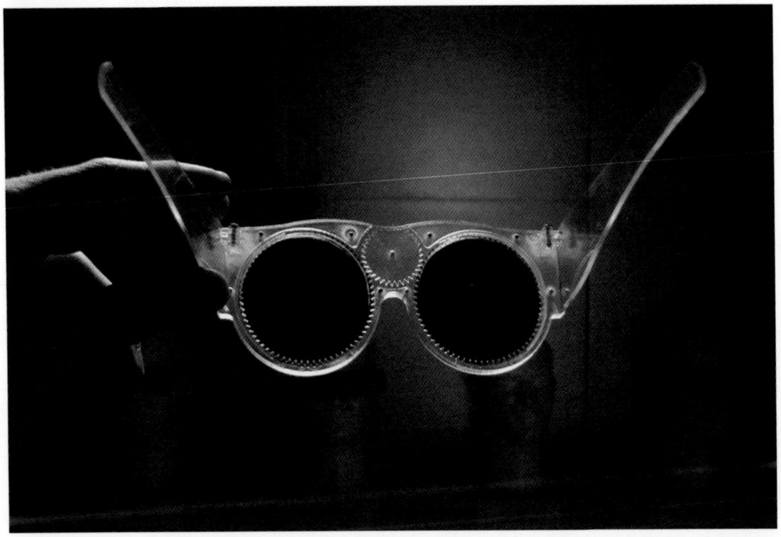

图 2.17-3 选择器
Selectors photograph

图 2.17-4 偏振图。在空间中保持光的偏振对发展空间状态至关重要。当光被某些材料或几何形状反射时，光容易失去它的偏振性。控制这一现象的两种方法是：使用吸光材料完全消除反射；或者使用与入射到其表面的波长方向相同的反射材料
Polarization diagram. Maintaining the polarization of light in a space is paramount to developing spatial states. Light can easily lose its polarization if it reflects off certain kinds of materials or geometries. Two means to control this are to either eliminate reflections completely using light absorbing materials, or to use material which reflect the same orientation of wavelength that was incident on its surface

图 2.17-4

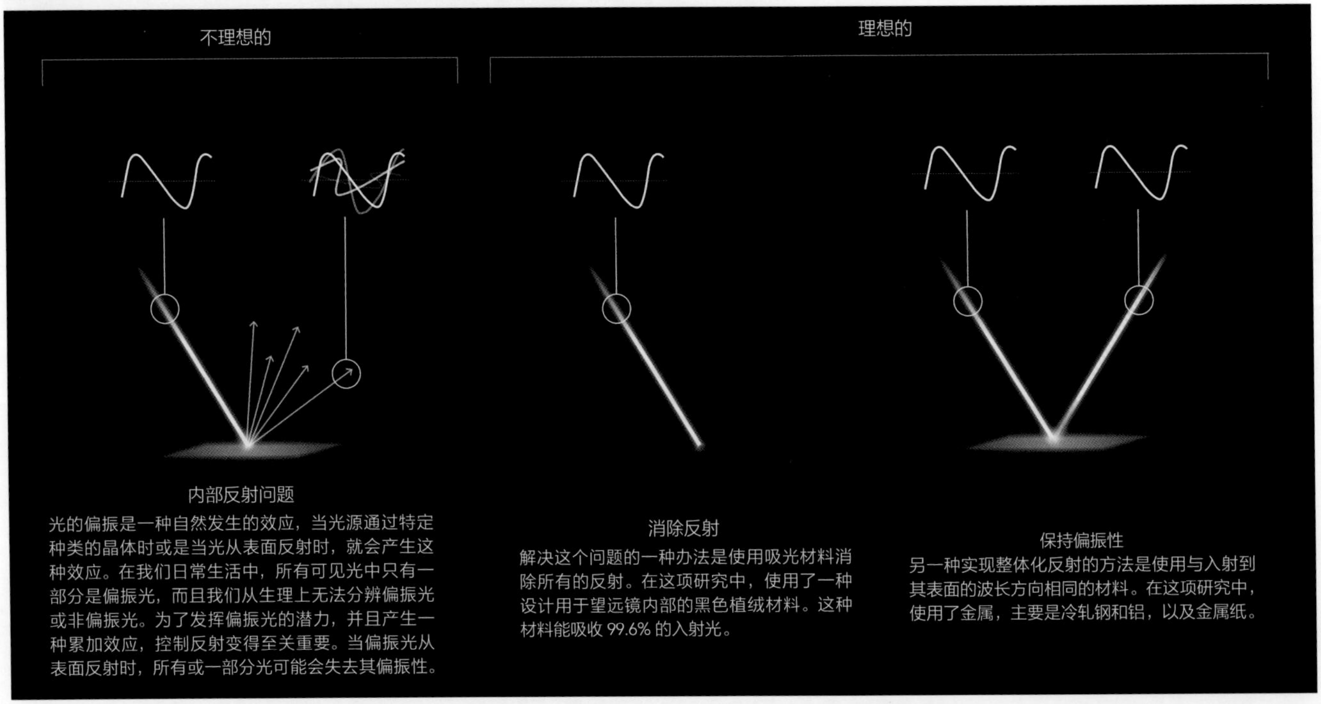
不理想的
理想的
内部反射问题
光的偏振是一种自然发生的效应，当光源通过特定种类的晶体时或是当光从表面反射时，就会产生这种效应。在我们日常生活中，所有可见光中只有一部分是偏振光，而且我们从生理上无法分辨偏振光或非偏振光。为了发挥偏振光的潜力，并且产生一种累加效应，控制反射变得至关重要。当偏振光从表面反射时，所有或一部分光可能会失去其偏振性。
消除反射
解决这个问题的一种办法是使用吸光材料消除所有的反射。在这项研究中，使用了一种设计用于望远镜内部的黑色植绒材料。这种材料能吸收 99.6% 的入射光。
保持偏振性
另一种实现整体化反射的方法是使用与入射到其表面的波长方向相同的材料。在这项研究中，使用了金属，主要是冷轧钢和铝，以及金属纸。

图 2.17-5

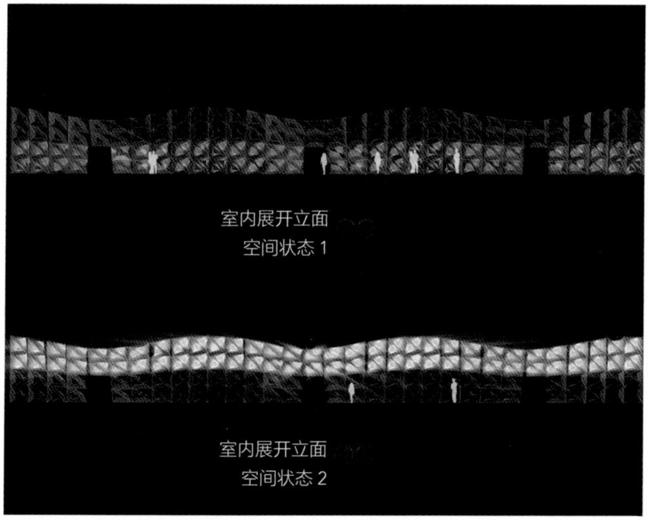

图 2.17-6

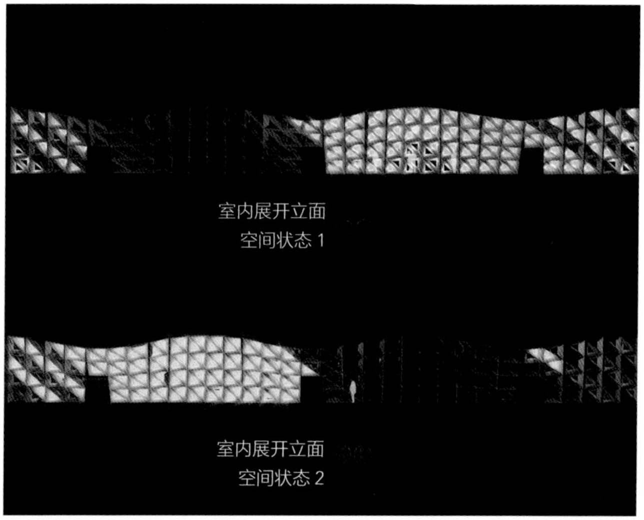

图 2.17-7

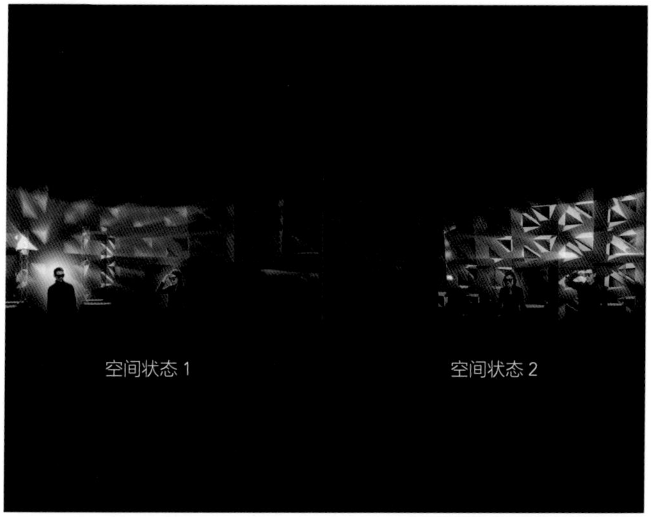

图 2.17-5 室内展开立面 1。这种多光圈的概念用于探索采用支持性原理的投机亭。在某种情况下，物体被体验为沐浴在红光和低高度中，同时它也存在沐浴在从高窗透进室内的白光的空间中
Unfoled interior elevations 1. A concept of multiple apertures was explored a speculative pavilion which employ enabling principles. One state could be experienced as bathed in red light and of low height, were as simultaneously existing in that space could be another bathed in white light emitted from a clerestory ribbon window

图 2.17-6 室内展开立面 2。这种多光圈概念也为空间状态的发展提供了更大的灵活性和可编程性。在这些实例中，空间中的不同凹口可以依靠观察者的镜头方向而被强化
Unfoled interior elevations 2. The concept of multiple apertures could also permit great flexibility and reprogrammability into the development of these spatial states. In this example, different bays of the space could be highlighted depending on the orientation of the observer's lenses

图 2.17-7 室内效果
Interior

嵌入式系统的实时传感与驱动力能够产生一个分散与网络化的组织，在这种组织中，单体建筑构件可以根据现场环境压力进行调整与适应，从而产生即时的整体配置。分布式智能可产生更加有效与更加弹性的系统，并能为管理环境动力提供更好的配置。以下项目是将建筑单元重新考虑为一个数字化的建筑量子，能够对动态的内部和外部压力做出动力学响应。

The real-time sensing and actuating capacities of embedded systems enables a decentralised and networked organisation where individual building components adjust and adapt locally to environmental pressures, generating emergent global configurations. The distributed intelligence yields a more efficient and resilient system that is better equipped for managing environmental dynamics. The following projects rethink the building unit as an architectural quantum that is augmented digitally and capable of kinetic responses to dynamic internal and external stresses.

3

能动构件

KINETIC COMPONENTS

3.1 可变隔热体 Variable Insulation

隔热体从墙体构件的限制中释放出来，使其能够适应热环境的改变。❶

克莱奥 · 巴斯特

Insulation is released from the confines of the wall assembly, freeing it to adapt to varying thermal conditions.❶

Cleo Buster

By relocating insulation from within the wall assembly to a flexible interior curtain, Variable Insulation is able to address shifting thermal needs as well as changing programmatic uses.

Two curtain prototypes were fabricated using two different materials. The first curtain is made of many individual aluminised Mylar pillows attached together. Each pillow is connected by polyethylene tubing to a centralised air pump and fitted with two-way valves. This enables the curtain to be inflated for increased insulation values, and deflated for decreased insulation values. The curtain responds to both exterior and interior temperatures, giving the inhabitant control over their environment.

The second prototype consists of individual cells constructed from one continuous piece of polyethylene tubing. Thus, the infrastructure becomes the structure, eliminating the need for two separate systems.

通过将隔热体从墙体构件中移出，并置入灵活柔性的室内窗帘中，可变隔热体就能够满足不断变化的热需求以及相应的空间规划改变。

两个窗帘原型是用两种不同的材料制作的。第一个窗帘是由许多单独的镀铝聚酯薄膜的枕头相互连接而组成。每个枕头都通过聚乙烯管连接到中央空气泵，并配有双向阀连接。这使得窗帘可以通过充气来提高绝热值，通过放气来降低绝热值，以此响应室外与室内的温度变化，让居住者进行自我环境控制。

第二个窗帘原型是由一个连续的聚乙烯管组成的独立单元构成，因此，基础设施变为结构，

❶ Guided by Christos Marcopoulos and Carol Moukheiber.

❶ 由克里斯托 · 马尔科普洛斯与卡罗尔 · 穆海贝尔指导。

消除了对两个独立系统的需求。每个单元有一个透明的乳胶气球连接着一个双向阀门，并与系统的中央矩阵相连。气球窗帘具有多重功能：同时起到绝热、调节私密性与控制光线的作用。

Each cell has a clear latex balloon attached to a two-way valve, which is connected to the central matrix of the system. The balloon curtain has multiple functions: it simultaneously acts to insulate, mediate privacy and control light.

图 3.1-1

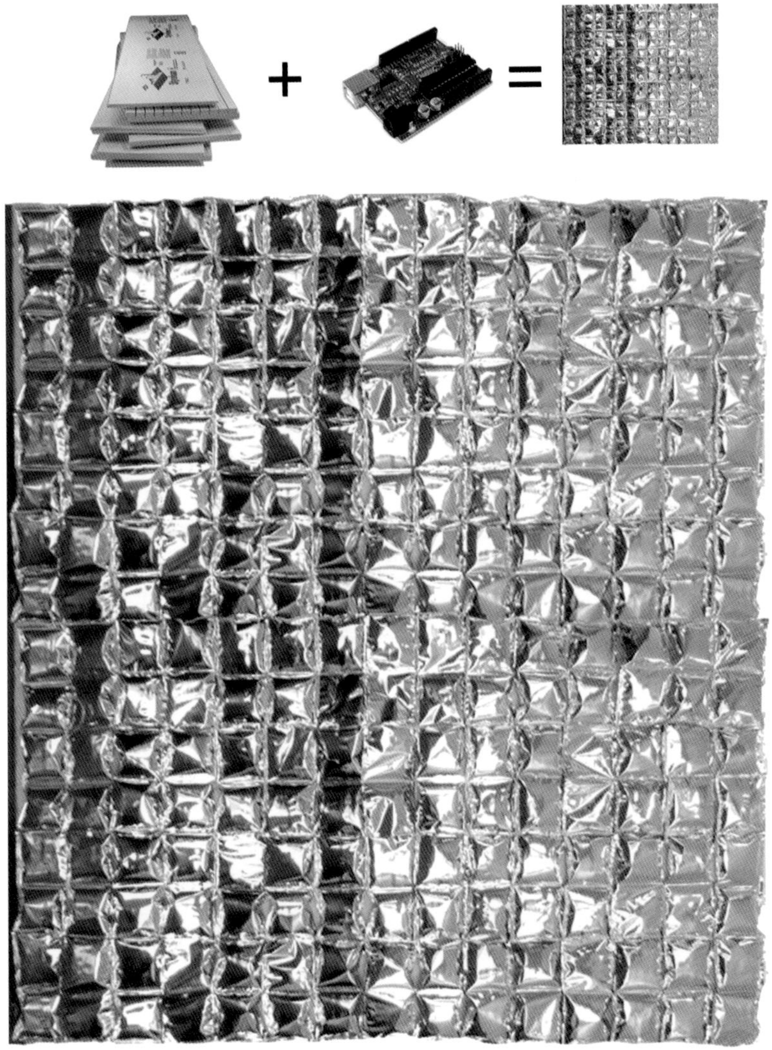

图 3.1-1 窗帘能够通过改变在每个聚酯薄膜枕头内的空气所占空间的大小来达到多个 R 值，每个枕头都是独立地响应它周围的环境参数
The curtain can achieve multiple R-values by altering the air space within each Mylar pillow. Each pillow is independently responsive to its surrounding ambient parameters

图 3.1-2

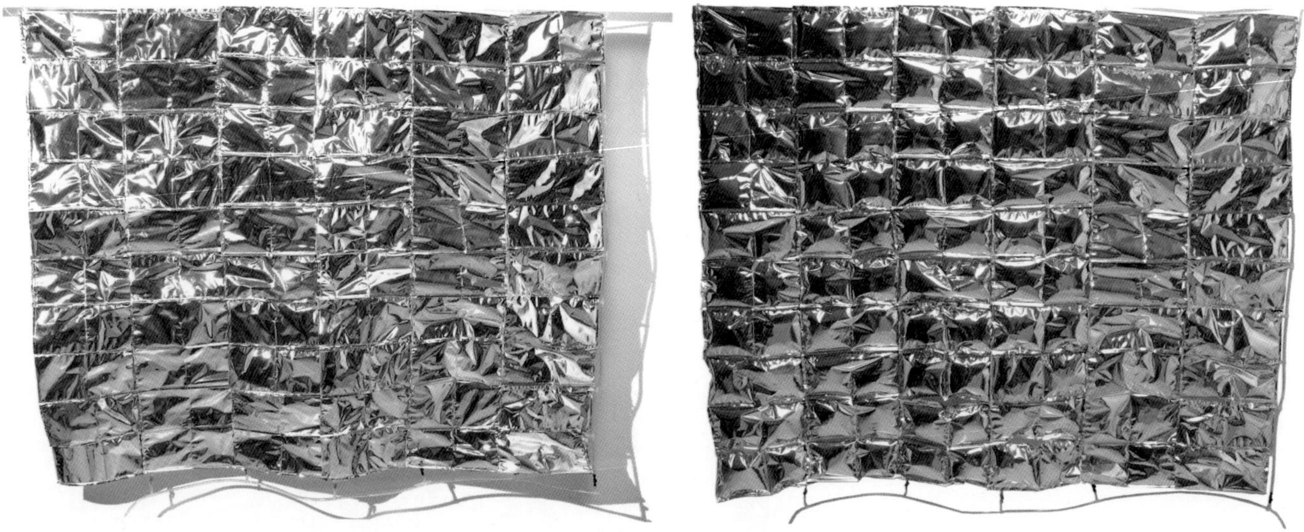

图 3.1-3

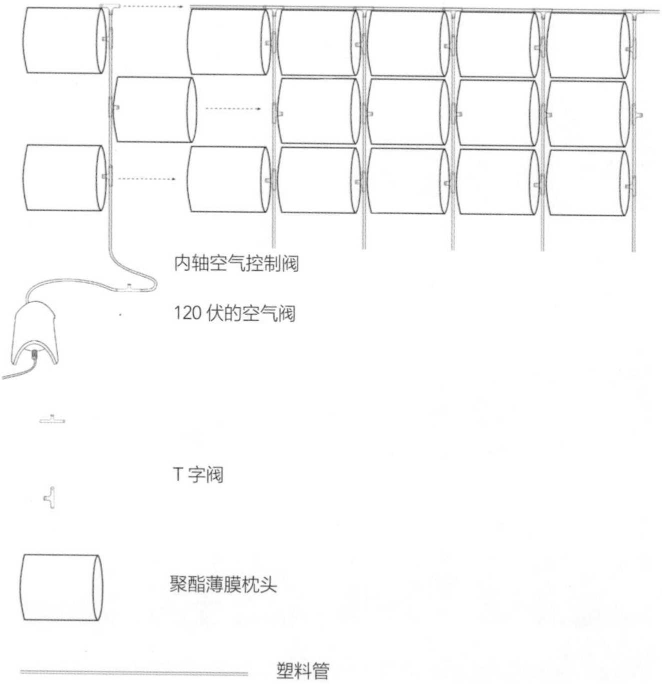

图 3.1-2 窗帘的充气与放气
The curtain deflated and inflated

图 3.1-3 每一个聚酯薄膜枕头插入空气供给设施
Each Mylar pillow plugs into the air supply infrastructure

图 3.1-4

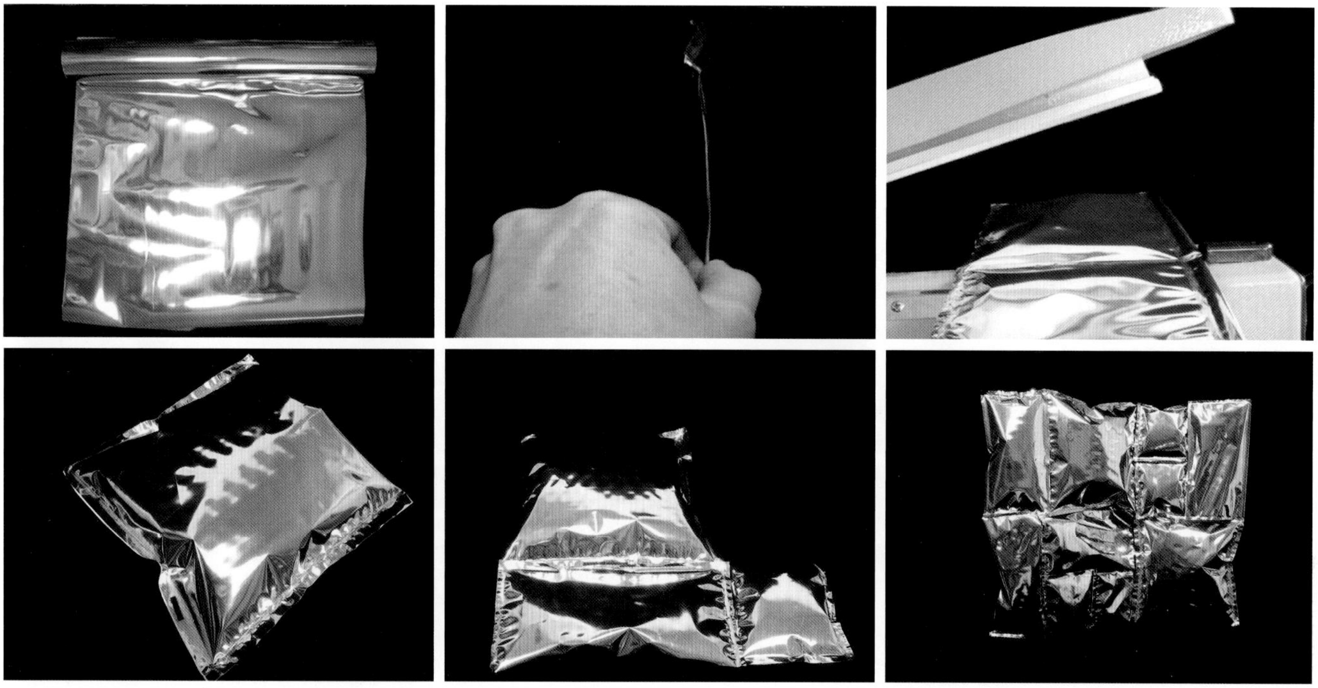

图 3.1-4 窗帘原型是用商业食品包装厂里的铝箔与一台封口机制成的
The prototype was constructed using foil and a sealing machine from a commercial food packaging plant

图 3.1-5

图 3.1-5 项目原型以气球与空气管作为结构与基础配置，探索改变材料表面性质（不透明、绝热与开放）的潜力
Using balloons and air tubing as both structure and infrastructure, this prototype explores the potential to change a ‘surface’s qualities: opaque and insulating vs. open

3.2 新风窗户 Fresh-Air Window

密闭的玻璃幕墙模块可提供新风。

NMinusOne 工作室，瓦伦蒂娜 · 梅勒

A sealed glazed curtain wall module is capable of supplying fresh air.

Studio NMinusOne with Valentina Mele

“新风窗户”是一个密封的幕墙单元，在双层玻璃窗中嵌入了热回收通风元件，它允许进入的新风被排出的室内空气加热或冷却。

从传统意义上讲，热回收通风设备一般位于中央，但相反的是，新风窗户却将热回收通风设备分散在整个建筑中。通过传感装置，每扇窗户能够测定室内空气品质，当室内空气品质下降时，相关的新风窗户便会通过引入新鲜空气来响应，所需设施就嵌在建筑的砌块内。

当你的家容纳了许多复杂的管道，如烟道、输气道、电线、灯具、入口、出口、烤炉、水槽、垃圾处理器、高保真音响、天线、导管、冰箱、热水器——当家里容纳了如此多的设施以至于这些硬件设施可以自行运转，无需借助房子的帮助，那么为什么还要一个房子来支撑它？因而，雷娜 · 班哈姆说“家不是房子”。

The ‘Fresh-Air Window’ is a sealed curtain wall unit with a Heat Recovery Ventilation element embedded in the double glazing, allowing incoming fresh air to be heated or cooled by the outgoing interior air.

Traditionally, heat recovery ventilators are centrally located. By contrast, the Fresh-Air Window decentralizes the Heat Recovery Ventilation throughout the building. Through sensing devices, every window has the capacity to gauge interior air quality. When local air quality declines, the respective Fresh-Air Windows respond by introducing fresh air. Infrastructure is now embedded within the building block.

When your house contains such a complex of piping, flues, ducts, wires, lights, inlets, outlets, ovens, sinks, refuse disposers, hi-fi reverberators, antennae, conduits, freezers, heaters — when it contains so many services that the hardware could stand up by itself without any assistance from the house, why have a house to hold it up? -Reyner Banham, ‘A Home Is Not A House’.

图 3.2-1

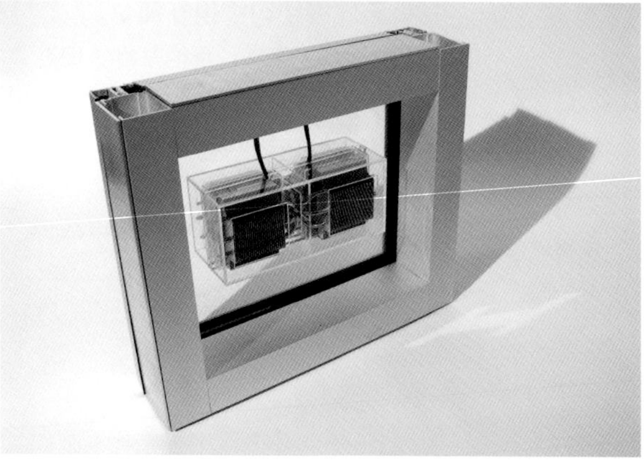

图 3.2-2

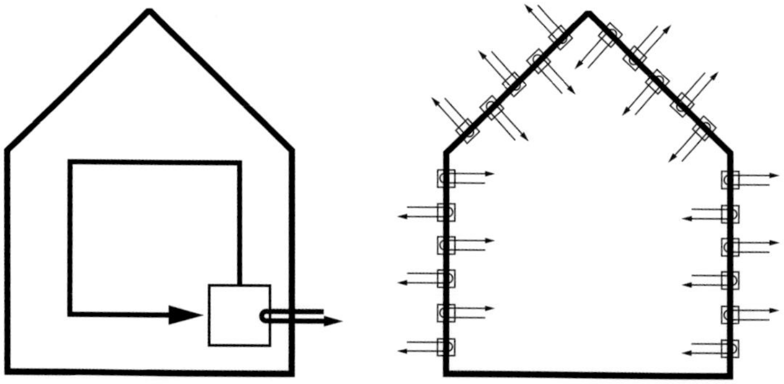

图 3.2-3

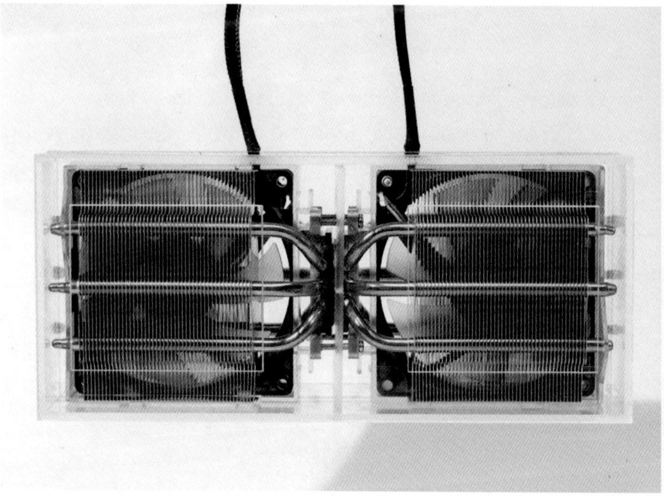

图 3.2-1 新风窗户采用了两个背对背的中央处理单元冷却槽，这种配置创造了一个即时性的热回收通风器
The Fresh-Air Window employs two back-to-back Central Processing Unit cooling sinks. This configuration creates an instant Heat Recovery Ventilator

图 3.2-2 传统系统：集中式的热回收通风；拟建系统：将热回收通风分散在整个建筑的围护结构
Traditional system: centralized heat-recovery ventilation; Proposed system: distributed heat recovery ventilation over the entire building envelope

图 3.2-3 两个背对背的中央处理冷却单元形成一个热回收通风器
Two back-to-back CPU cooling units form a Heat Recovery Ventilator

图 3.2-4

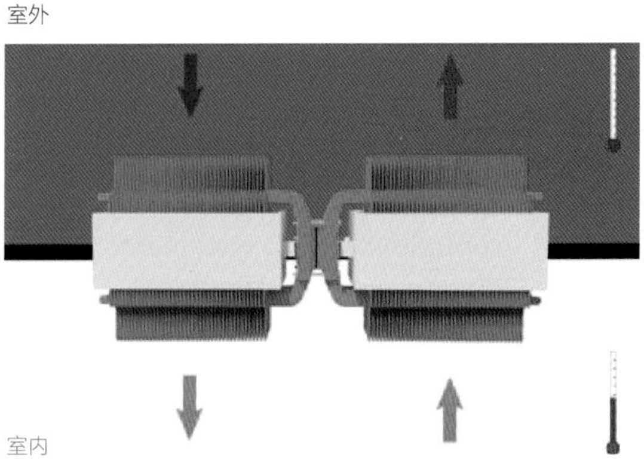

图 3.2-5

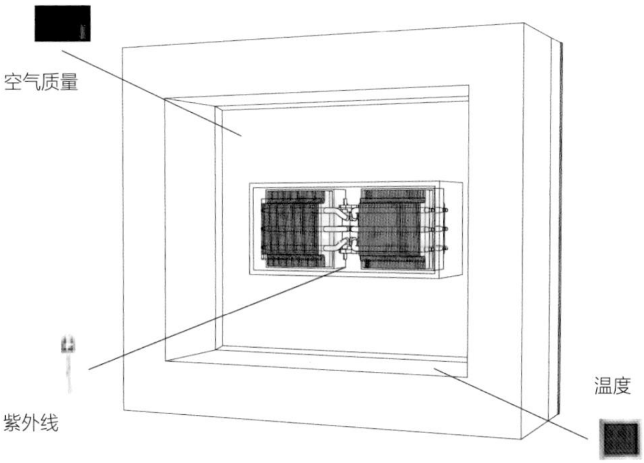

图 3.2-6

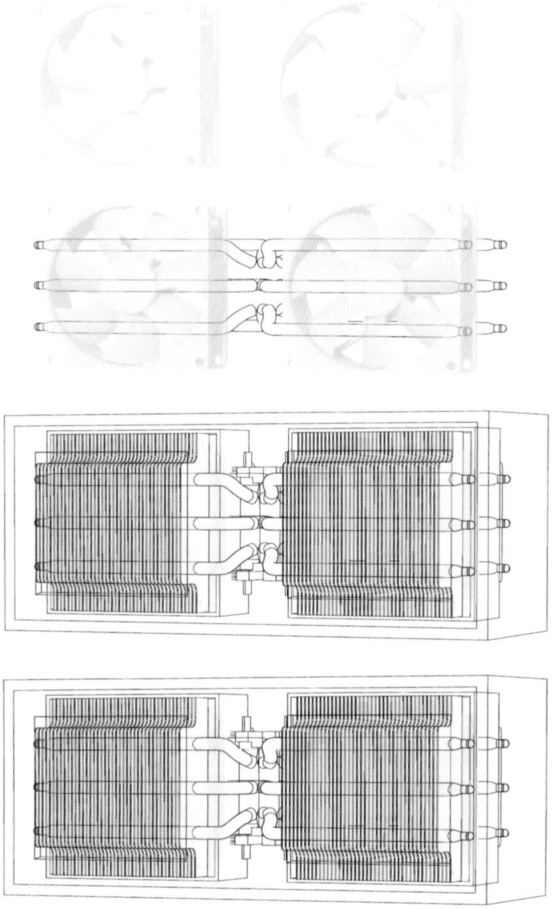

图 3.2-4 热回收通风器示意图：新鲜冷空气通过叶片从室外进入，被从室内排出的热而污浊的空气所加热
Diagram of the Heat Recovery Ventilator: fresh cold air comes in from the exterior through the fins and is heated by outgoing warm, stale air from the inside going out

图 3.2-5 作为空气质量监测器的窗户
The window as air quality monitor

图 3.2-6 中央处理单元热回收通风器组件：风扇、铜管与叶片
CPU HRV components: fans, copper tubing, and fins

3.3 供热与制冷玻璃 Heating and Cooling Glass

通过将先前的中央空调系统布置在玻璃系统中，玻璃就成了触觉传感器。❶

瓦伦蒂娜 · 梅勒

By distributing the previously centralised HVAC system into the glazing system, glass becomes a haptic transmitter.❶

Valentina Mele

'Heating and Cooling Glass' explores the potential for the migration of the central HVAC system into a decentralized modular surface.

The double glazed Heating and Cooling Glass unit is used opportunistically as the site of maximum difference in temperature. Mediating incoming and outgoing air at the envelope takes advantage of the 'building's skin as the interface between two different climatic zones. That difference is harnessed- through a Peltier device inserted in the glazing unit-to generate heating and cooling locally, as needed by the user, or as necessitated by weather conditions.

By distributing the previously centralised HVAC system into the glazing system, glass becomes not only a visual transmitter of the outside but also a haptic transmitter. In a tactile way, it fulfils the quest of modernism to achieve continuity between inside and outside.

❶ Guided by Christos Marcopoulos and Carol Moukheiber.

“供热与制冷玻璃”探索了中央空调系统向分散的模块化表面转移的潜力。

双层的供热与制冷玻璃单元恰好被用于温差最大的场所，建筑表皮作为不同气候区间的界面，被用来调节建筑围护结构的进气与排气，其原理是将珀耳帖效应❷装置插入玻璃单元，根据使用者的需要与天气条件来局部供热与制冷。

通过将先前的中央空调系统分布在玻璃系统中，玻璃不仅成为外部视觉景观的传导器，而且也是触觉传导器。在触觉上，它实现了现代主义追求的内外之间的连续性。

❶ 由克里斯托 · 马尔科普洛斯与卡罗尔 · 穆海贝尔指导。
❷ 译者注：1834 年法国科学家珀耳帖发现了热电制冷和制热现象即温差电效应。珀耳帖效应是指当有电流通过不同导体组成了回路时，在不同导体的接头处随电流方向的不同分别出现吸热、放热现象。

图 3.3-1

图 3.3-1 通过供热与制冷玻璃投射的阴影
Shadow cast by the Heating and Cooling Glass

图 3.3-2

图 3.3-2 微型供暖与空调装置由一个珀耳帖效应装置插入两个背对背安装的中央处理单元冷却槽之间，该装置是通过使电流反向流动到珀耳帖效应装置来实现制冷与供暖
The miniaturised heating and air-conditioning unit consists of a Peltier device inserted between two back-to-backmounted CPU cooling sinks. The device cools and heats by reversing the current flow to the Peltier device

图 3.3-3

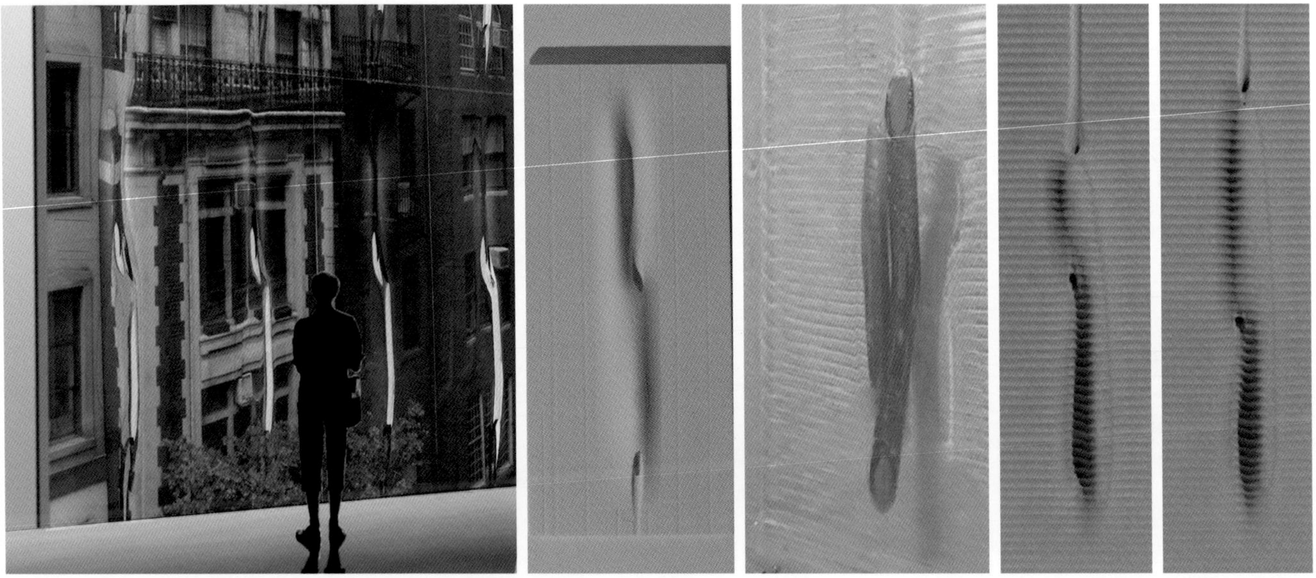

图 3.3-4

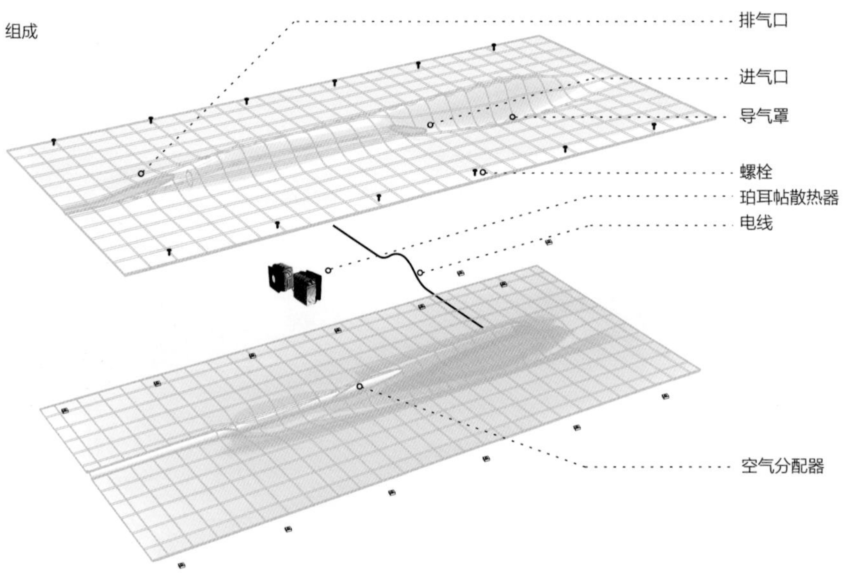

图 3.3-5

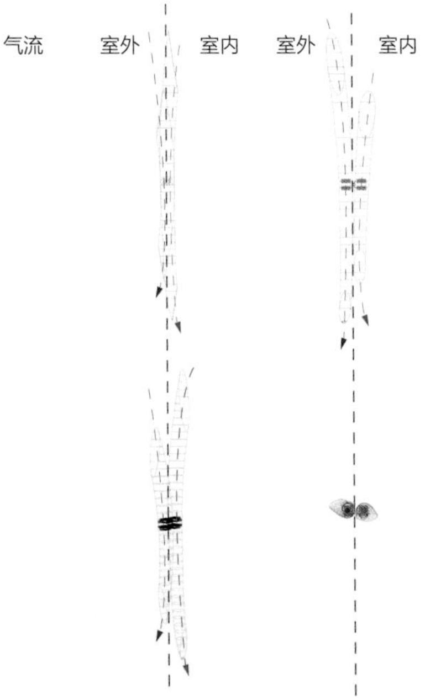

图 3.3-6

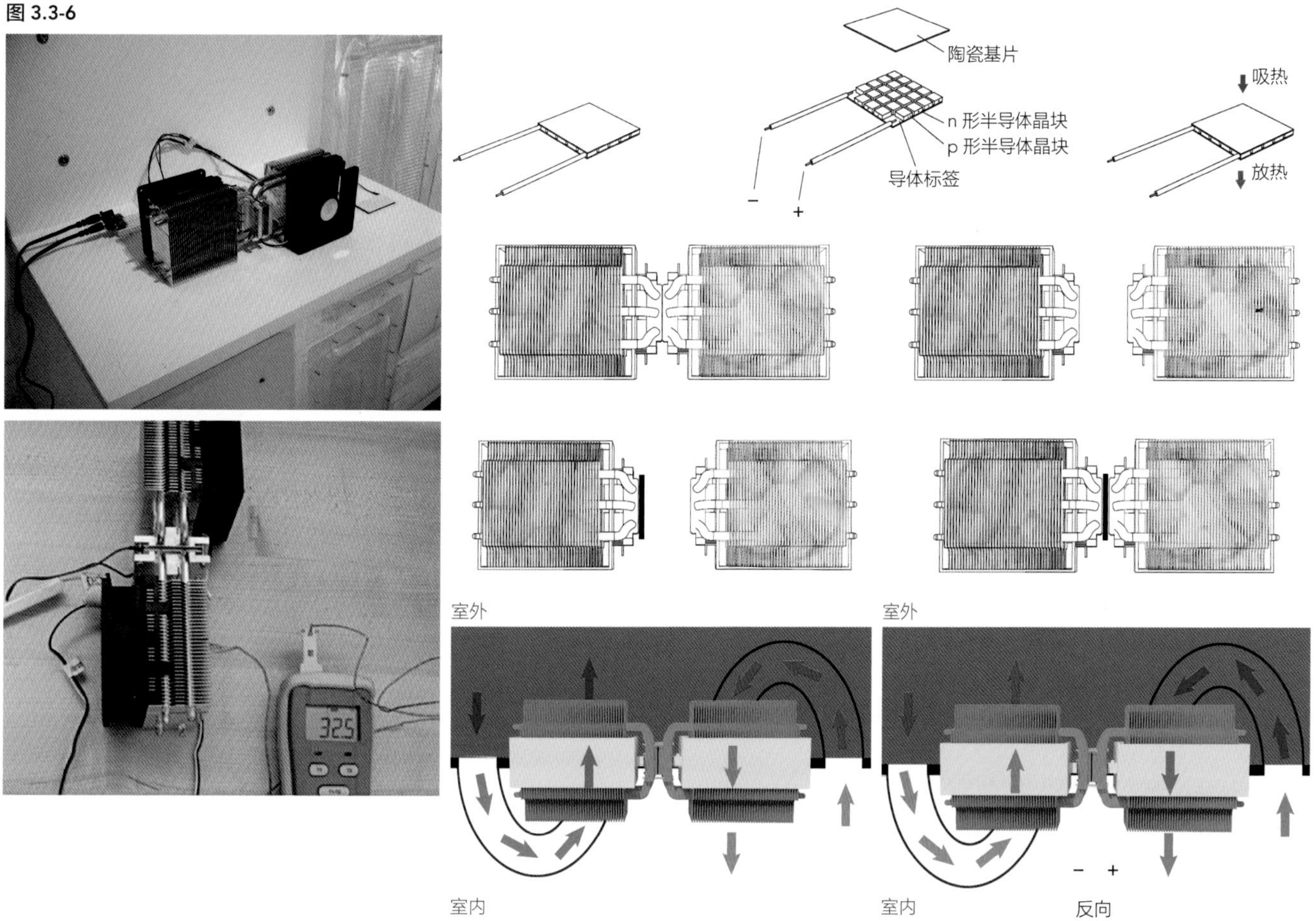

图 3.3-3 玻璃本身被弯曲以适应其新的性能
The glazing itself is slumped to accommodate its new performance

图 3.3-4 玻璃夹层。两片玻璃通过微型热泵使空气直接进出
The glazing sandwich. Two sheets of glass direct air in and out through the miniaturised heat pump

图 3.3-5 表示气流流动的平面与剖面的管道设施示意图
Diagram of the tubing infrastructure showing the airflow in plan section and elevation

图 3.3-6 珀耳帖效应设备实质上是一个微型热泵；通过将珀耳帖效应设备的电流反向流动，可根据居住者供暖与制冷的需求来调整气流流动
The Peltier device is essentially a mini-heat pump; by reversing the current on the Peltier device, airflows can be adjusted according to the heating and cooling needs of the inhabitants

3.4 可变几何桁架 Variable Geometry Truss

能够动态响应偏心荷载的可变几何桁架。❶

戴维 · 朗

A structural truss with a variable geometry that responds dynamically to eccentric loads.❶

David Long

Variable Geometry Truss contributes to the growing field of Intelligent Responsive Architecture by using a novel approach to assembly, construction, modularity, and intelligence. It references 'Swarm' behaviour observed in nature in flocking birds, fish, and bees.

Responsive architecture combines built forms with integrated systems that are capable of responding to changing conditions. Responsive architecture is not merely concerned with physicality but is more focused on behaviour. This emerging genre of architecture is at the forefront of an imminent paradigm shift in architectural design where buildings will no longer be static but rather dynamic objects which reflect the capabilities and requirements of the techno-environmentally conscious age.

可变几何桁架结构主要采用了一种新的集装配、建造、模块化与智能化于一体的新方式，带动了智能感应式建筑领域的新发展。它引用了在自然界中被观察到的群居的鸟、鱼与蜜蜂的"群"行为。

感应式建筑将建成环境与集成系统结合起来，以应对不断变化的环境。而且其不仅关注物质属性还注重行为，这一新兴的建筑类型是建筑设计即将到来的范式转换的前沿领域，在这种转变中，建筑不再是静态而是动态的对象，能够回应技术环保时代的能力与需求。

❶ Supervised by Rodolphe el-Khoury.

❶ 由鲁道夫 · 埃尔 - 库利监制。

图 3.4-1

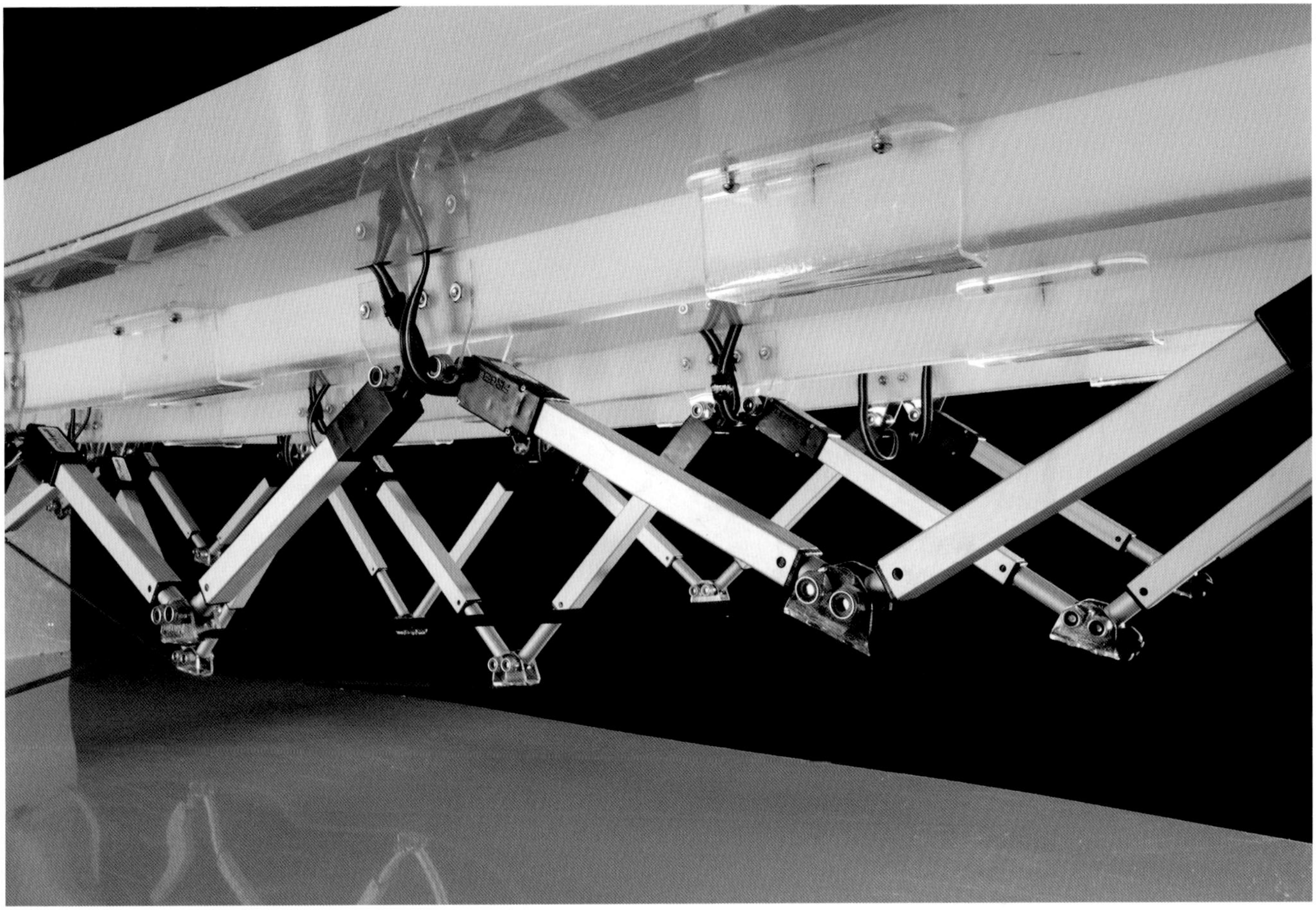

图 3.4-1 该设计使用了“群”逻辑，允许每个驱动桁架杆件能够连续和独立地适应由传感器阵列测量和传递的局部压力
The design uses a ‘Swarm’ logic allowing each actuated truss member to continuously and independently adapt in response to local pressures measured and relayed by an array of sensors

图 3.4-2

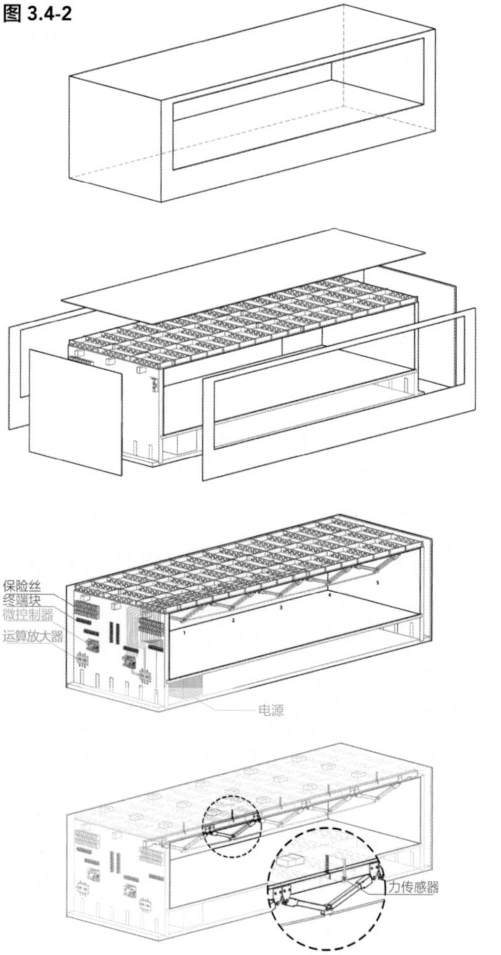

图 3.4-3

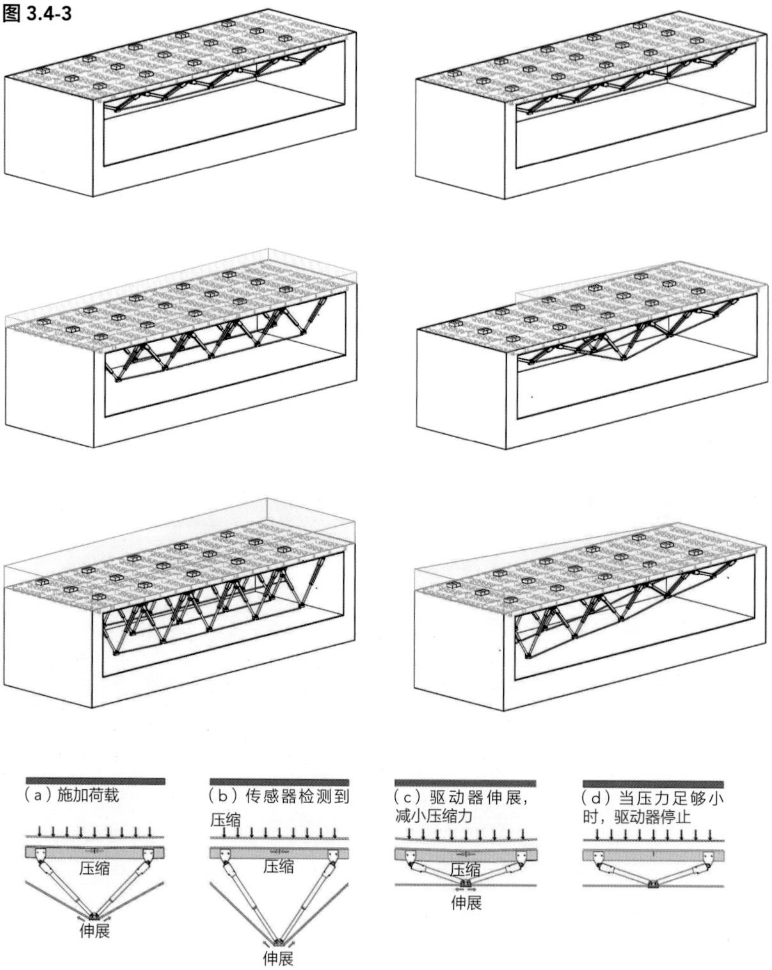

图 3.4-2 如果给定了上弦点荷载，在荷载下的两个网状骨架（驱动器）应延伸到相同的长度以分配荷载，为了实现这一目标，驱动反应桁架是以“群”的方式执行的。群中的单个单元是由两个驱动器与荷载感应器（置于上弦两个驱动器之间）构成

Given a point load on the top chord, both webs (actuators) below that load should extend to the same length to distribute the load. To achieve this, the actuated responsive truss is implemented as a ‘Swarm’; a single cell in the swarm consists of two actuators and a load sensor (positioned between the two actuators, on the top chord)

图 3.4-3 施加荷载 // 传感器检测到压缩 // 驱动器伸展，减小压缩力 // 当压力足够小时，驱动器停止

Load is applied // Sensor detects compression // Actuators extend, decreasing compressive force // Actuators stop when force is sufficiently small

图 3.4-4

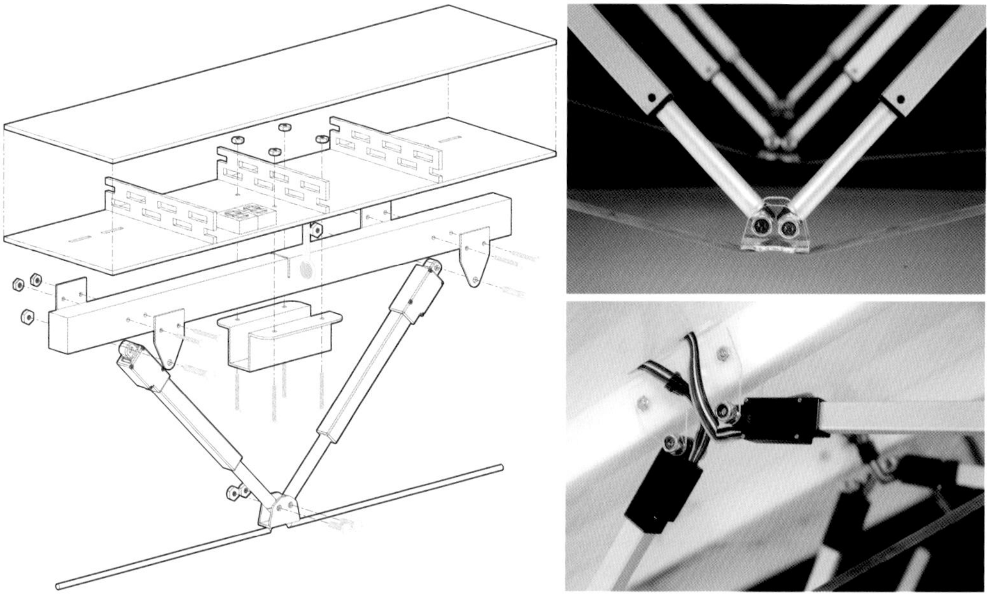

图 3.4-5

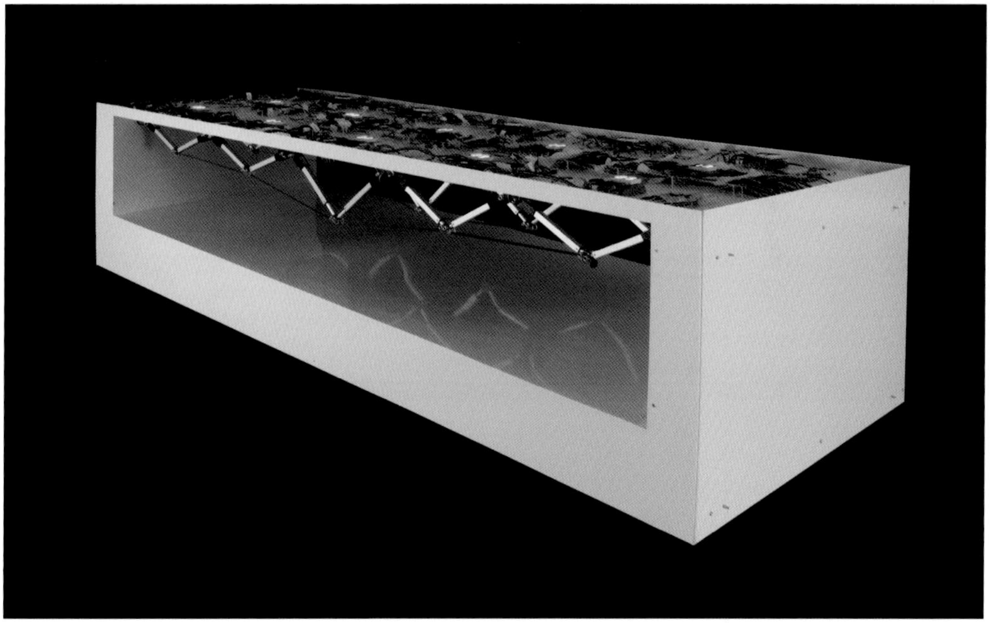

图 3.4-4 单一模块细部
Single cell detail

图 3.4-5 动态原型
Dynamic prototype

图 3.4-6

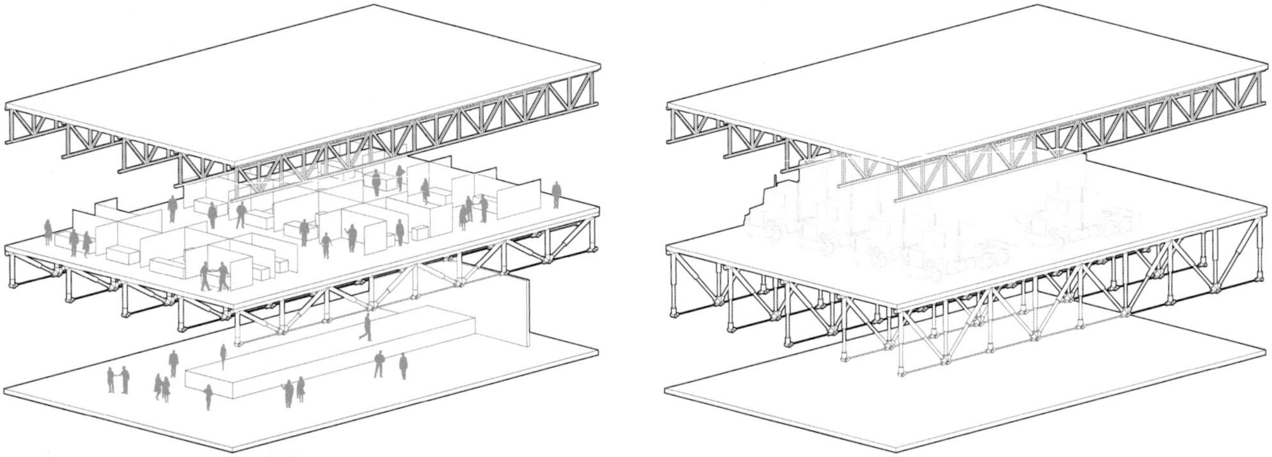

图 3.4-7

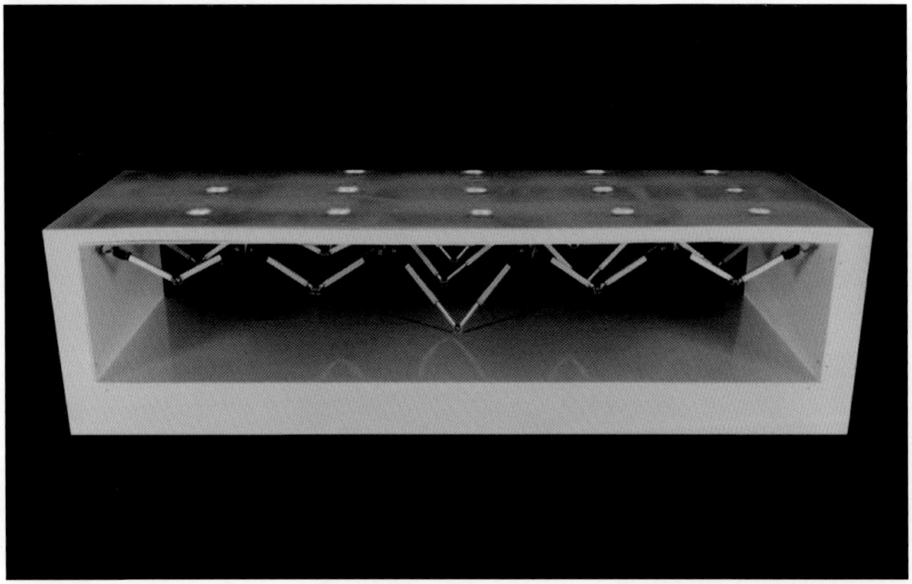

图 3.4-6 左图是装有能优化空间的感应式驱动桁架的会展中心，右图是桁架能够响应所承载的最大荷载
On the left, convention centre with responsive actuated truss that optimises space, and on the right, the truss responds to accommodate maximum loads

图 3.4-7 驱动感应桁架
Actuated responsive truss

3.5 寻光砖 Light-Seeking Brick

“寻光砖”是根据温度与光照条件而感应旋转，被赋予活力的砖墙。[1]

麦琪 · 纳尔逊

A brick wall is animated through responsive rotation based on temperature and lighting conditions.[1]

Maggie Nelson

“寻光砖”作为一个能动覆盖面系统，具有可变多孔性，能够创造基于环境识别的可变美学居家环境。

尽管砖在世界范围内仍然是最基本的建筑构件，但砖墙被看作本质上为静态的对象，砖砌体很少被视为是一种具有创新性的建筑材料。通过对砖墙单元体的再设计，植入感应式技术，将标准砖墙活化，使建立动态系统成为可能。

“寻光砖”通过旋转运动，它们的旋转度受温度和室外照明条件制约。基于环境原则，砖系统通过立面调节室温和采光，从而控制室内空间。

动态砖在交替的过程中运行，位于静态砖的

‘Light-Seeking Bricks’ act as a kinetic cladding system with variable porosity to create variable aesthetic conditions within the home based on environmental readings.

Though bricks remain a fundamental building component worldwide, the brick wall is viewed as a fundamentally static object. Brickwork is rarely seen as an innovative material. By redesigning the brick unit to house responsive technologies, it is possible to animate a standard brick wall to create a dynamic system.

The ‘Light-Seeking Bricks’ move through rotation. Their degree of rotation is conditioned based on temperature and exterior lighting conditions. By following an environmental logic, the brick system modulates heat and light gain through the façade for control of the interior space.

The kinetic bricks occur in alternating courses, sitting atop static bricks, within which are embedded light and temperature sensors that

---

❶ 由鲁道夫 · 埃尔 - 库利与纳希德 · 纳比安指导。

❶ Guided by Rodolphe el-Khoury and Nashid Nabian.

direct the brick rotation. Each section of static bricks is supported from the rear by a lightweight structure made of suspended cables or an alternative minimal system. The entire system acts as the rain screen portion of a double-skin façade, in front of a glass façade which functions as an air and water barrier and drainage plane.

Maintaining the brick form throughout the façade allows the interactive system to invoke the elements of surprise and wonder as a seemingly tradition material is operated in an extraordinary way.

顶部，其中嵌入了光和温度传感器，以指导砖的旋转。每部分静态砖通过由悬浮电缆或可变度最小的系统组成的一种轻型结构从后方支撑。整个系统作为一个双层立面的雨幕部分，置于玻璃立面前面，作为空气和水的屏障和排水平面。

在整个立面保持砖的形式，能够使交互系统激发一些惊奇和有效的元素，一个看似传统的材料在以一种非同寻常的方式运行。

图 3.5-1

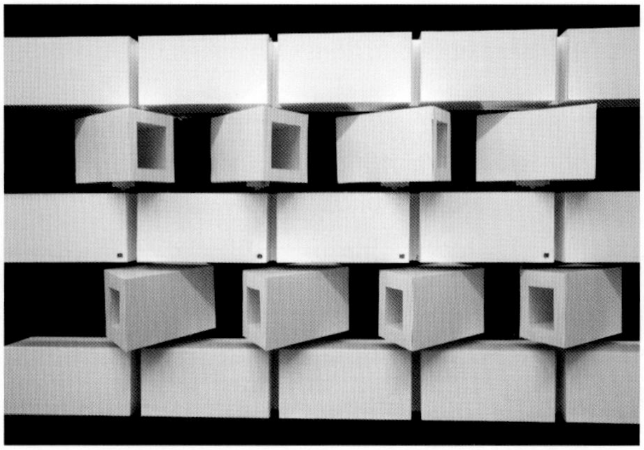

图 3.5-1 置于静态砖顶部的动态砖在交替过程中工作的原型。每块砖都嵌入了光和温度传感器，以指导砖的旋转
Prototype with kinetic bricks occurring in alternating courses atop static bricks. Light and temperature sensors to direct the brick rotation are embedded within each brick

图 3.5-2

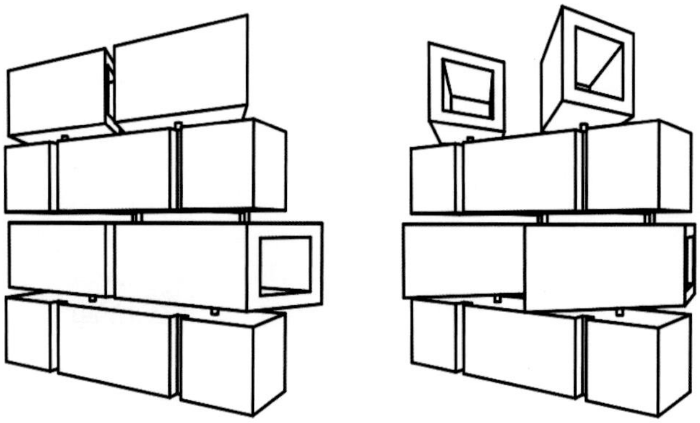

图 3.5-2 这种传统的建筑材料能被伺服电机推动
This traditional building material can be animated Servo Motors

图 3.5-3

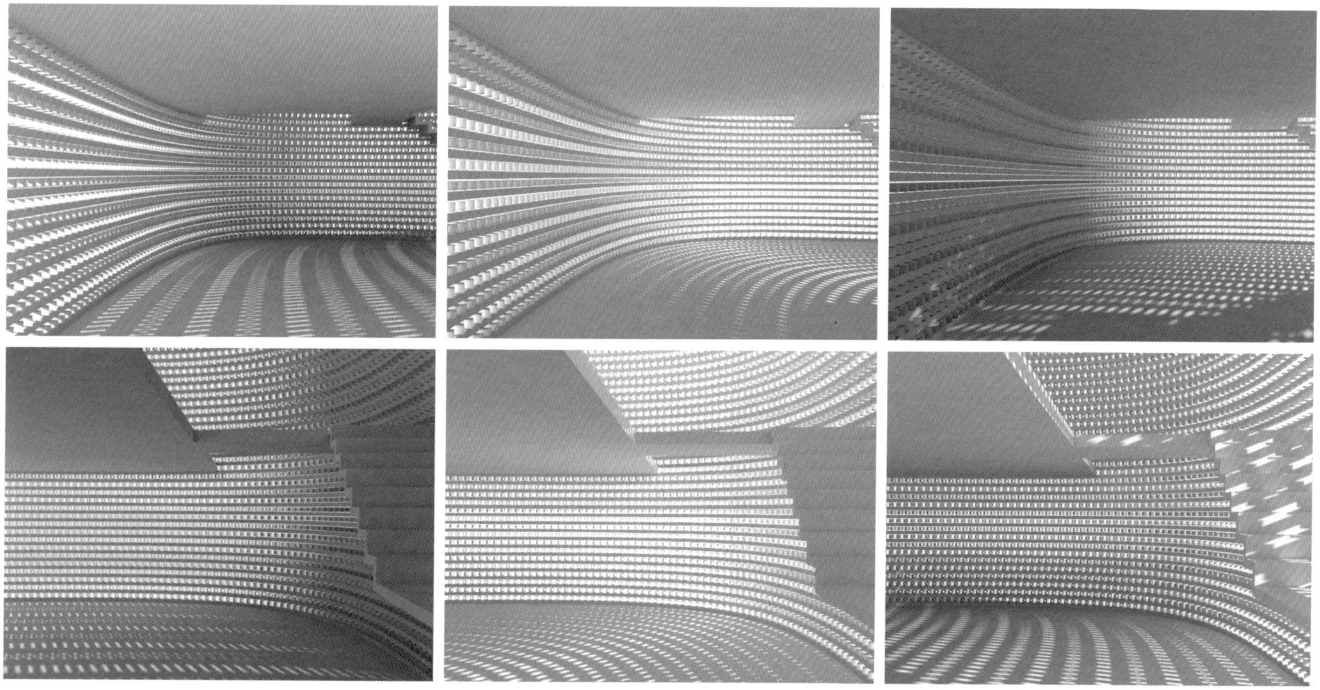

图 3.5-4

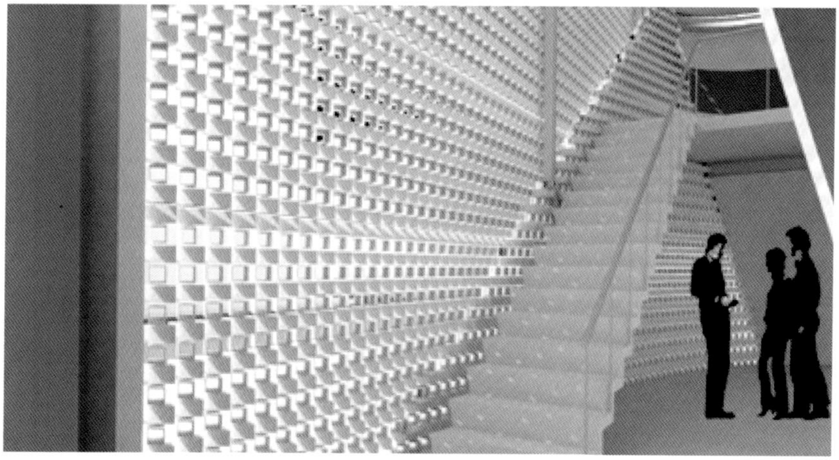

图 3.5-3 在一天的不同时间旋转砖立面效果图。感应式技术
Renderings of rotating brick façade at various times of day. Responsive technologies

图 3.5-4 旋转的砖立面在室内产生斑驳的光影效果
The rotating brick façade creates dappled light effects inside the home

图 3.5-5

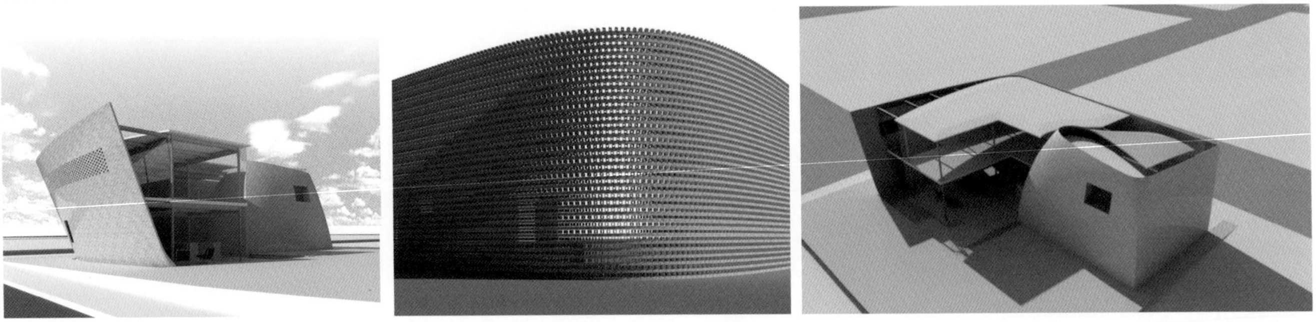

图 3.5-6

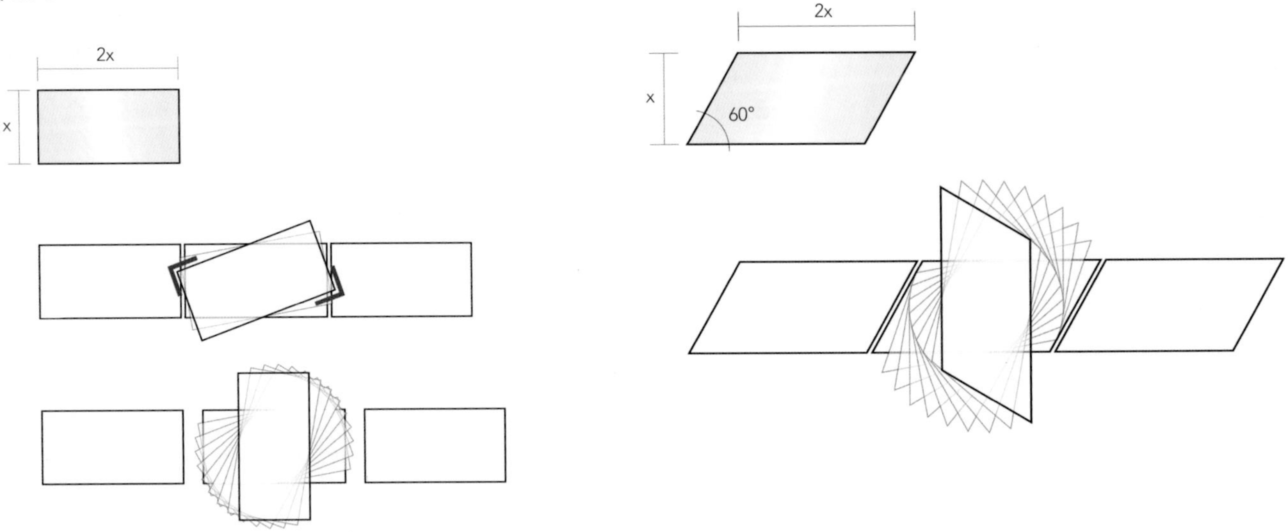

图 3.5-7

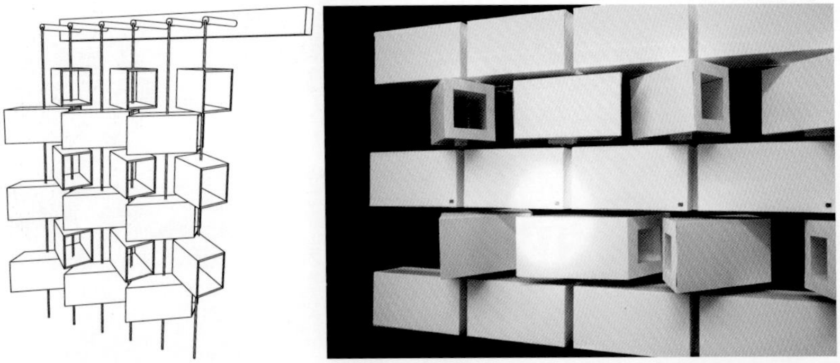

图 3.5-5 外部景象显示出砖构件阵列在这幢可感知住宅中的应用工作状态
Exterior view showing the array of brick applications at play in this sentient home

图 3.5-6 砖从矩形到平行四边形的形状转变可以支持其达到 90° 的旋转
The shift from rectangular to parallelogram-shaped bricks facilitates a rotation of up to 90 degrees

图 3.5-7 旋转砖系统是由悬挂在外围结构梁上的张拉索支承
The rotating brick system is supported by tension cables hung from a peripheral beam

3.6 智能墙体 Wallbot

将墙体计算机化，实现新空间的自主。[1]

奥托 · 吴

Computerising the walls for novel spatial autonomy.[1]

Otto Ng

“智能墙体”是一种智能化的墙体装置，能够根据居住者的行为或愿望做出反应，进行实时自我重新调配。

自 1914 年勒 · 柯布西耶推出“多米诺”住宅以来，墙体从承重结构的历史中解放出来。智能墙体是一种设计研究，试图进一步将墙体从过去的静态中解放出来。空间的边界除了能量和舒适度要求外，还应该由事件变化的压力和不可预测的物质流来不断塑造。

项目以波士顿的一个典型阁楼为主要研究场地，通过一组 21 米宽的墙壁来实现多重配置。从一种配置到另一种的转换需要墙体构件的平移、旋转和扩张。在此期间，墙体构件将被打散，产

[1] 由鲁道夫 · 埃尔 - 库利与纳希德 · 纳比安指导。

‘Wallbot’ is an intelligent wall unit, capable of reconfiguring itself in real time to respond to the behaviour or desire of the occupants.

Since Le Corbusier introduced the ‘Domino’ House in 1914, walls have been liberated from their historical role as loadbearing elements. Wallbot is a design investigation that attempts to even further liberate walls from their static past. The boundaries of space should be continuously shaped by the changing pressure of events and the unpredictable density of object flows, in addition to energy and comfort requirements.

Taking a typical loft in Boston as the principal site for the investigation, multiple configurations are achieved through a set of twenty-one metre wide wall sections. The transformation from one configuration to another requires translation, rotation and expansion of the wall pieces. In the interim, the wall sections will disperse, producing an exciting forest-like condition. Users communicate with Wallbots

[1] Guided by Rodolphe el-Khoury and Nashid Nabian.

through an intuitive graphic interface or through direct gestural control. A spectrum of radically different spatial configurations can be produced to suit a spectrum of domestic conditions.

生令人兴奋的森林般的状态。用户通过直观的图形界面或直接通过手势控制墙壁建造。这个系统可以产生完全不同空间配置下的光谱来适应家庭生活的需要。

图 3.6-1

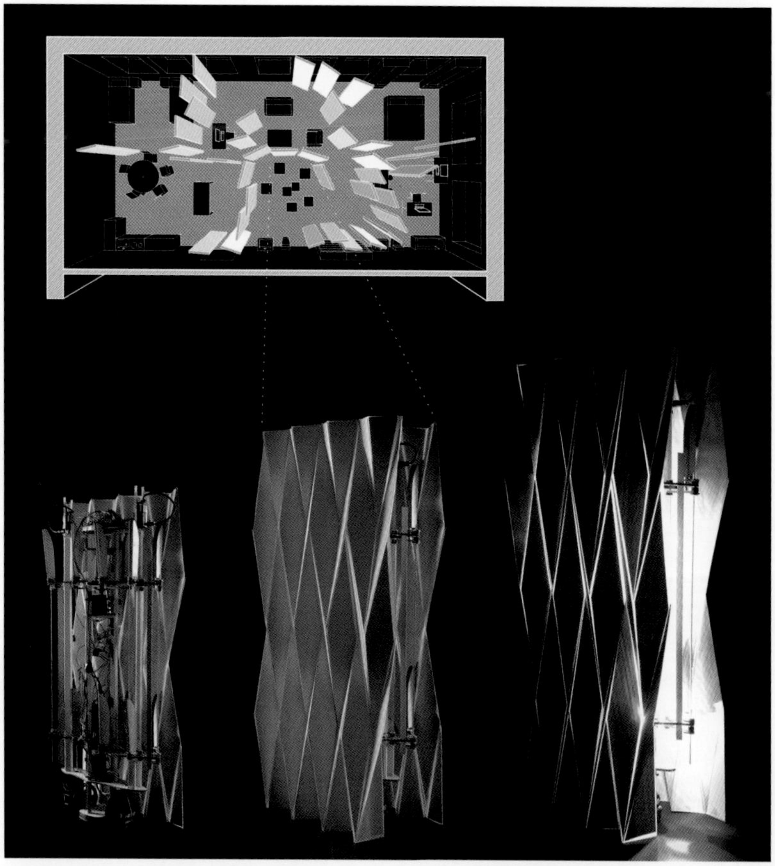

图 3.6-2

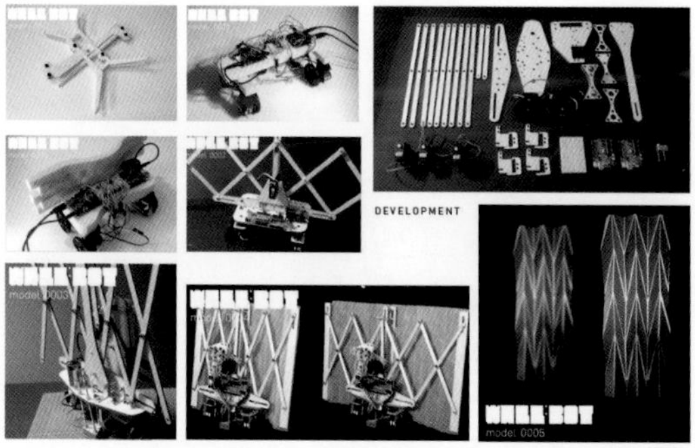

图 3.6-1 随着智能墙体的改变，空间配置也发生变化
Changing spatial conditions as Wallbots change configurations

图 3.6-2 墙体外壳附加到智能墙体的原型
Attaching the wall shell to a Wallbot prototype

图 3.6-3

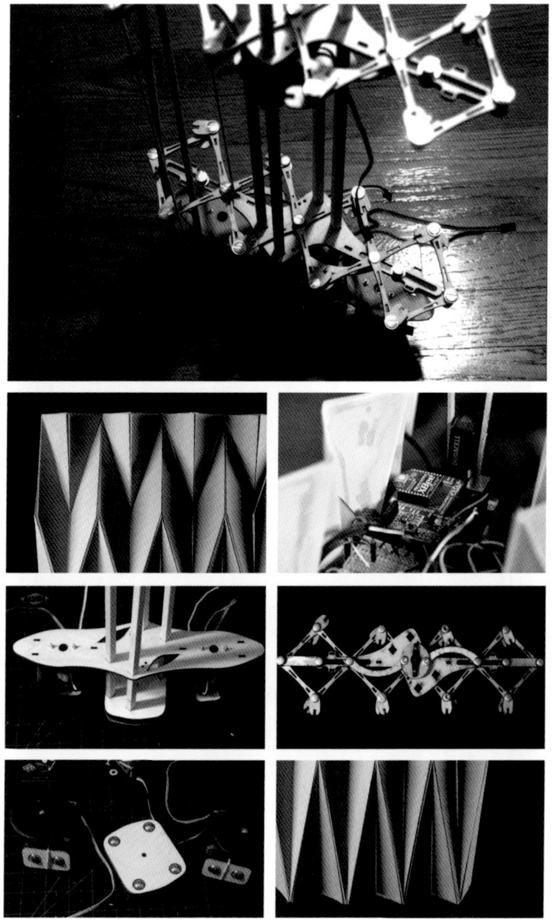

图 3.6-4

图 3.6-5

图 3.6-6

图 3.6-3 原型细节
Prototype details

图 3.6-4 扩展墙体的剖面部分：外壳材料允许刚性和柔性扩展。刚性外壳可通过打褶实现扩展和收缩的转换
Expanding wall section: the shell material is rigid and flexible to allow expansion. Pleating allows the rigid shell to transform through expansion and contraction

图 3.6-5 用户能够与智能墙体进行交流，通过直观的图形界面或直接的手势来控制
The user can communicate with Wallbots through an intuitive graphic interface or through direct gestural control

图 3.6-6 原型的群状况
Swarm condition of prototypes

图 3.6-7

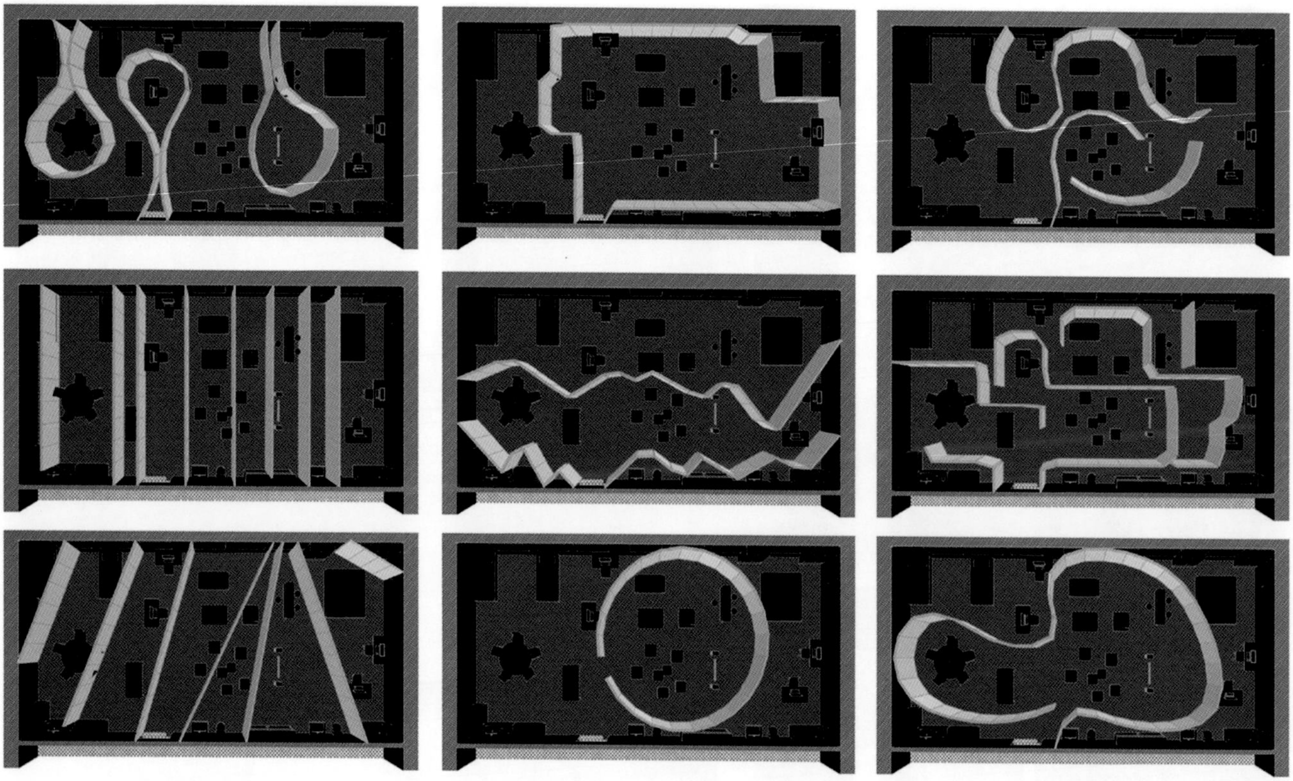

图 3.6-7　两种配置间的过渡
Transition between two configurations

3.7 窗帘 Curtain

“窗帘”需要对相互矛盾的因素做出反应，如隐私与光线，或视野与隔热。这些因素导致对织物的驱动，使之可在没有外部或硬件设施的情况下局部开启和关闭。❶

马赫塔卜·奥斯克

The curtain's need to respond to contradictory forces, privacy vs. light, or view vs. insulation results in an actuated textile that can open and close locally without an external or hard mechanical infrastructure.❶

Mahtab Oskuee

窗帘被理解为纺织品的一种特定表现形式，一种与建筑玻璃共同进化的柔软表面。从历史上看，窗帘同时扮演了多重角色，调节热动力、视觉、光线、颜色、气味和声音、取景活动和舞台气氛。随着 18 世纪和 19 世纪平板玻璃的广泛使用，鉴于窗帘具有移动性，能够调节热、光和观景，并可与日益增长的透明窗户相结合，窗帘得以从挂毯与帷幔中分离出来。尽管窗帘被现代主义斥为过于“柔弱”——女性化和装饰性，但窗帘的使用已经成为他们作品的关键与同义词。窗帘的灵活性和实用性是无与伦比的，可以依靠它来抵消开放平面（空间分异和分隔）和玻璃幕墙系统（热力学调节）的副作用。

密斯·凡·德·罗的范斯沃斯住宅（1951

The curtain is understood as a specific manifestation of textile, a soft surface that has co-evolved with architectural glazing. Historically, the curtain has taken on multiple and simultaneous roles, mediating thermodynamic flows, sight, light, color, smell and sound, framing activities, and staging atmospheres. With the widespread use of plate glass in the 18th and 19th century, the curtain differentiated itself from tapestry and drapery by gaining mobility, working as a thermal, light and view filtering membrane in conjunction with the increasingly large and transparent window. Though reviled as 'weak' — effeminate and decorative-by the modernists, its use became critical and synonymous with their work. The curtain's flexibility and utility was unmatched, relied on to counteract the side effects of open plans (space differentiation and partitioning) and glazed curtain wall systems (thermodynamic regulation).

Mies van der Rohe's Farnsworth House (1951) best epitomizes the dilemma of glass and psychological distress mitigated by the curtain —

❶ 由卡罗尔·穆海贝尔指导。

❶ Guided by Carol Moukheiber.

providing comfort to the inhabitant from the relentless exposure to the outside. With Lilly Reich in the Exposition 'de la Mode', Berlin (1927), the curtain comes off the wall to become a soft, dynamic, and sensual form of spatial enclosure. Building upon that work, the project embraces the curtain's association with the feminine, exploring its potential to provide insurmountable utility in the modulation of space through the dynamic entanglement of environmental, socio-cultural, experiential, and psychological elements.

What it does:

This iteration of the curtain prototype denies the view. It is equipped with proximity sensors, which upon sensing the presence of a viewer closes a cluster of modules providing privacy. Caressing the surface with one's hand would expand the zone of privacy. Alternative versions are being considered with reversible motion, along with environmental sensors.

Technique: Crochet — Topological Surface and Soft Actuation

The desire for light and privacy manifested itself in more porous textiles — for example traditional crocheted lace curtains filtered the light imbuing the space with atmospheric qualities. The membrane's need to respond to contradictory forces, e.g., privacy vs. light, or view vs. insulation resulted in an actuated textile that can open and close locally without the aid of an external or hard mechanical infrastructure. Crochet, as a quintessential domestic technique is harnessed for its capacity to model mathematically complex topological surfaces.

A cylindrical hyperbolic paraboloid module is constructed using wool fiber for its insulating value. Shape Memory Alloy (SMA) wire is woven into the outer expanded edge, stabilizing the form with added stiffness. The modules are stitched together into a holey 3D surface. The shape's crenelated geometry — the result of an exponential increase in stitches with each additional concentric row-allows the module to close and open without deforming the entire surface. The desire for a non-deformable surface is the formal driver for the hyperextended modules.

年）最成功地体现了玻璃的困境和由窗帘减轻的心理困扰——为居民提供舒适的环境，让他们免受外界无端的干扰。莉莉 · 赖希在 1927 年的柏林“风尚展览”上让窗帘从墙上脱落，成为一种柔软、动感、感性的空间隔断。该作品在窗帘与女性之间建立联系，通过环境、社会文化、体验和心理元素的动态影响，探索其在空间调节中不可逾越作用的潜力。

它做了什么：

窗帘原型的迭代否定了这种观点。它配备了近程传感器，一旦感知到观察者的存在，就会关闭一组能提供隐私的模块，用手轻抚其表面会扩大隐私区域。替代性样本的研究正在考虑结合环境传感器的可逆运动。

技术：钩针——拓扑表面和软驱动

在多孔的纺织品上表现出更多对光线和隐私的需要，例如传统的钩针花边窗帘过滤了光线，使空间充满了氛围。薄膜需要对相互矛盾的两方面作出反应，例如隐私与光线、视野与隔热，从而产生了一种驱动性的织物，在不借助外力或机械设备的情况下局部打开和关闭。用钩针进行编织作为一种典型的家庭工艺，是利用其能力为复杂的拓扑表面进行数学建模。

利用羊毛纤维的隔热值来构造一个圆柱形双曲抛物面模块。由形状记忆合金（SMA）线织成的外边缘能够增加刚度，使形状更稳定。这些模块被缝合在一起，成为一个多孔的 3D 表面。该形状的几何锯齿状纹理——每增加一排同心圆针数，缝线就会呈指数增长——允许模块在整个表面不变形的情况下关闭和打开。对非变形曲面的需求是超扩展模块的正式驱动。

拉伸和释放的张力被锯齿状褶皱所提供的额外长度所吸收。位于边缘的额外长度同样能容纳更长的形状记忆合金，通过不连续的脉冲电流增加其驱动，并将能耗降到最低。虽然所有的模块都基于同样的钩针图案或结构，但由于纤维性质和不同的个体钩针张力，不同模块的锯齿形状也不同。在收缩和释放时，每个模块的反应模式略有不同，导致在非均匀的活性表面产生变化和可感知的愉悦感。

下一个迭代将会解决其他形式的环境感知，包括光和温度。不同纤维特性的结合，可根据制造方法和能量使用调节吸收或转移热量与可伸缩性。

The tension of pulling and releasing is absorbed by the extra length afforded by the folds of the crenellation. The edge's extra length also accommodates a longer SMA, increasing its actuation which through a discontinuous, pulsing current minimizes energy use. Although the modules are all based on the same stitch pattern or structure, the crenellation varies from module to module due to the nature of the fiber and the varying individual stitch tensions. Upon contraction and release, each module reacts with a slightly different pattern, resulting in the variability and palpable pleasure of a non-uniform active surface.

The next iteration will address additional forms of environmental sensing, including light and temperature. The incorporation of various fiber properties could mediate (absorb or deflect heat), and scalability both in terms of fabrication methods and energy use.

图 3.7-2

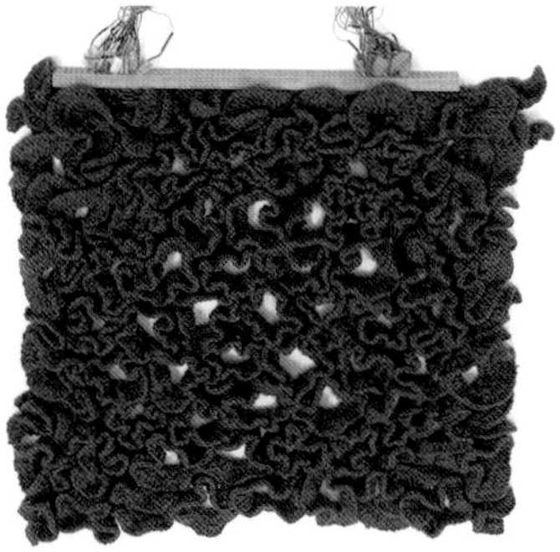

图 3.7-1

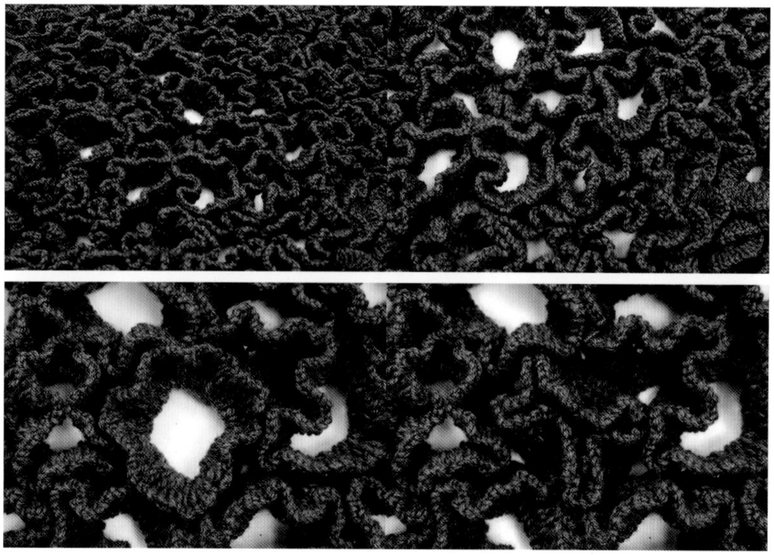

图 3.7-1 窗帘原型
Curtain proto

图 3.7-2 高动态范围导线
HDR wire

图 3.7-3

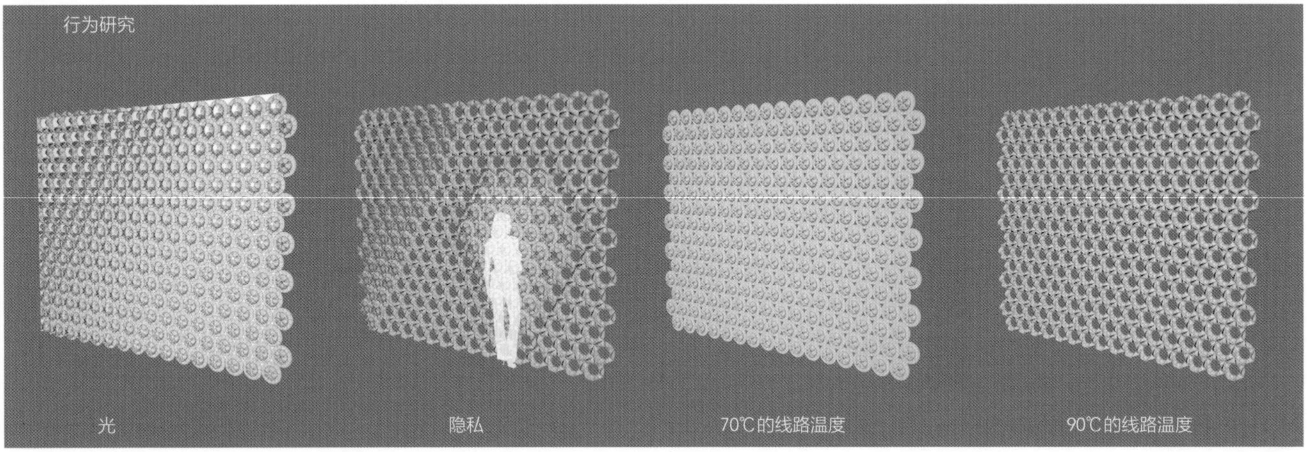

图 3.7-3 窗帘与人行为研究
Behaviour study

图 3.7-4

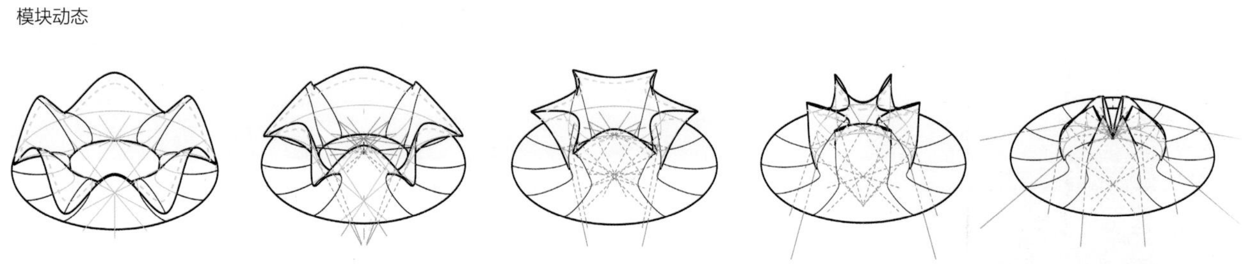

图 3.7-4 圆柱形双曲抛物面模块的动态性
Module dynamics

图 3.7-5

线路设计

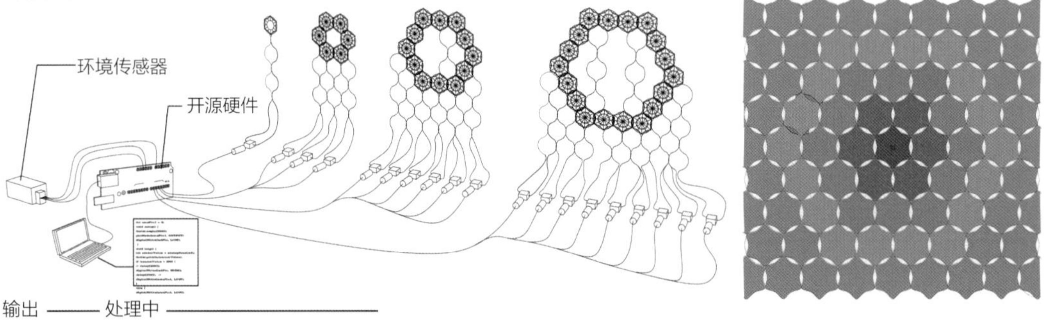

图 3.7-5 窗帘模块中线的设计
Wiring design

图 3.7-6

形状记忆合金供电要求

电阻（欧姆）=[电源（伏特）/ 所需电流（安培）] – 线的总电阻（欧姆）
欧姆定律：电压 = 电流 x 电阻

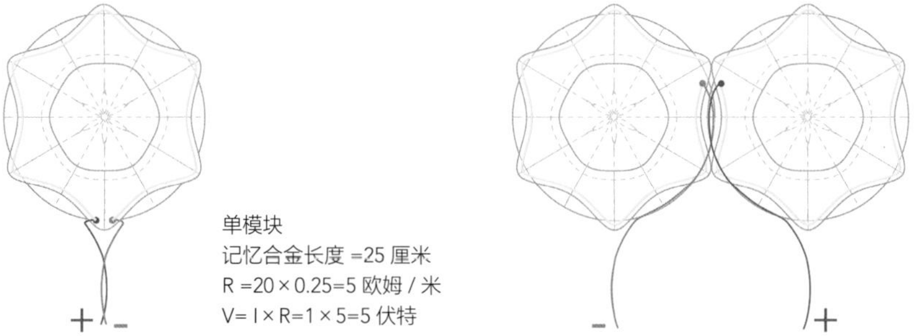

单模块
记忆合金长度 =25 厘米
R =20 × 0.25=5 欧姆 / 米
V= I × R=1 × 5=5 伏特

2 个系列模块
记忆合金长度 –50 厘米
R =20 × 0.5=10 欧姆 / 米
V= I × R=1 × 10=10 伏特

图 3.7-6 形状记忆合金的模块化设计
Shape memory Alloy

图 3.7-7

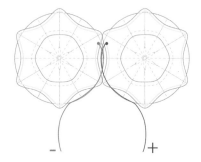

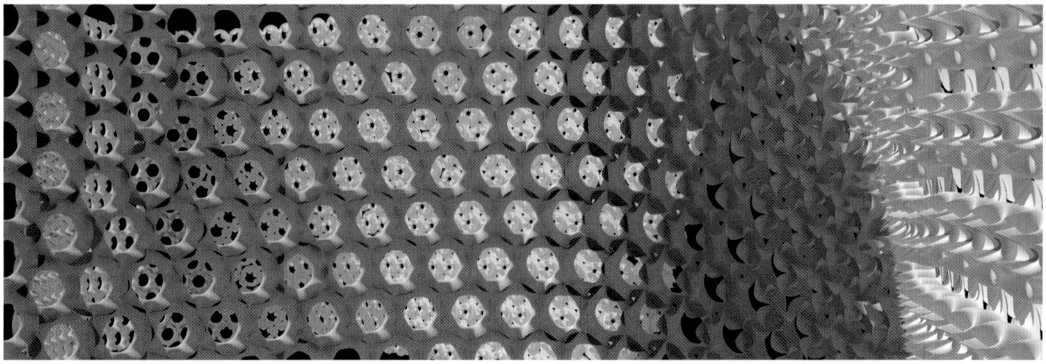

图 3.7-7 窗帘效果图
Render

3.8 感应窗帘 Sentient Curtain

“感应窗帘”配备了由“XYZ 交互”提供技术支持的感应式技术。它可以精确地定位和跟踪空间中的物体；它通过精确地调整材料的透明度来做出回应。

设计：KLF 事务所；技术支持：XYZ 交互

‘Sentient Curtain’ is equipped with sensing technology that is powered by ‘XYZ Interactive’. It can locate and track a body in space with great precision; it responds by finely adjusting the transparency of its material.

Designed by: KLF,
Powered by: XYZ interactive

What

States and modes of interaction (Summer mode)

1. Maximum opacity: screen

When no human presence is sensed in the room — the system can distinguish between pets and humans based on size, the curtain assumes maximum opacity to limit solar heat gain to a minimum. The opacity value is not fixed; it varies in response to the intensity of direct sunlight measured by built-in photo-sensors.

2. From opacity to transparency: screen and view

The sensor's detection of human presence initiates a gradual deformation of the fabric to decrease the opacity of the curtain. The longer the presence is detected the more transparent the curtain.

做什么

交互的状态和模式（夏季模式）

1. 最大不透明性：屏幕

当房间内没有人时（系统可以根据大小区分宠物和人类），窗帘采用最大不透明度来限制太阳能热增益，使其达到最小。不透明度值不是固定的，它随内置的光传感器测量到的直射光强度而变化。

2. 从不透明到透明：屏幕和视野

传感器对人存在的检测使织物逐渐变形，以降低窗帘的不透明度。检测到人存在的时间越长，窗帘就越透明。使用者一旦离开房间，窗帘就会恢复到原来的不透明状态。使用者可以在窗帘的初始编程中调节窗帘“清除”的速度和透明度的范围，以实现速度和感应灵敏度达到适当的平衡，例如，你不希望窗帘对路过的人做出反应，而且也不想让它花太长时间来适应刚刚进入房间的人。

3. 更透明：视野

当使用者非常靠近窗帘时，“清除”过程会加速，因为这个位置表明使用者对视野很有兴趣。在这种情况下，直接面对使用者的区域逐渐呈现出更大的透明度，以提供更好的视野。此处响应的灵敏度和感应速度同样取决于使用者的校准。

4. 最大透明度：视野，立即

该视野可以覆盖渐进的过程，并用一个“划过”手势激发瞬时的“清除”，从而达到立即需要的最大的透明度。

怎么做

输入、处理和输出
1. 输入

A. 光传感

窗帘上有两个内置在上部机械外壳中的光电传感器：

As soon as the user leaves the room the curtain reverts to its original opaque condition. The speed at which the curtain ‘clears’ and the extent of the transparency can be adjusted by the user in the initial programming of the curtain to achieve the right balance in speed and sensitivity of responsiveness. For instance, you don’t want the curtain to respond to someone who is just passing by but also you don’t want it to take too long in adjusting to the presence of someone who has just entered the room.

3. More transparency: view

The ‘clearing’ process accelerates when the user is very close to the curtain, in a position that indicates an active interest in the view. In this situation the area immediately facing the user gradually assumes greater transparency to deliver a better possible. Here again the sensitivity and speed of response is subject to user calibration.

4. Maximum transparency: view, immediately

The view can override the gradual process and provoke an instant ‘clearing’ with a ‘sweeping’ gesture where maximum transparency is needed immediately.

How

Input, processing, and output
1. Input

A. Photo-sensing

The curtain features two photo-sensors that are embedded in the upper mechanism casing:

• 一个朝外的光电传感器测量太阳光的强度。

• 一个朝内的光电传感器测量室内环境光的强度。

B. 运动、位置和手势感应

窗帘的 XYZ 手势感应 -ZX 传感器具有多种配置，具体配置取决于窗户的尺寸和所需的敏感度。

• 标准 XYZ 手势感应 -ZX 传感器被集成到每个面板层中心的设计中，具有一个典型的手势控制高度。
• 远程 XYZ 手势感应 -ZX 传感器仅安装在遮阳板顶部的控制模块中。
• 两个传感器条（顶部和底部，或在每个面板的侧面）的组合，结合了反射和主动式技术，更像是 3D 触摸屏。
• 嵌框式的安装方式（传感器条位于面板的四周）实现了更好的性能。

C. 使用者输入

使用者通过遥控装置校准窗帘的感应度，该装置允许调整以下设置：

• 延迟：调整对存在的响应速度
• 不透明度：调整最大不透明度水平
• 亮度：调整最大亮度水平
• 夏季 / 冬季模式：从被动冷却模式切换到被动加热模式
• 内存：使用者创建或预设的程序

• A photo-sensor facing outwards measures the intensity of sunlight.
• A photo sensor facing inwards measures the intensity of the ambient light in the room.

B. Motion, position and gesture sensing

The curtain features XYZ Gesturesense-ZX sensors in a variety of configuration depending on the dimensions of the window and desired sensitivity

• Standard XYZ Gesturesense-ZX sensor integrated into the design at the center of each panel level with a typical hand gesture control height.
• Long-range XYZ Gesturesense-ZX sensor mounted in the control module in the top of the shade only.
• A combination of two sensor bars (top and bottom, or at the sides of each panel) which incorporate both a reflective and active technology for more of a course 3D touchscreen-like capability.
• A Bezel type of implementation (sensor bars all around the panel) for even better performance.

C. User input

The user calibrates the responsiveness of the curtain by means of a remote control device that would allow the adjustment of the following settings:

• Delay: adjusts the speed of response to presence
• Opacity: adjusts maximum opacity levels
• Brightness: adjusts maximum brightness levels
• Summer / Winter mode: witches from passive cooling to passive heating mode
• Memory: user-created or preset programs

2. 处理

这个原型使用了 Arduino 开源硬件微控制器，将为窗帘的控制模块设计或定制一个微处理电路。

3. 输出

驱动器是记忆合金电缆（肌线）。由控制模块触发的长度的微小变化会使窗帘的结构变形，从而增加或降低其透明度。

在哪里

感应窗帘适用于有大面积阳光照射的环境，南面和西面的玻璃幕墙是理想的场所。玻璃幕墙可以使人们在高层建筑中看到壮观的景色，但在夏季也容易受到太阳过度热增益的影响。在白天，经常无人居住的家庭环境将从存在 一 响应遮阳中获益最大，仅在需要时提供视野。但是，办公室和玻璃公共空间也可以充分利用感应窗帘带来的好处。

2. Processing

The prototype uses an Arduino micro-controller. A micro-processing circuit will be designed or customized for the control module of the curtain.

3. Output

The actuators are memory-alloy cables (muscle wire). Minute changes in their lengths triggered by the control module deform the structure of the curtain to increase or decrease its transparency.

Where

Sentient curtain is suitable for environments with extensive exposure to sunlight. South and west-acing glass curtain walls are ideal sites. They usually allow for spectacular views in high rise buildings but also are susceptible to excessive solar heat gain in the summer months. Domestic environments that are often left unoccupied during the day would benefit most from presence-responisve shading, delivering the view only when needed. But offices and glass public rooms could also capitalize on the benefits of a properly programmed Sentient Curtain.

图 3.8-1

图 3.8-1 如果你的窗帘挡住了阳光怎么办……当房间没有人的时候，景色无关紧要。但是当窗帘被松开变得更加通透，只有当有人在房间里的时候才能享受到景色。想象一下如果窗帘变得愈发松弛，它的不透明度就越好，人在房中的视野也会越好
What if your curtain blocked the sun rays out… indifferent to the view when no one is around. But loosened up to become more see-through, delivering the view only when someone is in the room. Imagine this curtain getting increasingly looser, trading its opacity for a better view the longer the room is occupied

图 3.8-2

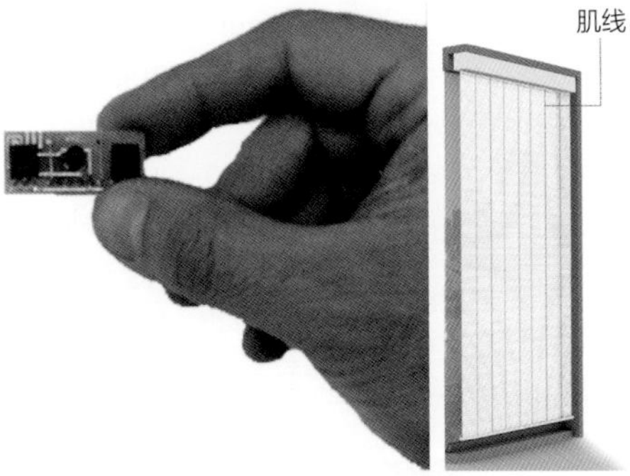

图 3.8-2 “感应窗帘”配备了由“XYZ 交互”提供技术支持的感应式技术。它可以精确地定位和跟踪空间中的物体；它通过精确地调整材料的透明度来做出响应。窗帘的透明性随嵌入的记忆合金电缆（肌线）的轻柔动作而变化，这些电缆可以张开或闭合织物的编织结构
‘Sentient Curtain’ is equipped with sensing technology that is powered by ‘XYZ Interactive’. It can locate and track a body in space with great precision; it responds by finely adjusting the transparency of its material. Transparency varies with the gentle action of embedded memory-alloy cables (muscle-wire) that splay open or close the woven structure of the fabric

图 3.8-3

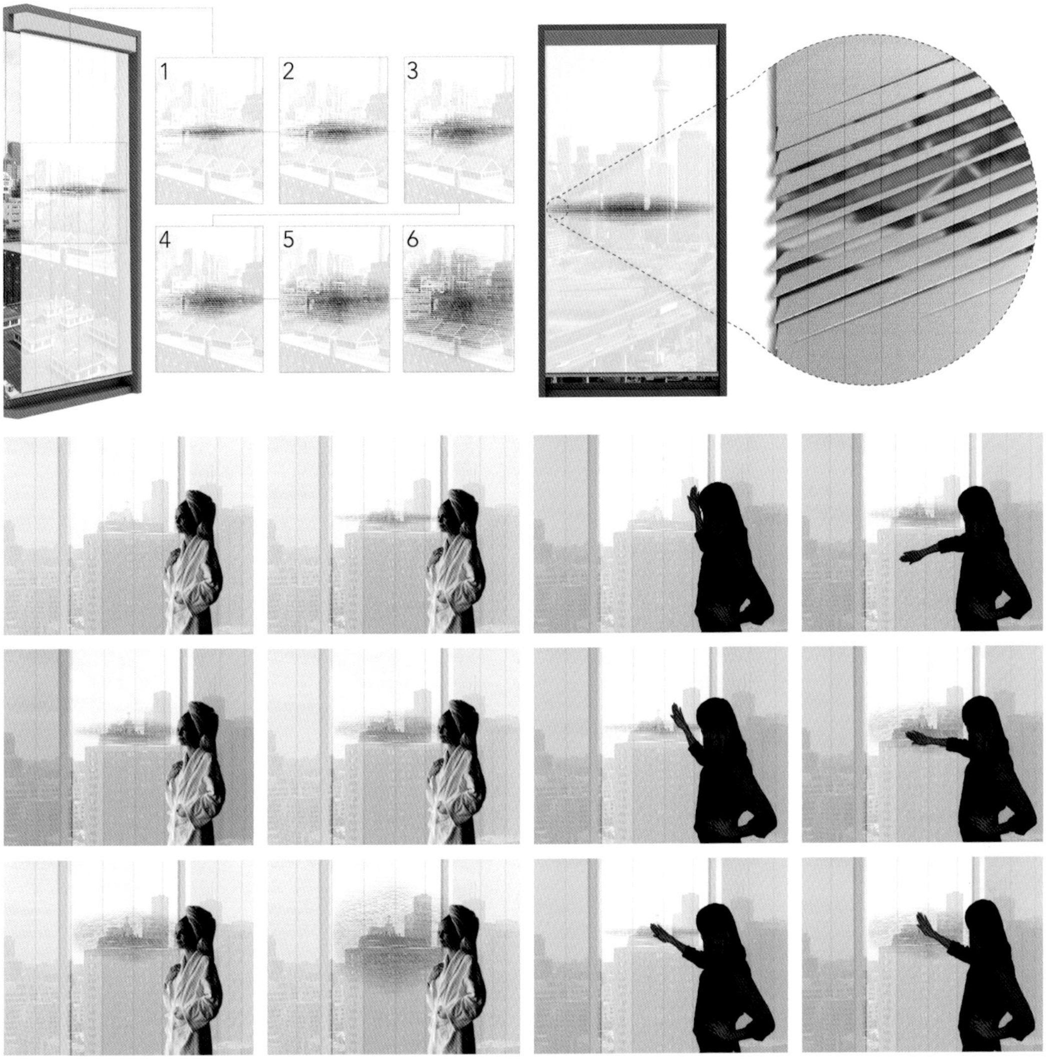

图 3.8-3 如果窗帘知道什么时候暂停它的屏蔽功能，以便在你需要的时间和地点给你最好的视野，那会是一种怎样的体验？“感应窗帘”会对靠近窗户的人做出反应，变得更加透明，但只是在人直接面对窗帘的区域。他们逗留的时间越长，这个区域就越开阔和通透。如果你想尽快透过窗帘看得尽可能清楚该怎么办？“感应窗帘”的手势读取传感器可以立即响应，用一个“划过”的手势可以迅速把你指定的区域变成一个透明的薄膜

What if this curtain knows when to suspend altogether its screening function to give you the best view where and when you want it... ‘Sentient Curtain’ responds to people approaching the window and becomes even more transparent but only in the area directly facing them. The longer they linger the more expanded and transparent that area. What if you wanted to see through the curtain as clearly as possible and immediately? ‘Sentient Curtain’ responds instantly thanks to its gesture-reading sensor. It promptly turns the area you designate with a ‘sweeping’ gesture into a transparent film

图 3.8-4

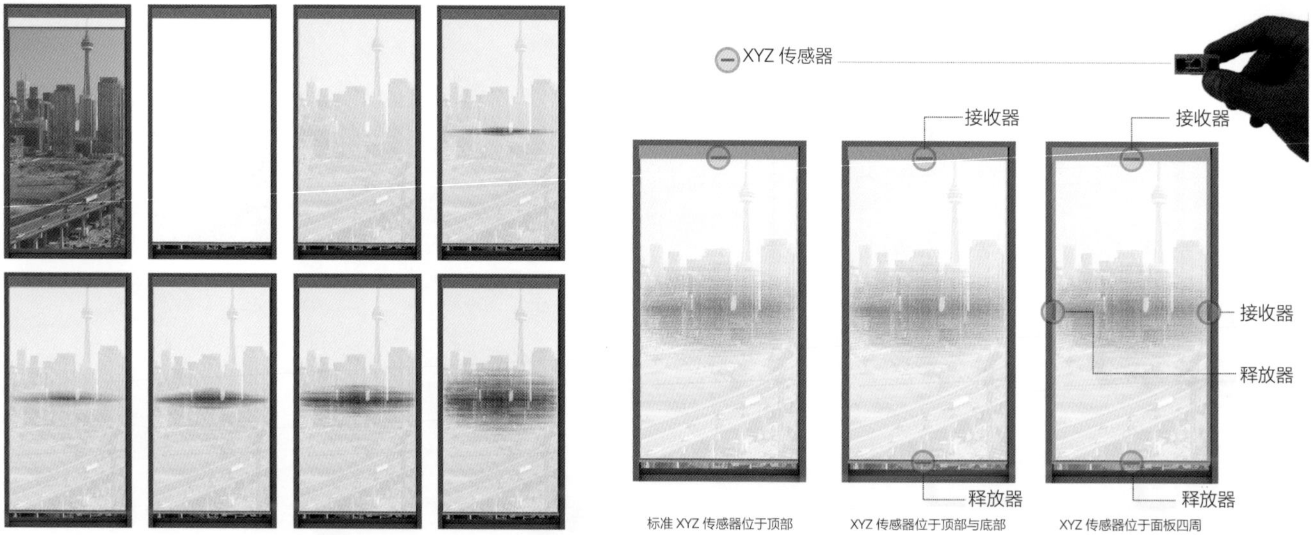

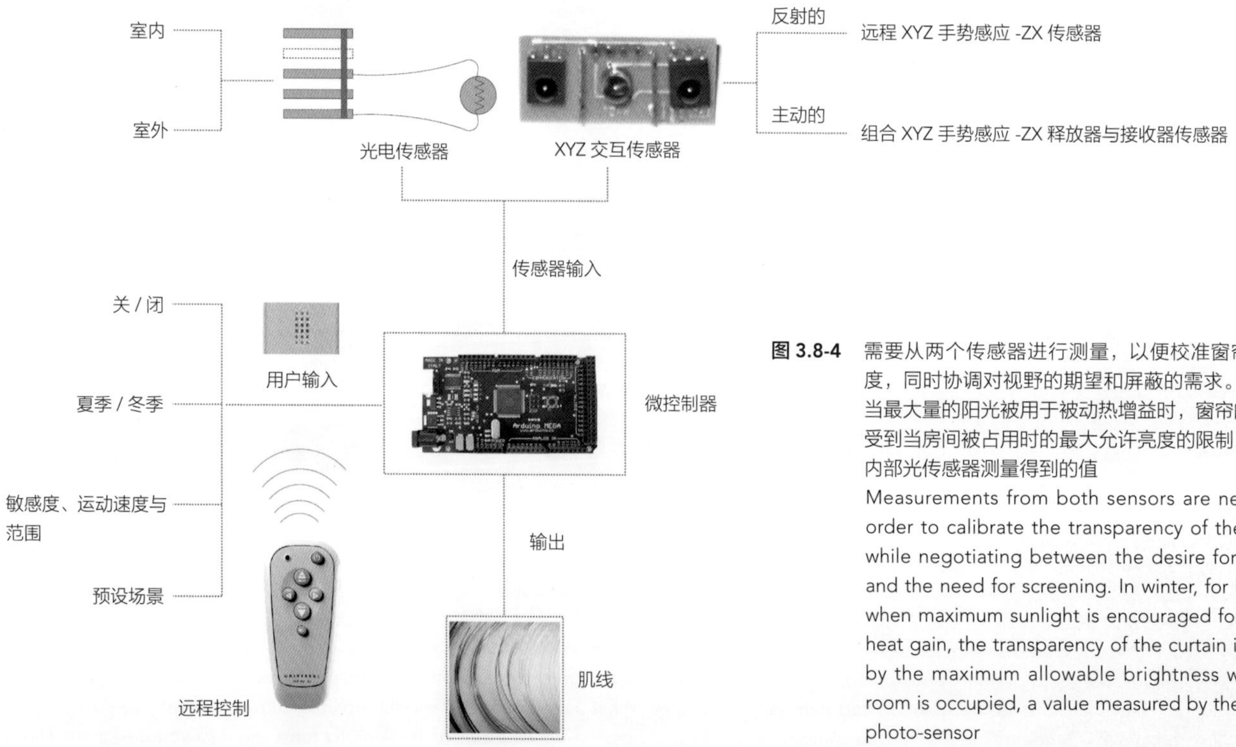

图 3.8-4 需要从两个传感器进行测量，以便校准窗帘的透明度，同时协调对视野的期望和屏蔽的需求。在冬季，当最大量的阳光被用于被动热增益时，窗帘的透明度受到当房间被占用时的最大允许亮度的限制，这是由内部光传感器测量得到的值

Measurements from both sensors are needed in order to calibrate the transparency of the curtain while negotiating between the desire for viewing and the need for screening. In winter, for instance, when maximum sunlight is encouraged for passive heat gain, the transparency of the curtain is limited by the maximum allowable brightness when the room is occupied, a value measured by the internal photo-sensor

建筑对可变环境的反应与适应能力在室内与室外的临界面上的应用是最有价值的。尽管传统建筑结构倾向于气密界面，但是感应式系统却能使可渗透的围护结构有效地与建筑协调一致，而非仅仅阻隔界面两端极端状况的接触。下面的项目探讨了一些场景和技术，这些场景和技术使新潮的覆层要素，例如表皮细胞等，能够通过传感和感应式表皮智能地调节能量、光、空气和水的动态交换。

A building's capacity to respond and adapt to varying conditions is most useful at the threshold between the interior and exterior. Whereas traditional structure favoured an airtight boundary, responsive systems enable permeable envelopes that have a capacity to productively negotiate rather than simply prevent contact between varying extremes. The following projects explore scenarios and technologies that fashion cladding elements like epidermal cells with a capacity to intelligently modulate the dynamic exchanges of energy, light, air, and water through a sensing and responsive skin.

4

可变覆盖面

VARIABLE CLADDING

4.1 群落瓷砖 Swarm Tile

交互式模块化瓷砖系统。[1]

沙迪 · 拉莫斯

Interactive Modular Tile System.[1]

Shadi Ramos

'Swarm Tile' is an interactive modular tiling system. The tiles react to their surrounding environment through sensing technology, or respond to more global conditions of their context of use and the occupancy patterns of the tiled space through network connections among all the tiles.

With a very simple and straightforward architecture, the proposed electronically enhanced, responsive system of Swarm Tiles examines the spatial and conceptual ramifications of a very interesting concept: since so much of the physical world is based on modular units of materials, what would happen if these modules possessed some of the attributes of the biological cell network that constitutes a living creature? What if they were able to sense their context and augment their surrounding environment with programmable behaviours?

Each cell or module is composed of a sensor, multiple light emitting diodes (LEDs), wireless communication, and an embedded micro-

“群落瓷砖”是一个交互式的模块化瓷砖系统。瓷砖通过感应技术应对其周围环境，或通过所有瓷砖间网络化的连接，对瓷砖使用环境和瓷砖空间的占用模式做出积极的响应。

以下电子增强系列瓷砖的感应式系统是一个非常简单而直接的体系结构，它验证了非常有趣的空间与概念性的结果：既然如此多的物质世界是基于材料的模块化单元，如果这些模块具有某些构成一个生物体的生物细胞网络属性时会发生什么事情？如果他们通过编程行为感受并增强他们周围的环境又会怎么样？

每个单元或模块由一个传感器、多个发光二

---

❶ Guided by Rodolphe el-Khoury and Nashid Nabian.

❶ 由鲁道夫 · 埃尔 - 库利与纳希德 · 纳比安指导。

极管（LED）、无线通信和一个嵌入式微控制器组成。出于原型设计的目的，一个简单的开关可以记录瓷砖被踩踏的信息，并将此信息发送至Arduino开源硬件[1]板。在一种场景下，硬件板会通过调节嵌入式LED来激活当时的条件，以应对记录下的环境状态；在另一场景下，当时被记录的条件是通过安装的紫蜂无线电[2]与其他瓷砖进行无线通信。这使得瓷砖间形成网络化，在此每块瓷砖能够通过单元自动机模式来响应整体评估状况：当受到刺激时，信息被即时处理，并适时送回单元本身与系统内的其他单元。

controller. For prototyping purposes, a simple push button registers that the tile is stepped on and sends this information to an Arduino board. In one scenario, the board activates the local condition by regulating embedded LEDs in response to the registered context. In another scenario, the locally registered condition is wirelessly communicated with other tiles via a mounted ZigBee radio. This allows for the creation of a network condition among the tiles where each responds to a collectively evaluated context based on cellular automata logic: if a stimulus is present, the information is processed and an appropriate response is sent back to the cell itself and to other cells in the system.

---

❶ 译者注：Arduino是一款便捷灵活、方便上手的开源电子原型平台，包含硬件（各种型号的arduino板）和软件（arduino IDE），适用于艺术家、设计师、爱好者和对于“互动”有兴趣的人。Arduino是一个基于开放原始码的软硬体平台，构建于开放原始码simple I / O介面版，并且具有使用类似Java、C语言的Processing / Wiring开发环境。Arduino能通过各种各样的传感器来感知环境，通过控制灯光、马达和其他装置来反馈与影响环境。

❷ 译者注：ZigBee，也称紫蜂，是一项新型的无线通信技术，适用于传输范围短、数据传输速率低的一系列电子元器件设备之间。ZigBee无线通信技术可在数以千计的微小传感器间，依托专门的无线电标准达成相互协调通信，主要特色有：低速、低耗电、低成本、低复杂度、快速、可靠、安全。

图 4.1-1

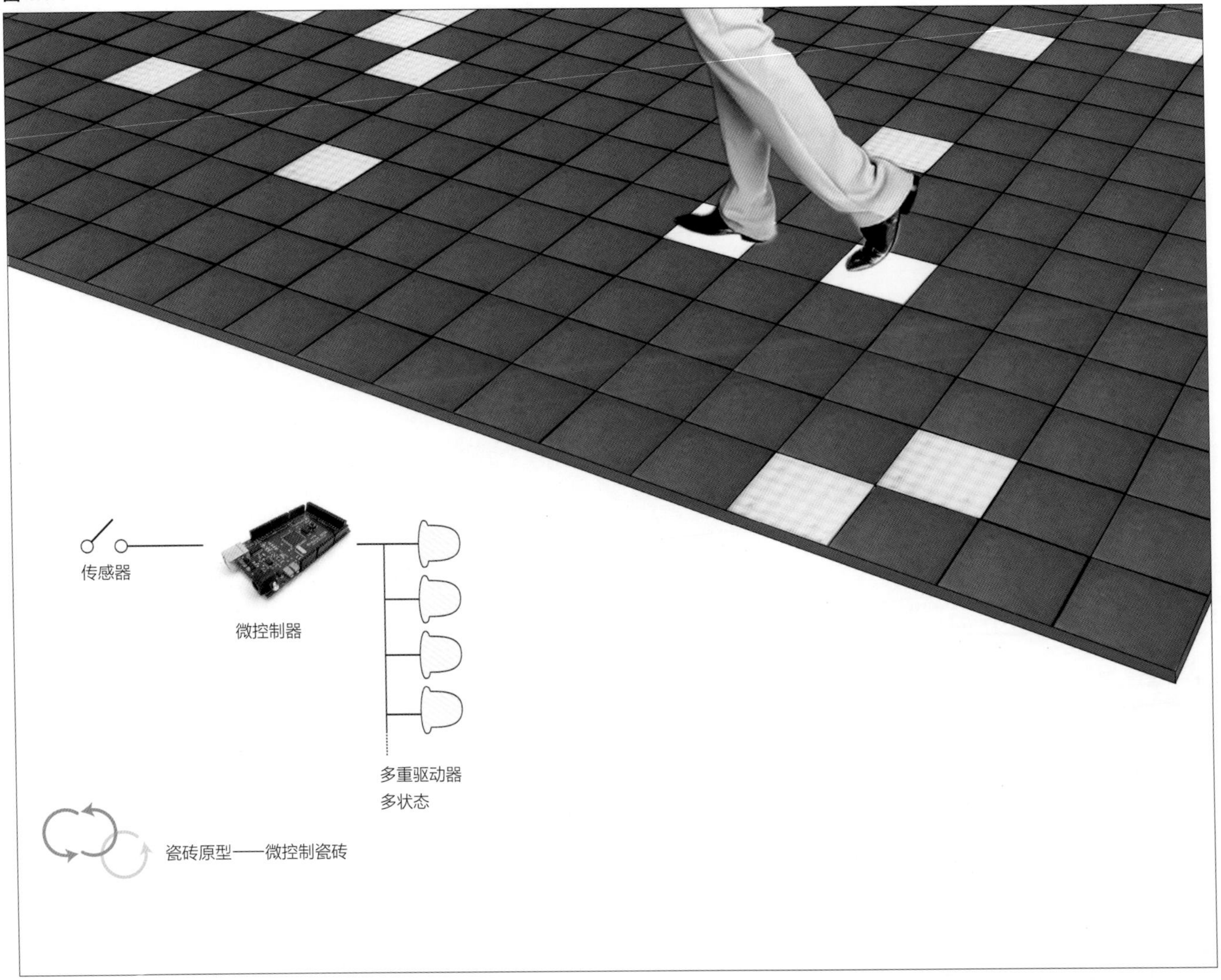

图 4.1-1 微控制瓷砖。传感器与一个能够激活单个或多个 LED 传感器的 Arduino 开源硬件微控制器相联系
Microcontrolled Tile. The sensor communicates with an Arduino microcontroller which then activates single or multiple LED sensors

图 4.1-2

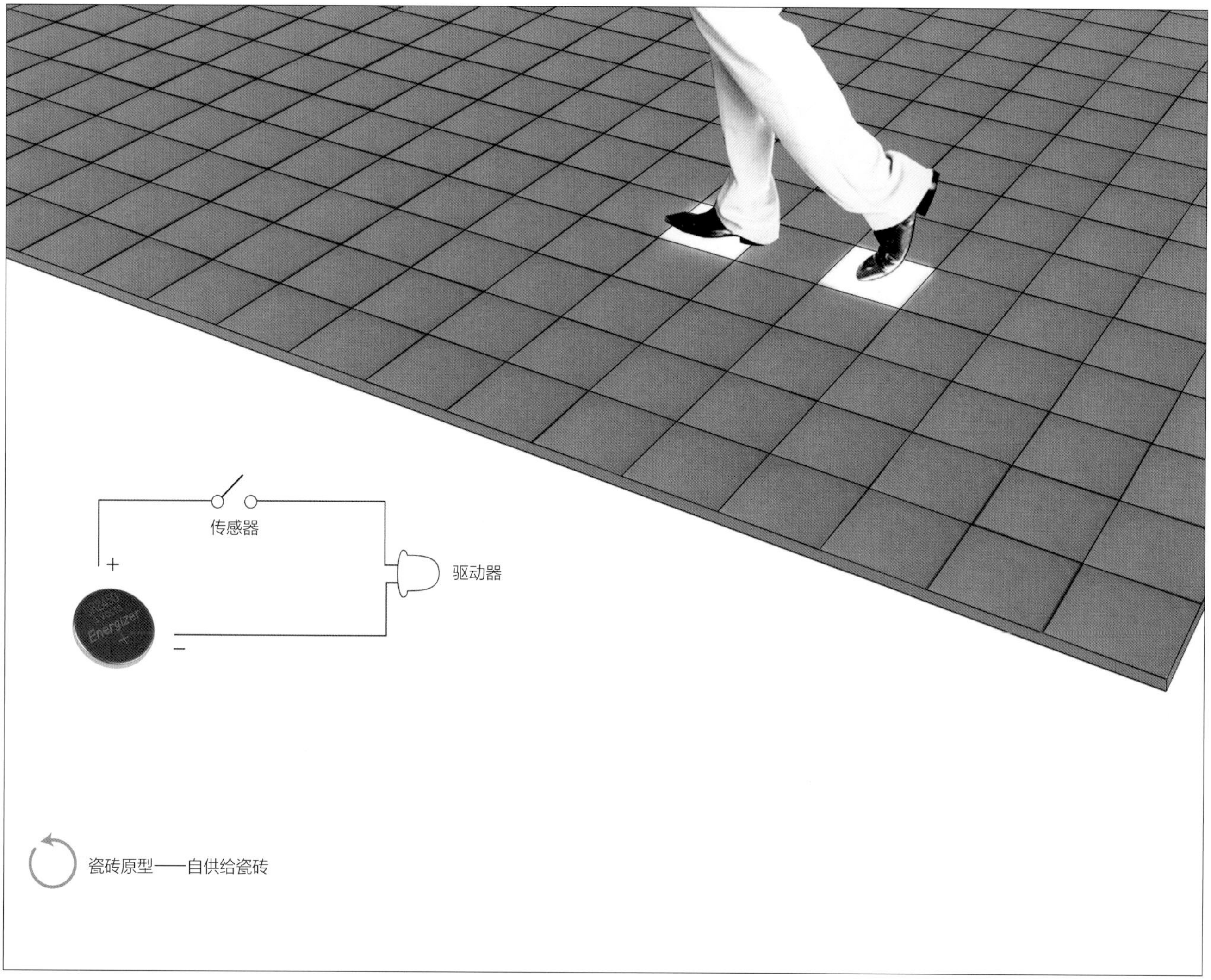

图 4.1-2 自供给瓷砖。传感器直接与 LED 驱动器相连接。当按下按钮（瓷砖被踩踏）时，电路闭合，LED 发光

Self-sufficient Tile. The sensor communicates directly with the actuator LED. When the button is pressed (tile is stepped on), the circuit is completed and the LEDs turn on

图 4.1-3

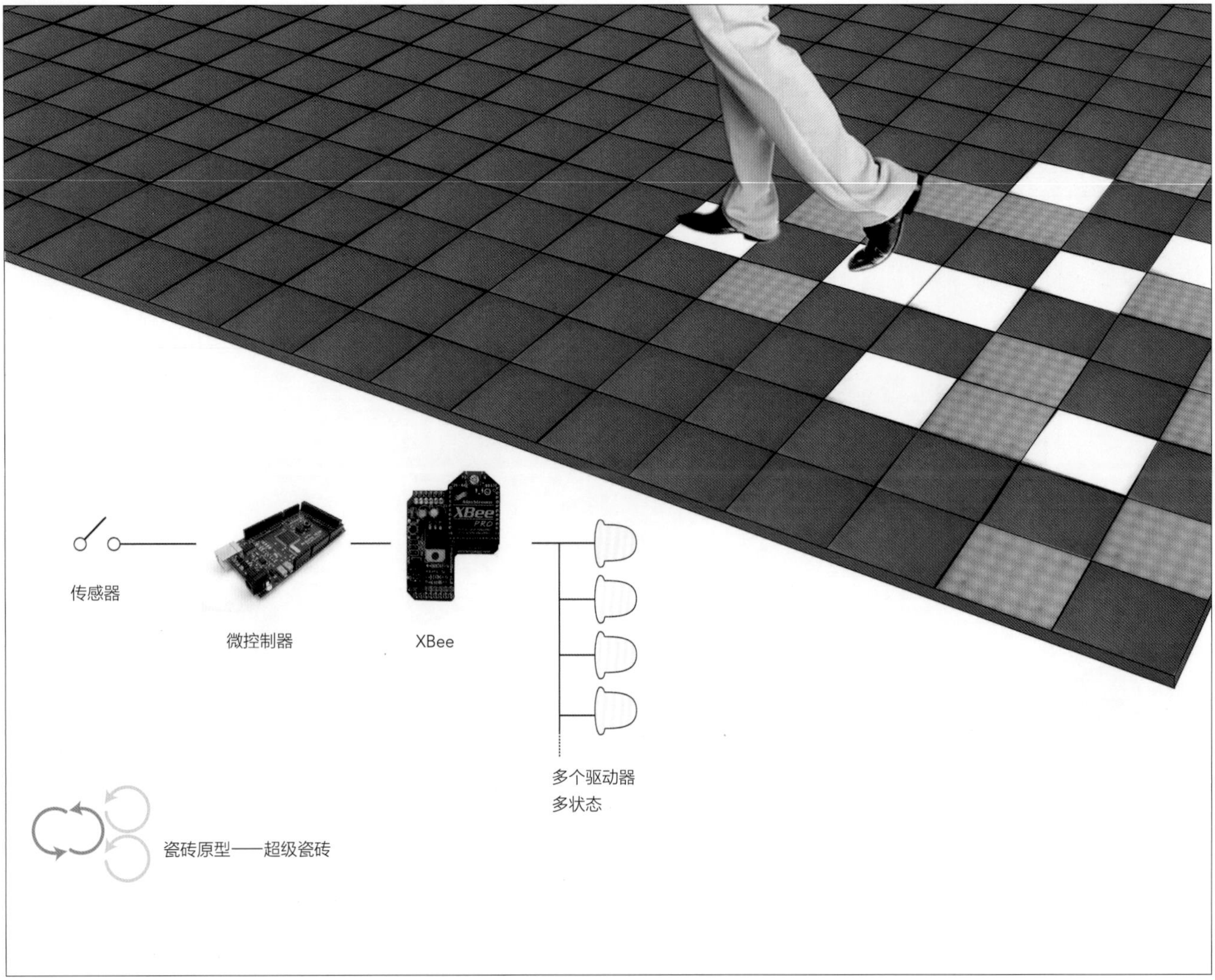

图 4.1-3 超级瓷砖。该模式采用了与微控制器瓷砖相似的方式。但这种瓷砖能够通过 XBee（一种无线网络适配器）与其他瓷砖或电脑进行通信联系。这使得远程数据的虚拟收集与分析得以实现，数据可以反馈给处理软件和可视化，以便更好地使用

Super Tile. This model operates in a similar manner to the microcontrolled tile. However, this tile is capable of communicating with other tiles or a computer through Xbee, a wireless network adaptor. This allows for virtual collection and analysis of data from a remote source. The data can be fed to the processing software and visualised for better usage

图 4.1-4

图 4.1-4 “群落瓷砖”演示
‘Swarm Tile’ demonstration

图 4.1-5

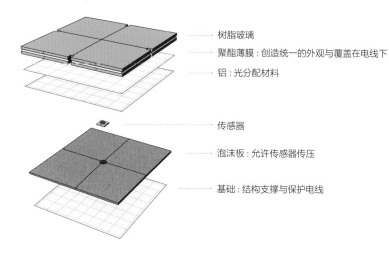

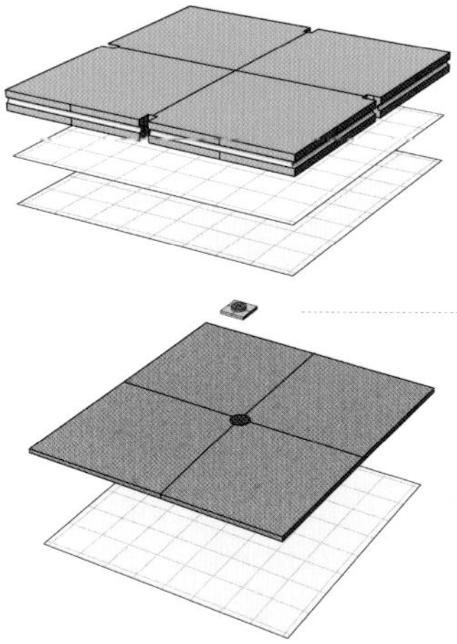

结构说明：瓷砖结构是基于电线连接需要与开关传感器驱动。

图 4.1-5 “群落瓷砖”组件显示了材料支撑、光漫散元件与压力感应机制
‘Swarm Tile’ components showing material support, light diffusing elements and a pressure sensing mechanism

图 4.1-6

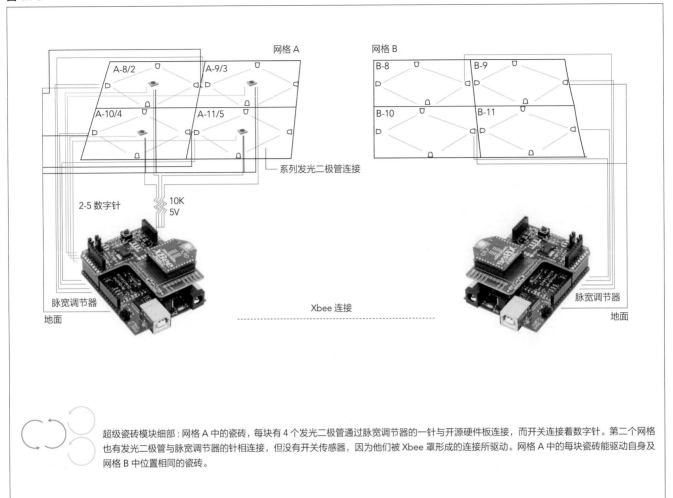

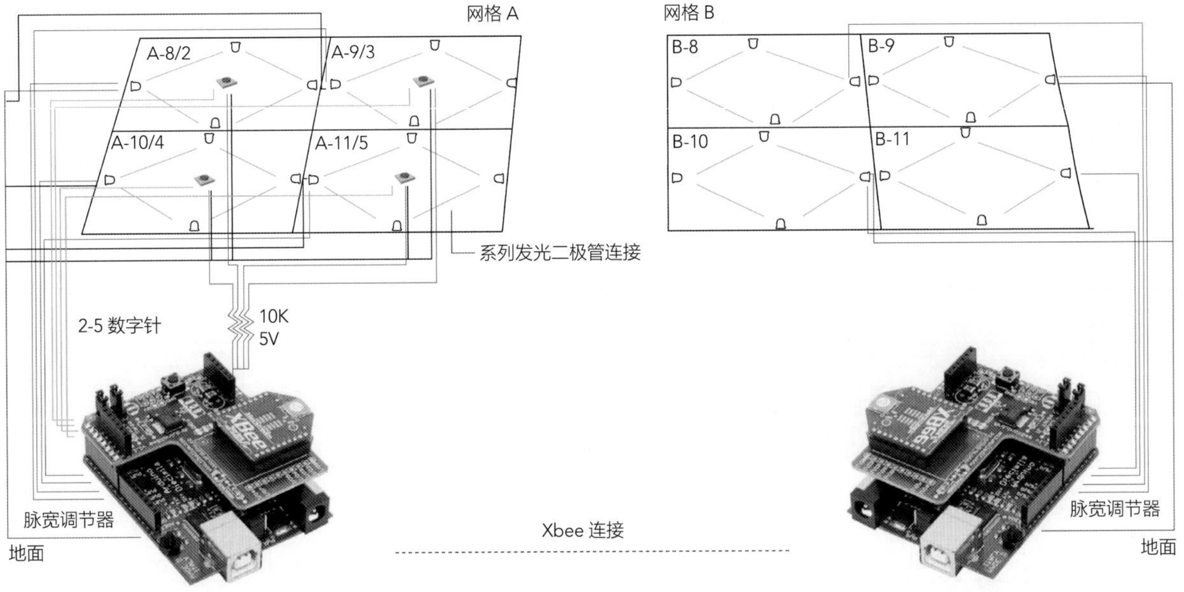

超级瓷砖模块细部 : 网格 A 中的瓷砖，每块有 4 个发光二极管通过脉宽调节器的一针与开源硬件板连接，而开关连接着数字针。第二个网格也有发光二极管与脉宽调节器的针相连接，但没有开关传感器，因为他们被 Xbee 罩形成的连接所驱动。网格 A 中的每块瓷砖能驱动自身及网格 B 中位置相同的瓷砖。

图 4.1-6 超级瓷砖细部。网格 A 的每块瓷砖能驱动自身及网格 B 中位置相同的瓷砖
Super Tile detail. Each tile in grid A activates itself and its equivalent tile in grid B

图 4.1-7

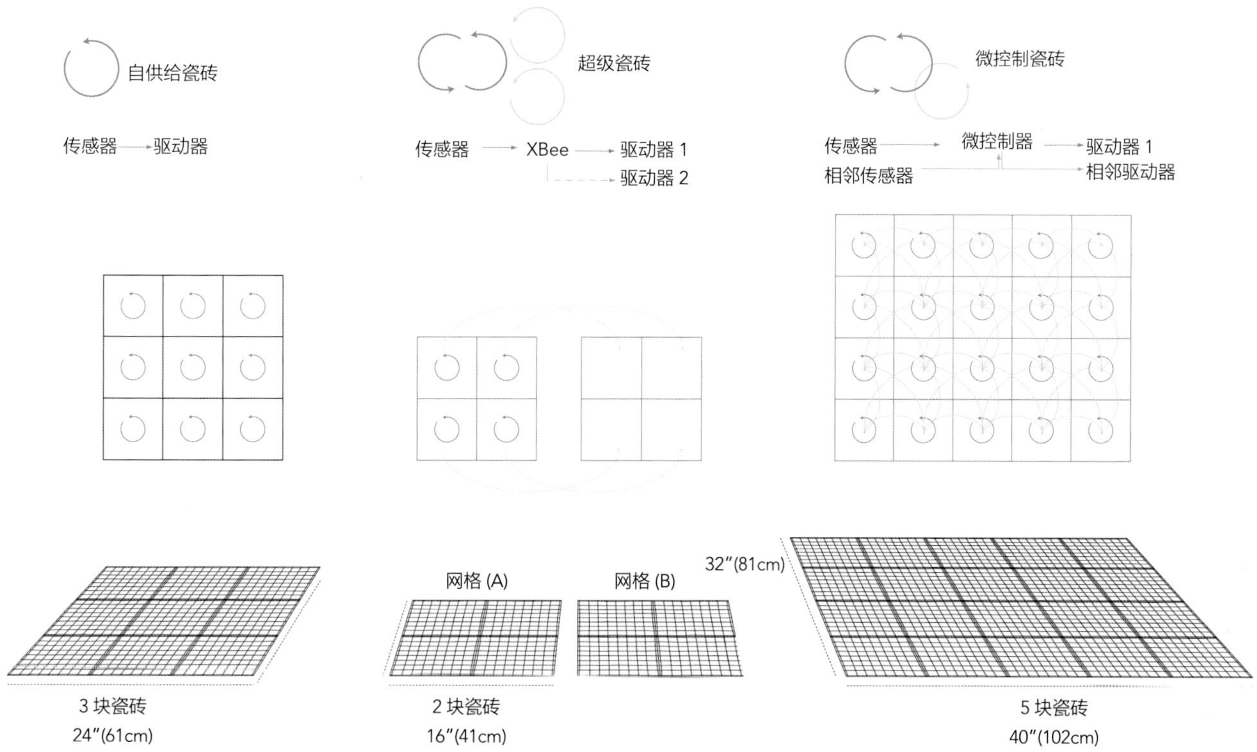

模式建立 : 工作模式将包括三个独立的原型。第一个原型是 3 × 3 的自供给瓷砖网络网格，当它们驱动时就发光。第二个原型包括两个 2 × 2 的超级网格。当网格 A 中的瓷砖被驱动时，网格 B 中相同位置的瓷砖通过 Xbee 无线连接而发光。最后第一个原型是一个 5 × 4 的微控制瓷砖网格，这些瓷砖证明了细胞自动机模式，通过影响相邻瓷砖的状态而驱动。

图 4.1-7 “群落瓷砖”的建造说明
Construction specifications for the‘Swarm Tile’

4.2 充气围护结构 Pneumatic Envelope

具有可变透明度的热敏隔热幕墙。❶

瑞克 · 索莱

A thermally responsive insulating curtain wall with variable transparency.❶

Rick Sole

The 'Pneumatic Wall' is a thermally responsive insulating curtain wall with variable transparency, oscillating in the gradient between wall and window.

The 'Pneumatic Wall' has a cellular structure, with each cell or air-pocket consisting of three membranes forming two separately controlled pneumatic pockets. Through sensor data processed in heat transfer analysis software, individual pneumatic pockets are controlled to inflate or deflate as required by user preference, or by changing climatic conditions.

Passive convective air flows are created based on thermal gradients on the interior side. By dynamically altering the thermal properties of a wall in section, a variety of air circulation patterns are achieved within the space. Operable windows take the form of air pockets on a central revolving axis which can be rotated to provide fresh air. Each membrane

“充气墙”是一个具有可变透明度的热敏隔热幕墙，在墙与窗之间以一定速率变化。

“充气墙”具有蜂窝结构，其中每个单元或空气囊由三个膜构成两个独立控制的充气袋。通过传热分析软件处理传感器数据，根据使用者的需要，或者改变气候条件来控制每个充气袋充气或放气。

被动式空气对流的产生是基于室内的热梯度。通过动态改变墙体剖面的热工性能，室内多样化的空气循环模式得以实现。可控窗户采用能够绕中心旋转轴旋转的充气袋形式，从而为室内提供新鲜空气。每层充气膜都具有透明与不透明的模

❶ Guided by Christos Marcopoulos and Carol Moukheiber.

❶ 由克里斯托 · 马尔科普洛斯与卡罗尔 · 穆海贝尔指导。

式，通过调节空气袋，可以实现不同的透明度。幕墙的外观可以根据所选择的室内条件与隔热程度发生变化。

has its own pattern of opacity and transparency. By modulating the air pockets, different transparencies are achieved. The curtain wall changes in appearance based on the selected interior condition, or level of insulation.

图 4.2-1

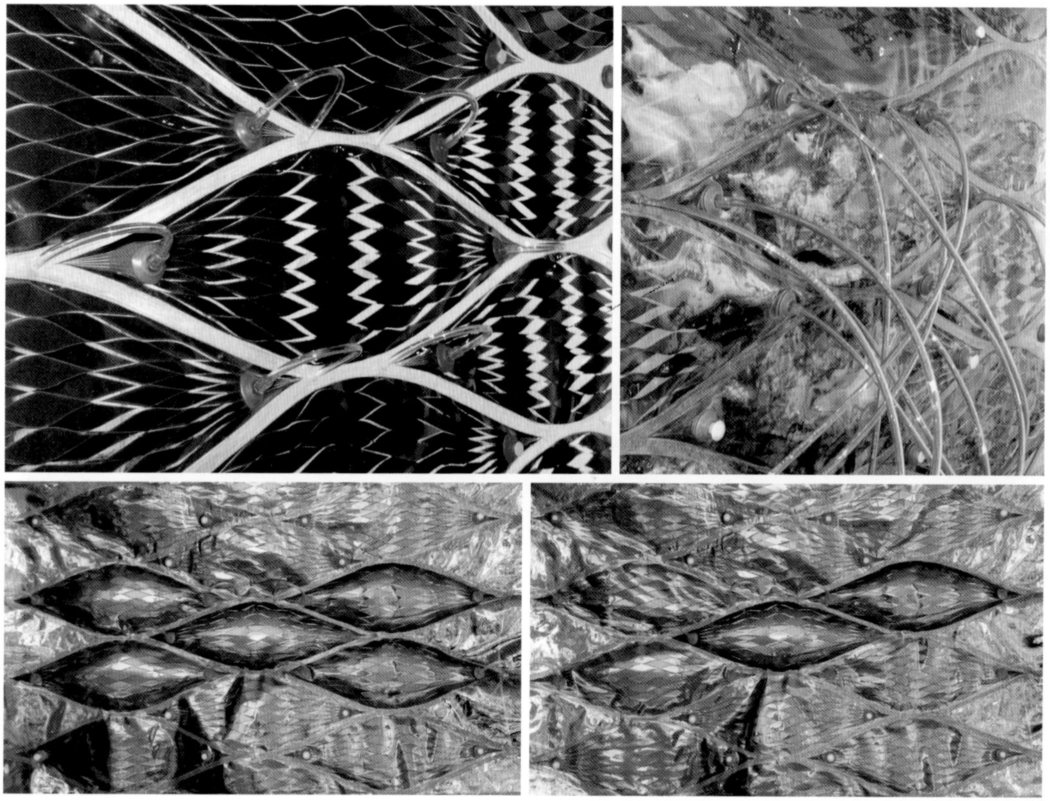

图 4.2-1 充气围护结构充气膨胀。越膨胀越隔热，且墙体变得更透明
The Pneumatic Envelope fully inflated. The more inflated, the more insulating, and the more transparent the wall becomes

图 4.2-2

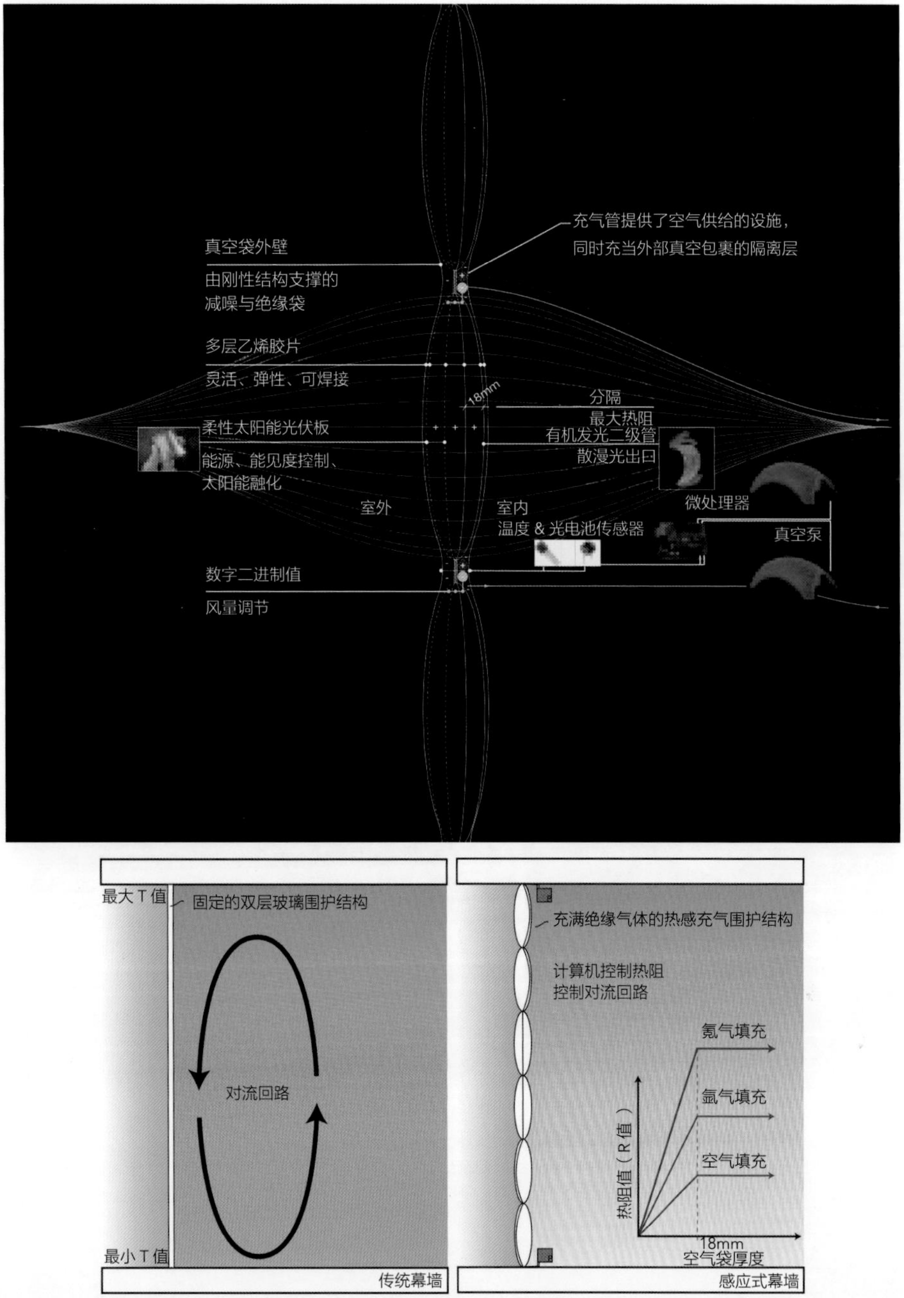

图 4.2-2 每个“枕头”充气或放气以提供局部的热舒适度。墙体垂直截面隔热水平的变化使得空间内产生空气对流，以减轻垂直热滞留问题

Each ‘pillow’ inflates or deflates as necessary to provide local thermal comfort. A variation in insulation levels across the wall’s vertical section allows for the generation of convective flows within a space, alleviating the problem of vertical heat entrapment

图 4.2-3

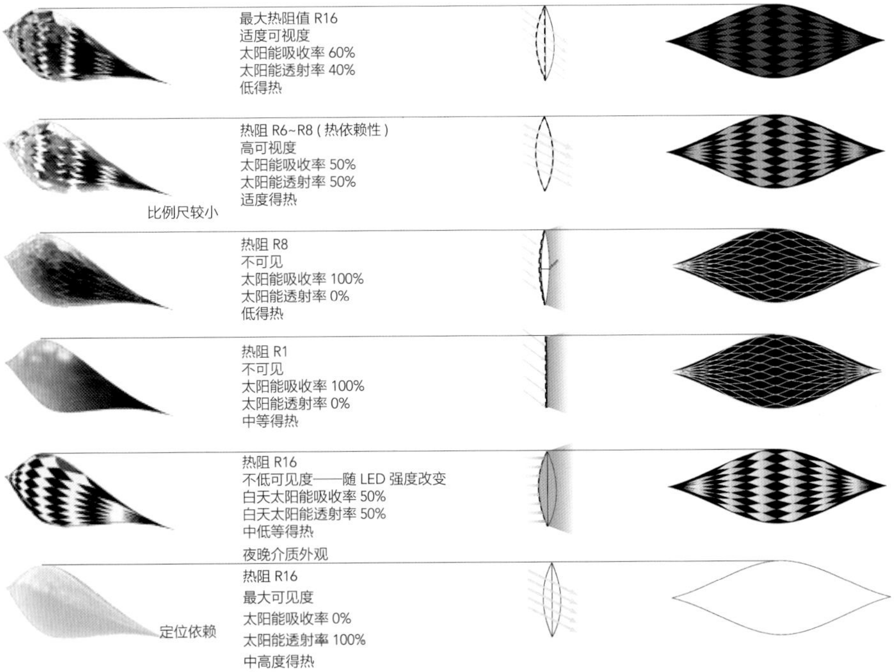

图 4.2-3 使用者可与墙体进行远程对话，并随时激活其热工性能

Users can speak to the wall remotely and activate its thermal properties at will

图 4.2-4

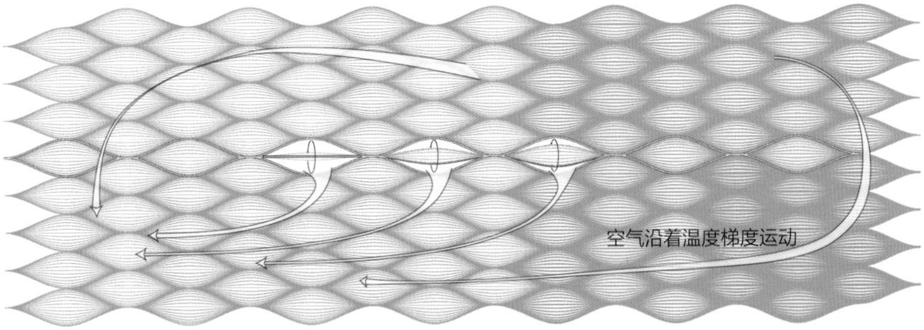

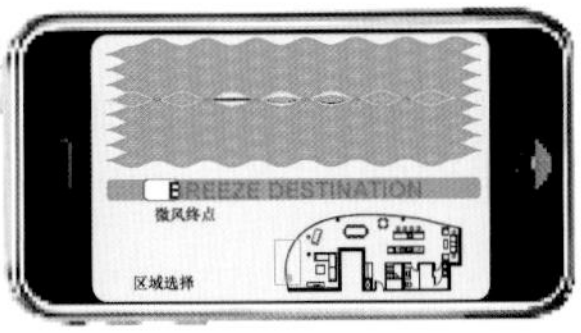

图 4.2-4 可控气袋能沿其中心轴旋转

Operable cells pivot along their central axis

4.3 聚光幕墙 Poly-Glazed Curtain Wall

多层玻璃系统能收集能量与塑造室内氛围。[1]

埃利 · 雅巴哈提

A multi-layered glazing system that harvests energy and sets up interior mood.[1]

Ali Yarbakhti

The 'Poly-Glazed Curtain Wall' is an interactive cladding system that responds intelligently to changing external circumstances and internal performance requirements. Each panel is capable of projecting the diversity inside the building to the outside while maintaining optimal visual and thermal conditions indoors.

The proposed electronically enhanced architecture of the 'Poly-Glazed Curtain Wall' cladding system consists of multi-layer striped glass sheets that transform the architectural facade to a dynamic mechanism capable of changing colour and transparency in response to interior and exterior circumstances. Since the proposal includes the deployment of a transparent solar panel as an embedded layer, the panel is self-sufficient in terms of energy consumption.

Each layer in the panel consists of alternating transparent and coloured translucent strips, and is placed on a geared shaft connected to a micro servo motor. By changing the rotational degree of each servo depending on the pattern in which the sliding layers are placed in relation to each other, the coloured translucent strips are juxtaposed to create a colour composition that is unique to a particular setting. For prototyping

“聚光幕墙”是一个交互式的覆层系统，能够智能响应变化的室外环境和室内性能要求。每个面板能够展现建筑从室内到室外的差异，同时保持最佳的室内光热条件。

该“聚光幕墙”覆层系统的电子增强结构是由多层条纹玻璃板构成，将建筑立面转化为能够根据室内外环境改变颜色和透明度的动态机制。由于该方案配置了一个透明的太阳能电池板作为嵌入层，电池板在能源消耗方面是自给自足的。

每层面板是由交替的透明和彩色半透明条组成，并放置在与微型伺服电机相连的齿轮传动轴上。根据滑动层彼此相关的图案来改变伺服电机的旋转程度，将半透明的色带并置，以创建一种针对特定设置的独特色彩构成。为了制作原型，每个面板都有内嵌的开源硬件板，可通过控制伺

[1] Guided by Rodolphe el-Khoury and Nashid Nabian.

[1] 由鲁道夫 · 埃尔 - 库利与纳希德 · 纳比安指导。

服电机的角度来动态改变面层的排列。通过一个传感器来记录周围环境，辨识室内的活动级别，使用者也能够通过直接输入来改写自动颜色设置。在其他可能实现的场景中，面层能够响应温度或光照强度，并应用由计算机设定的最佳透明度或颜色组合的运算法则，来响应环境条件的变化。

purposes, each panel has an embedded Arduino board that dynamically changes the arrangement of layers by controlling the servo angles. The context is registered via a presence sensor that identifies the level of activity inside. The users also can override the automated colour setting via direct input. In other possible implementation scenarios, the panels can respond to temperature or light intensity and apply algorithms that computationally decide on the optimum transparency/ colour combination in response to changing environmental circumstances.

图 4.3-1

图 4.3-1 分层玻璃和彩色过滤细部
Detail of layered glazing and chromatic filtering

图 4.3-2 多层玻璃系统和伺服马达激活的滑动系统细部
Diagrams detailing multi-layered glazing system and mechanism of servo-activated sliding

图 4.3-2

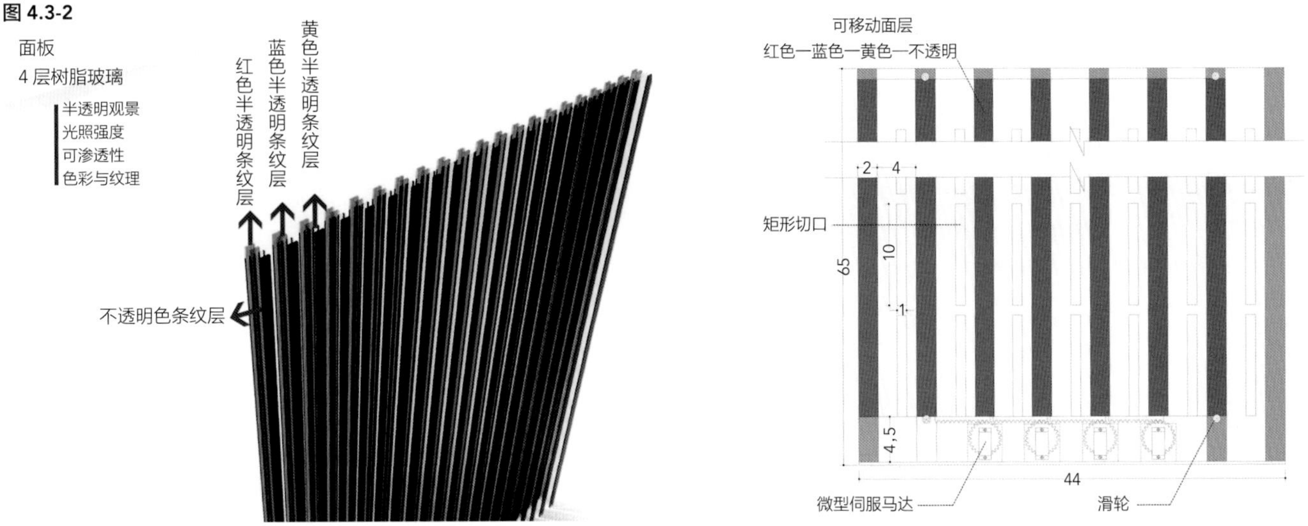

图 4.3-3

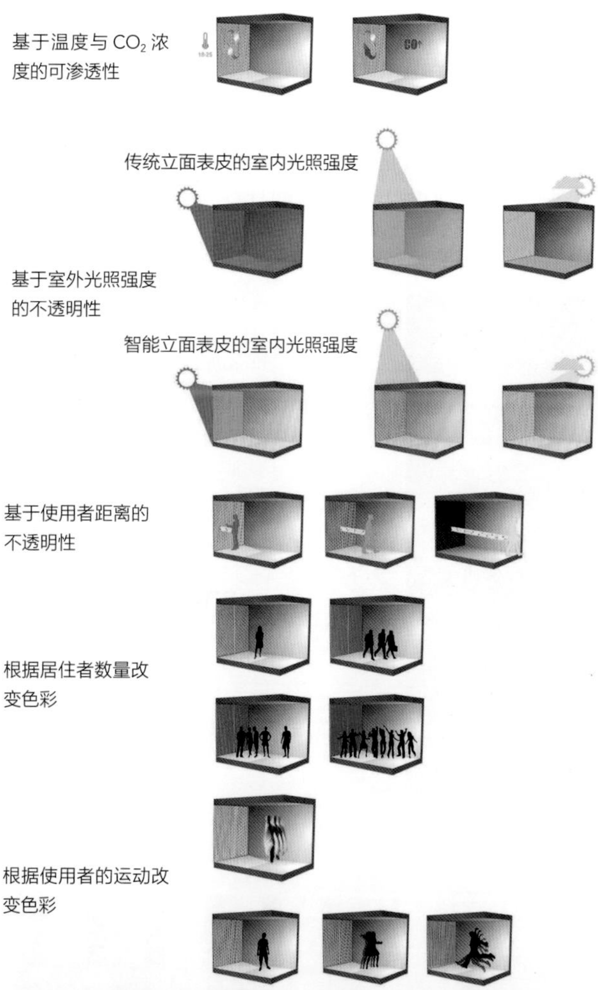

图 4.3-4

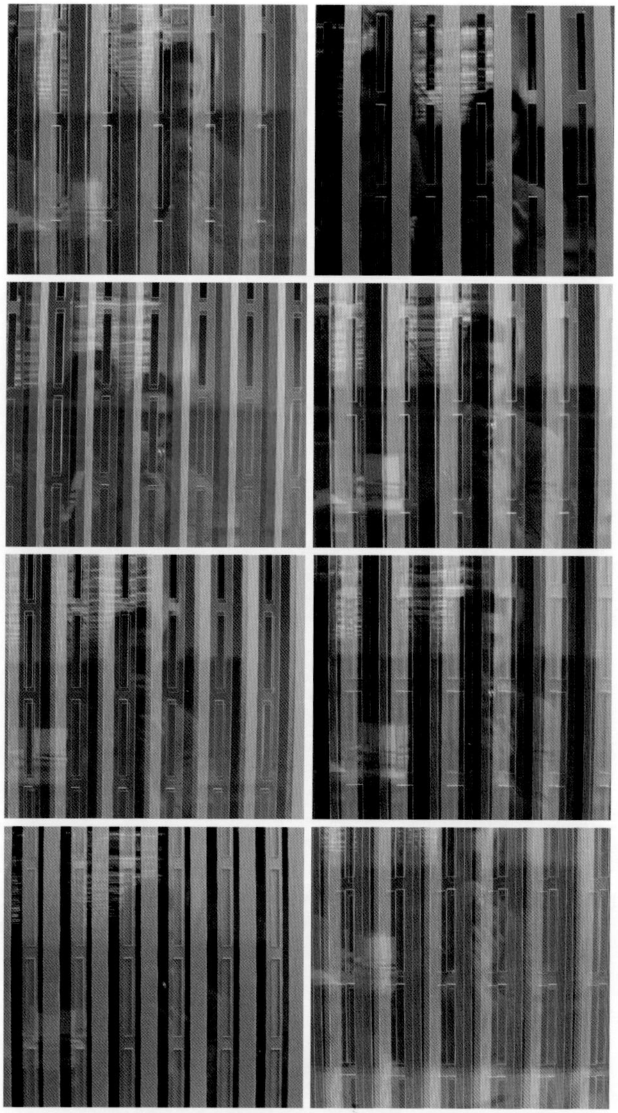

图 4.3-3 在不同活动的预期或反应下，光过滤和情绪设置的各种应用场景
Various scenarios of light-filtering and mood-setting configurations in anticipation of or in response to different activities

图 4.3-4 聚光幕墙
Poly-glazed curtain wall

4.4 修道院的微气候 Iviron Microclimate

个体环境可被定制，来响应居住者。[1]

朱利安纳 · 萨斯曼

Individual environments are customized to respond to their inhabitants.[1]

Julianna Sassaman

当居住者通过修道院数排的感应式地板与顶棚吊顶板时，会形成一系列可定制的微气候。

修道院使用触感制动器，包括灯光、供热器和扬声器。当居住者穿过排列的制动器时，它们被激活，修道院吊顶板里的传感器激活了这些嵌入式技术。驱动器被配置为可读取居住者所携带的个人身份信息，居住者通过该信息传达其对气候的偏好。

修道院吊顶板的大小尺寸取决于嵌入其中的特定制动器的特性与制动器的影响范围。

Iviron forms a series of customisable microclimates, created when inhabitants pass between arrays of responsive floor and ceiling tiles.

Iviron uses haptic actuators, including lights, heaters and speakers, which are activated as the inhabitant passes through the array. Sensors in the Iviron tiles trigger the activation of these embedded technologies. The actuators are configured to read a personal identity profile carried by the inhabitant, who communicates his or her climate preferences.

Iviron tiles vary in size, depending on the nature of the particular actuator embedded within them, and on the actuator's sphere of influence.

❶ 由鲁道夫 · 埃尔 - 库利与纳希德 · 纳比安指导。

❶ Guided by Rodolphe el-Khoury and Nashid Nabian.

图 4.4-1

微气候 : 光

60° 光锥

微气候 : 声

90° 声锥

微气候 : 热

微气候 : 热 - 光 - 声

60° 光锥

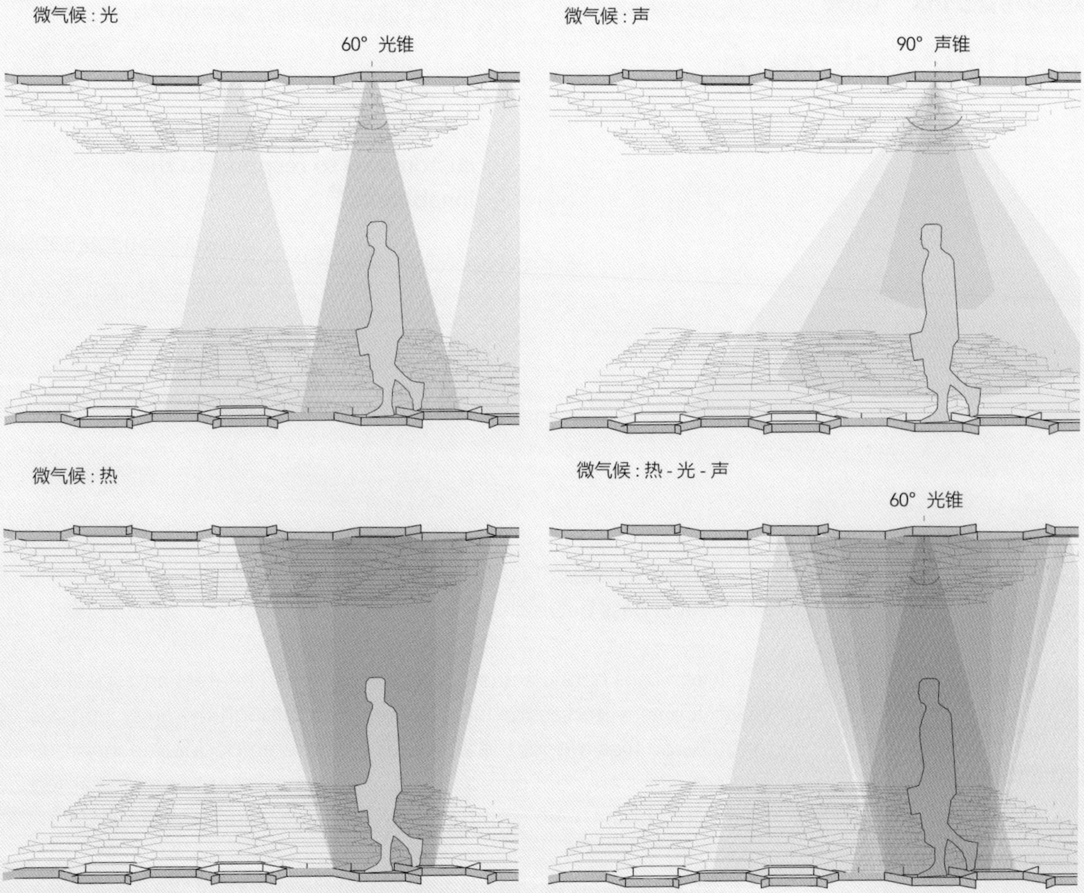

图 4.4-2

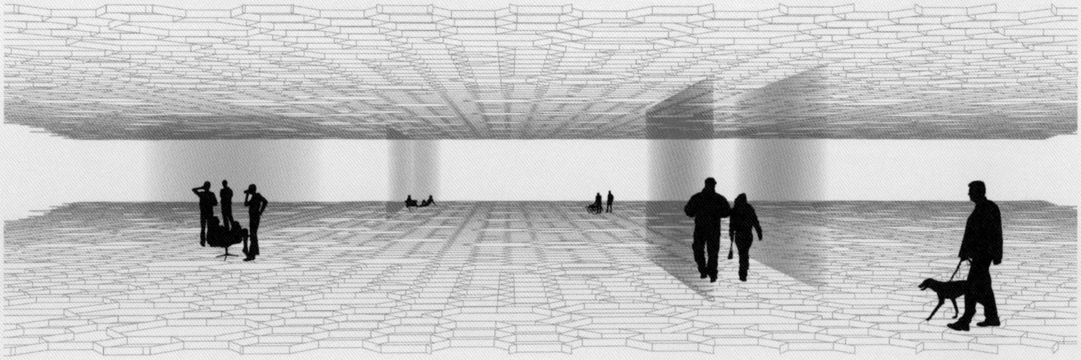

图 4.4-3

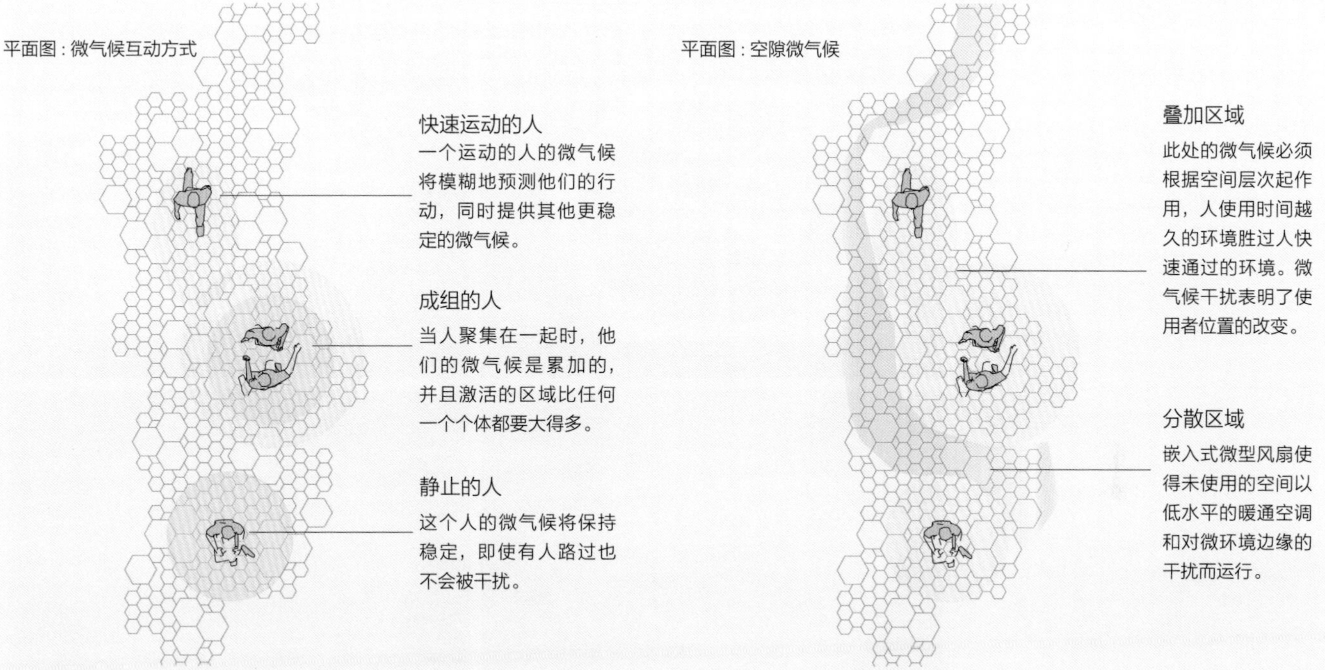
平面图 : 微气候互动方式
快速运动的人
一个运动的人的微气候将模糊地预测他们的行动，同时提供其他更稳定的微气候。
成组的人
当人聚集在一起时，他们的微气候是累加的，并且激活的区域比任何一个个体都要大得多。
静止的人
这个人的微气候将保持稳定，即使有人路过也不会被干扰。
平面图 : 空隙微气候
叠加区域
此处的微气候必须根据空间层次起作用，人使用时间越久的环境胜过人快速通过的环境。微气候干扰表明了使用者位置的改变。
分散区域
嵌入式微型风扇使得未使用的空间以低水平的暖通空调和对微环境边缘的干扰而运行。

图 4.4-1 包括灯光、音响和供热的嵌入式技术
Embedded technologies include lighting, sound and heat

图 4.4-2 动态开放的空间效果
The result is a dynamic open space

图 4.4-3 微气候相互作用和间隙气候
Microclimate interaction and interstitial climates

图 4.4-4

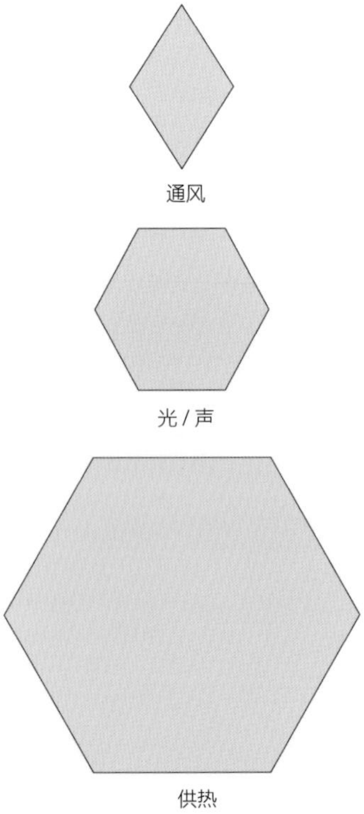

图 4.4-5

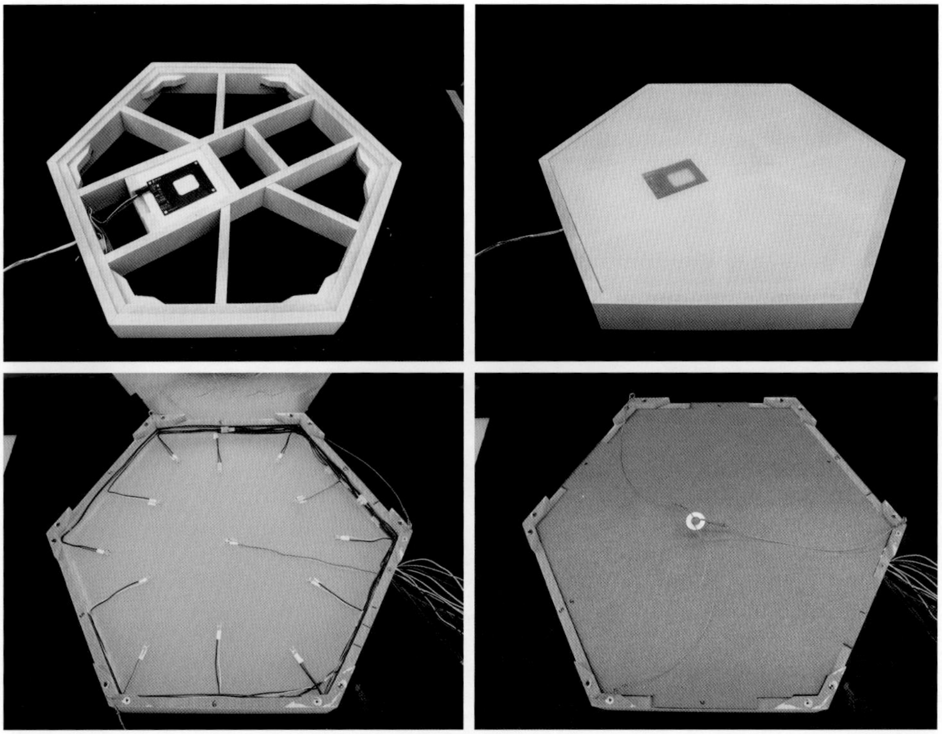

图 4.4-4 包括三种不同类型的嵌入式技术的吊顶板设计对策
The tile strategy includes three types for the various embedded technologies

图 4.4-5 地板原型。无线射频识别（RFID）读取器是嵌入地板瓷砖内的
Floor prototype. The RFID reader is embedded within the floor tile

图 4.4-6 三种实施场景：俱乐部、日常工作社区空间与家庭空间
Three scenarios for application. Club, live-work community space and domestic space

图 4.4-7 包括所有组件的全电路
Full circuit including all components

图 4.4-6

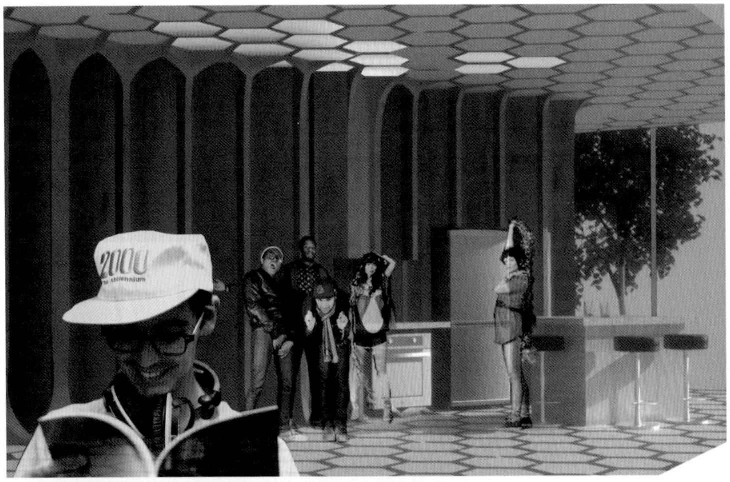

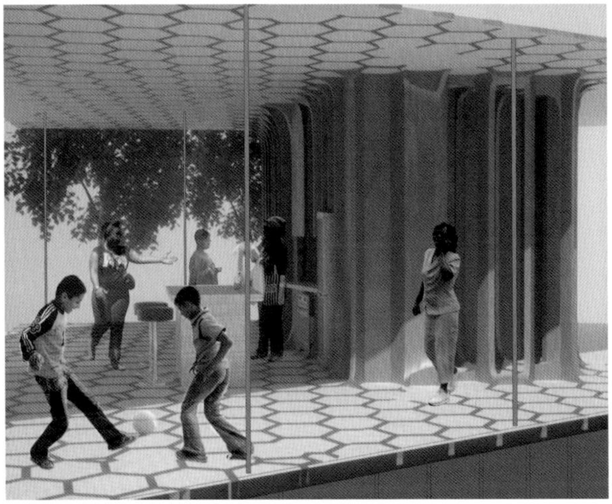

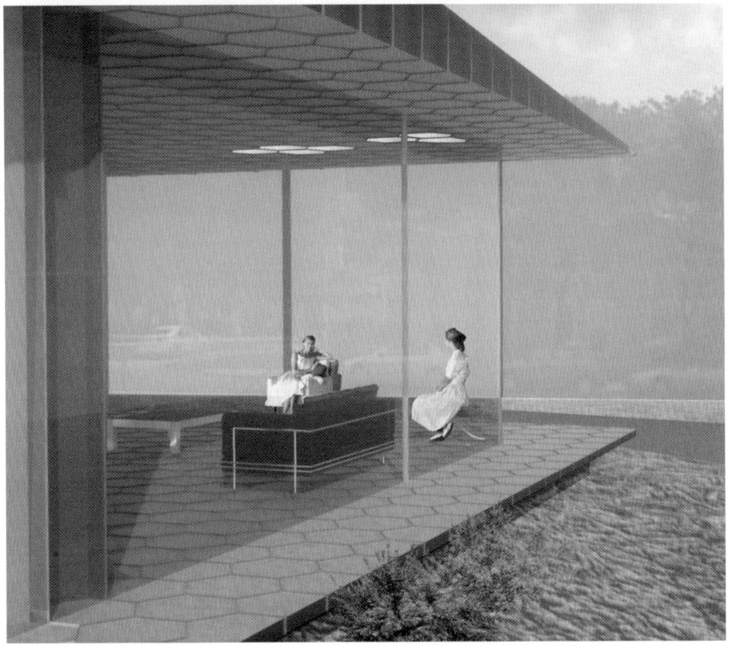

图 4.4-7

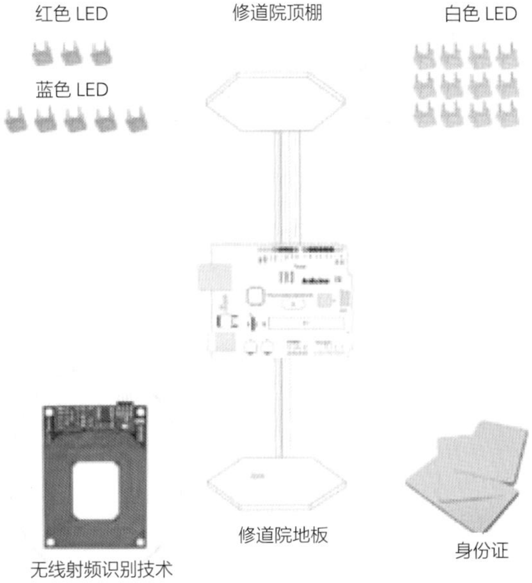
红色 LED
修道院顶棚
白色 LED
蓝色 LED
无线射频识别技术
修道院地板
身份证

4.5 可调节的天气 Mediating Weathers

> **在建筑围护结构内创建了第三种具有性能的天气空间。❶**
>
> 娜迪亚 · 沃勒斯
>
> **Performative third weather spaces are created within the building envelope.❶**
>
> Nadya Volicer

Taking the glass block as a starting point, 'Mediating Weathers' explores the potential of this material to optimize daylight, adjust for solar gain, and convert to a privacy screen by night.

This project is about the 'third weather' present within the home. This is the weather found in the inbetween spaces: between panes of glass, in wall cavities, in root cellars and greenhouses. These spaces are conditioned to have environments distinct from both inside and outside. How can we create 'third weather' spaces which continue to mediate between the exterior and interior, while offering physiological benefits to inhabitants, and spectacle for viewers both inside and out? The glass block has long been one of the most common third weather building materials, housing air in a hollow block of glass which insulates while acting as glazing and structure.

“可调节的天气”是以玻璃块为出发点，探讨了此种材料的潜力，以优化采光和获取太阳能，并在夜间转换为隐私屏幕。

该项目是关于在家中存在的“第三种天气”，此种天气能在以下中空的空间被发现：玻璃窗格、空心墙、地窖和温室。这些空间有条件形成与室内外截然不同的环境。我们如何创造一种能够不断调节室内外差异的“第三种天气”空间，同时又为居住者提供有益的生理影响，为室内外提供好的景观？玻璃块一直是最常见的“第三种天气”建筑材料之一，隔热中空玻璃块内装有空气，同时作为玻璃与结构。

❶ Guided by Rodolphe el-Khoury and Nashid Nabian.

❶ 由鲁道夫 · 埃尔 - 库利与纳希德 · 纳比安指导。

树叶墙原型在阳光下工作，利用光电池来驱动风扇，创造出树叶的颤动以过滤阳光，并将斑驳的光影投射到室内。当没有阳光直射时，内部的 LED 灯就会照亮玻璃块。夜间模式是追踪居住者的运动方式，而不是追踪阳光。当人被给定的玻璃块检测到时，风扇将被激发而使树叶颤动，从外部，人可以观察到居住者的运动，并将它看作是隔着墙体追踪颤动的树叶。

第二个原型——雾墙，是利用雾达到类似的效果。它依赖玻璃块内部的薄雾屏蔽太阳能，并在夜晚提供私密性。 第三个原型——雨墙，它是基于蒸发与冷凝在自给自足的系统原理下运作。如果玻璃块能够通风，夜间冷凝水将有助于为起居空间制冷。夜晚，通过允许室内空气向室外流动并穿过“第三种天气”，“第三种天气”将冷却家居环境。在此种模式中，内置的隐形屏幕就是树叶。最后两个原型（日光墙和风墙原型）通过将捕风器与双层表皮立面相结合，利用风能和太阳能调节“第三种天气”的空间。在一年中口光不充足的日子里，通过凹透镜风格根据季节开启和关闭光圈来补充阳光的自然路径，将阳光引入室内而增加采光。

The foliage wall prototype works in the sunlight by using a photocell to trigger a fan, creating a flutter of leaves to filter the sunlight and cast dappled light on the interior. When out of direct sunlight, an internal LED illuminates the block. Night mode tracks the movements of the inhabitant rather than the sun. When a presence is detected by a given block, the fans will be triggered and the leaves flutter. From the exterior, one would observe the movements of the inhabitants as a fluttering of leaves tracking across the surface of the wall.

A second prototype, the fog wall, uses fog to achieve similar outcomes. It relies upon mist within the block to shield from solar gain and provide privacy at night. A third prototype, the rain wall, operates on the principle of the self-sustaining system based on evaporation and condensation. Provided the block is able to ventilate, the nightly condensate would help to cool the living space. By allowing interior night air to pass through this ‘third weather’ as it moves to the exterior, it would cool the domestic environment. In this model, the built-in privacy screen is the foliage. Two final models harness wind and sun to mediate the third weather space by merging a wind catcher and a double skin façade, opening and closing apertures depending on the season through a grid of concave lenses which supplement the sun’s natural path to increase sunlight during the darker days of the year.

图 4.5-1

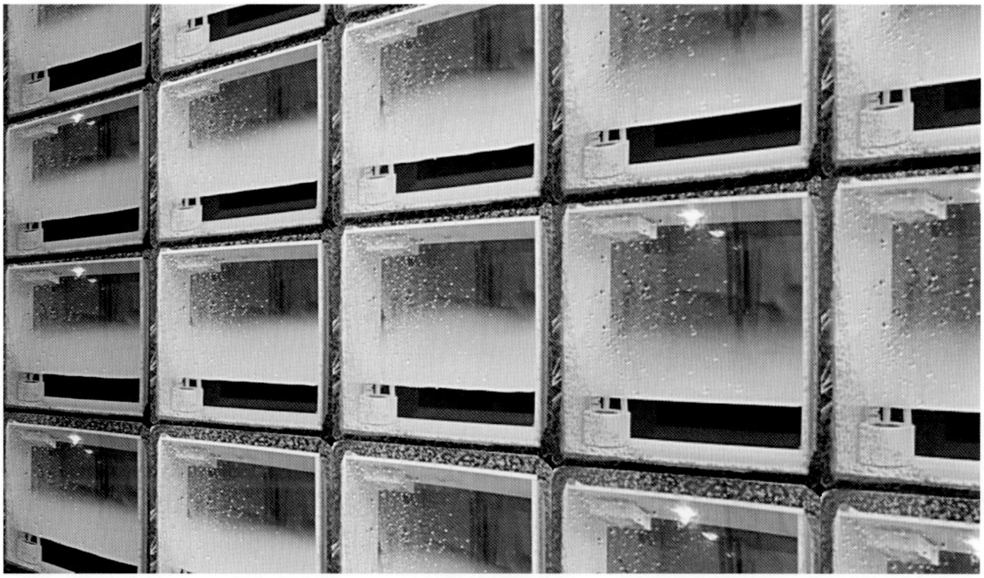

图 4.5-1 雾墙原型
Fog wall prototype

图 4.5-2

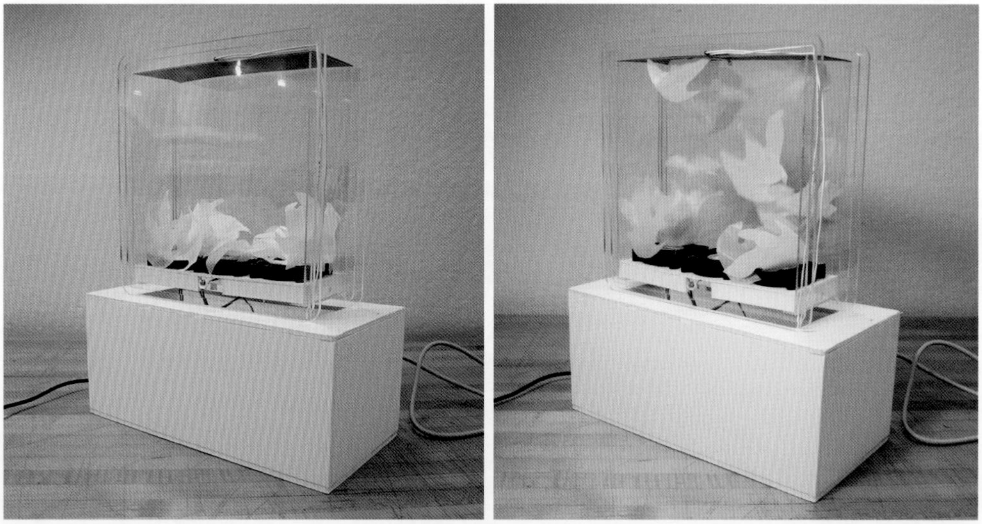

图 4.5-2 树叶墙原型
Foliage wall prototype

图 4.5-3

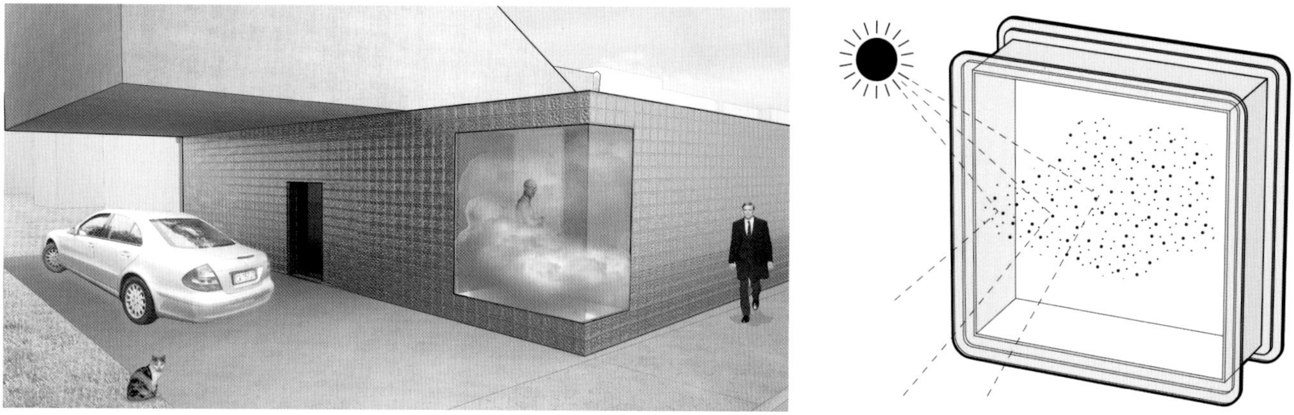

图 4.5-3　雾墙原型
Fog wall prototype

图 4.5-4

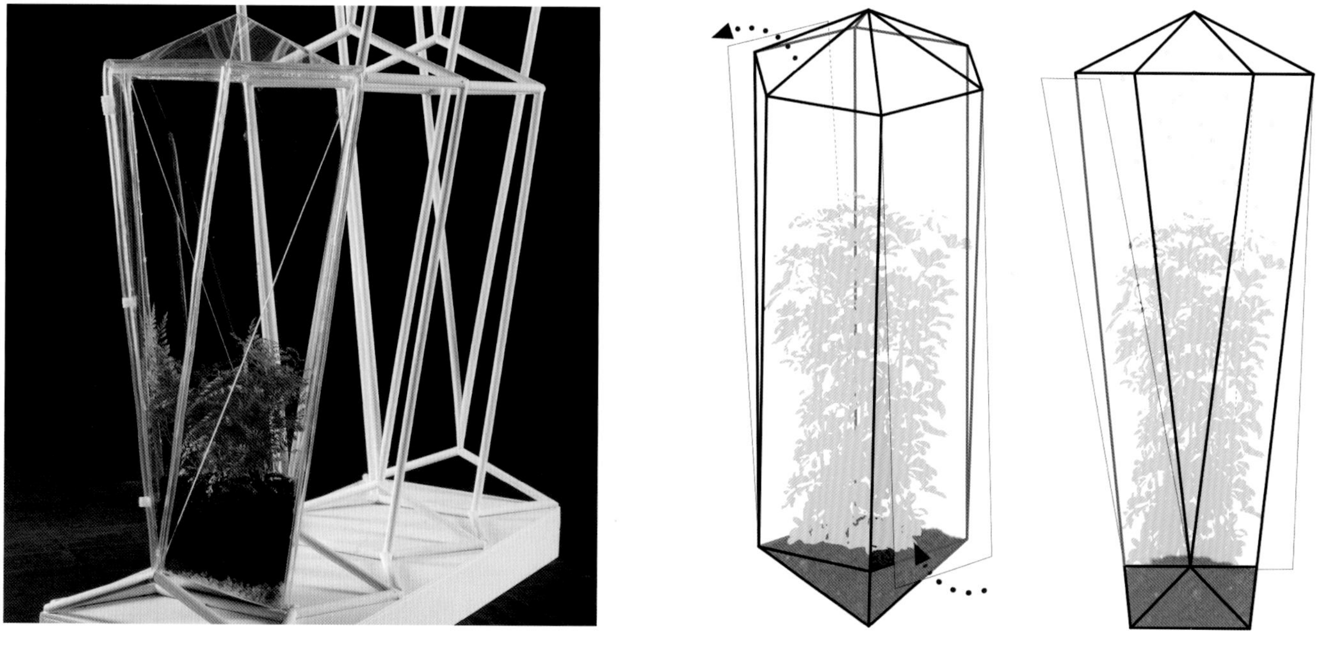

图 4.5-4　雨墙原型
Rain wall prototype

图 4.5-5

图 4.5-6

图 4.5-7

图 4.5-8

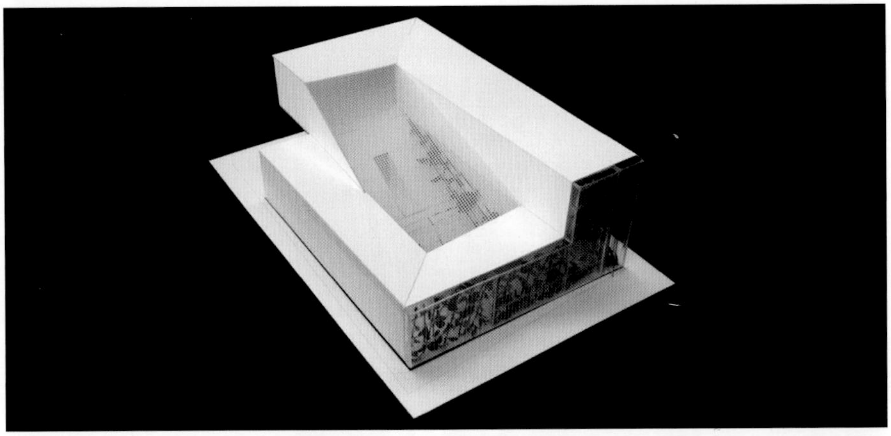

图 4.5-5 日光墙原型
Sun wall prototype

图 4.5-6 风墙原型
Wind wall prototype

图 4.5-7 雨墙效果图
Rain wall render cropped color

图 4.5-8 树叶雨墙在一定比例的建筑模型中
Foliage rain wall in a scale model house

4.6 日光建筑 The Sunlight House

一个利用建筑结构将光传输到室内黑暗空间的系统。❶

尤利娅 · 本茨瓦

A system which uses the structure of the house to transmit light into dark interior spaces.❶

Yuliya Bentcheva

“日光建筑”使用太阳能跟踪装置、光学纤维和传光 / 发光材料，利用自然光改变室内空间。

随着科技的进步，我们正在越来越远离自然环境，人们花费更多的时间在室内和空间上，试图重现自然环境，因此新技术正在探索将日光引入这些室内没有采光的区域。

“日光建筑”通过三个基本构件将阳光传输进室内黑暗空间：太阳能跟踪装置、光纤、传光 / 发光材料。太阳能跟踪装置采用双凸透镜，将收集的太阳光导入光纤。这些光纤通过由传光或发光材料制成的采光墙，将光传送进建筑内部深处。采光墙的材料可以是玻璃或塑料，并能被纳入建筑结构或作为灯具使用。

The ‘Sunlight House’ uses a solar tracking device, optical fibres, and light transmitting / emitting material to transform interior spaces using natural light.

With the advance of technology, society is becoming increasingly detached from the natural environment. People spend much more time indoors and in spaces that attempt to recreate the natural environment, and as such, new technologies are attempting to incorporate sunlight into these often dark areas.

The ‘Sunlight House’ brings sunlight into dark interior spaces through three basic components: a solar tracking device, optical fibres, and a light transmitting / emitting material. The solar tracking device uses a double convex lens in order to focus collected sunlight onto the fibres. These fibres transmit light deep into the house through a light wall, made of a light transmitting or emitting material. This may be glass or plastic and can be incorporated into the structure of the building or used as a light fixture.

---

❶ 由鲁道夫 · 埃尔 - 库利与纳希德 · 纳比安指导。

❶ Guided by Rodolphe el-Khoury and Nashid Nabian.

图 4.6-1

图 4.6-2

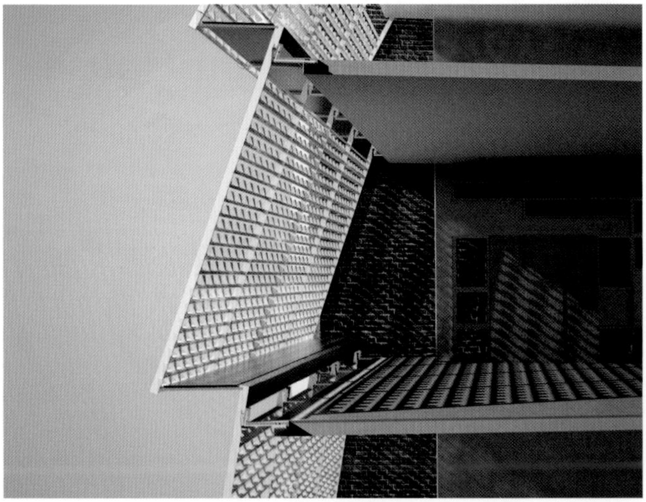

图 4.6-3

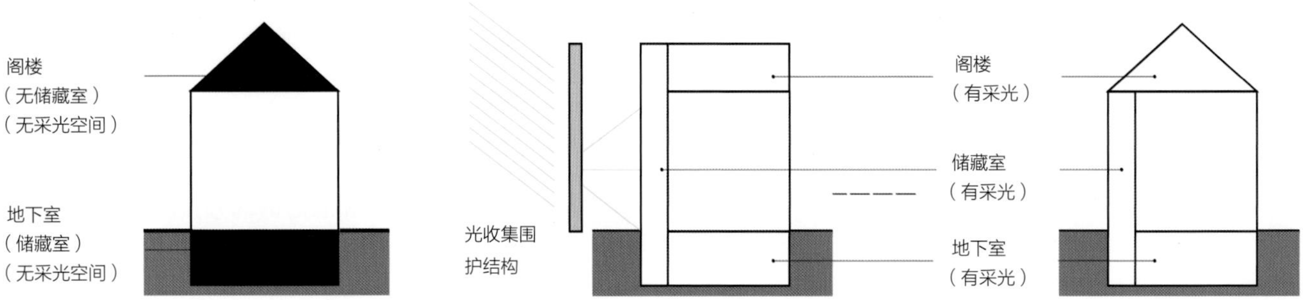

图 4.6-1 外立面是由太阳能跟踪装置构成
Exterior facade built from solar tracking devices

图 4.6-2 建筑剖断面表现出构造细部与开口
Section through facade, showing structural details and openings

图 4.6-3 利用光收集围护结构为阁楼和地下室采光
Reviving the attic and basement using a light collecting envelope

图 4.6-4

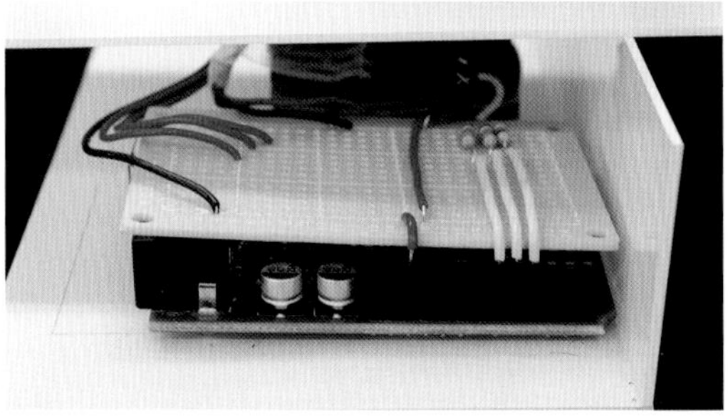

图 4.6-5

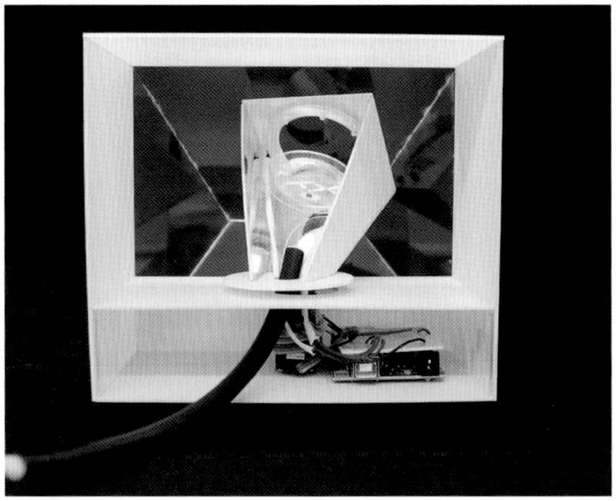

图 4.6-6

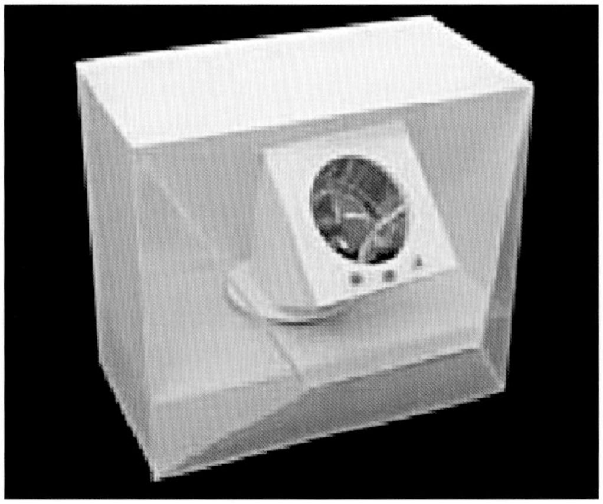

图 4.6-7

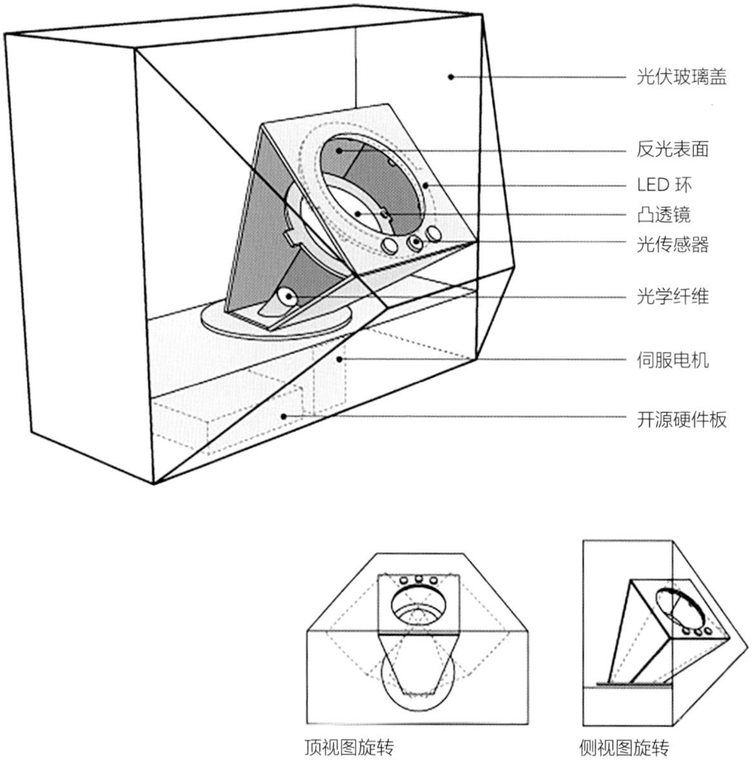

图 4.6-4 光纤和开源硬件装置
Optical fibre and Arduino setup

图 4.6-5 太阳能跟踪装置原型
Solar Tracking Device prototype

图 4.6-6 开源硬件和电路板
Arduino and circuit board

图 4.6-7 太阳能跟踪装置组件
Solar tracking device assembly

图 4.6-8

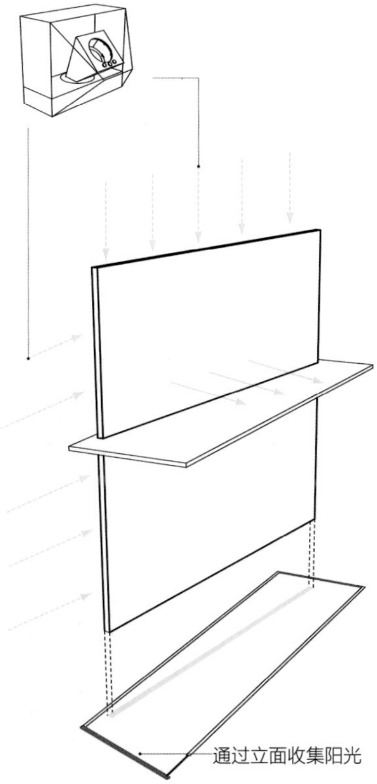

图 4.6-9

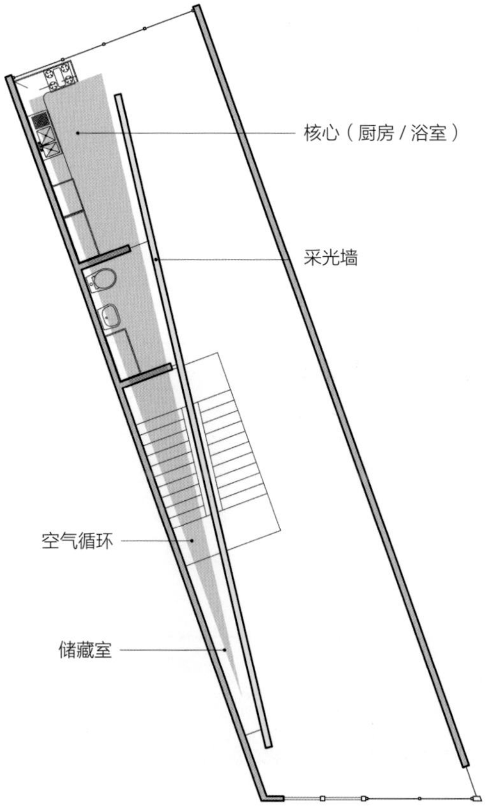

图 4.6-10

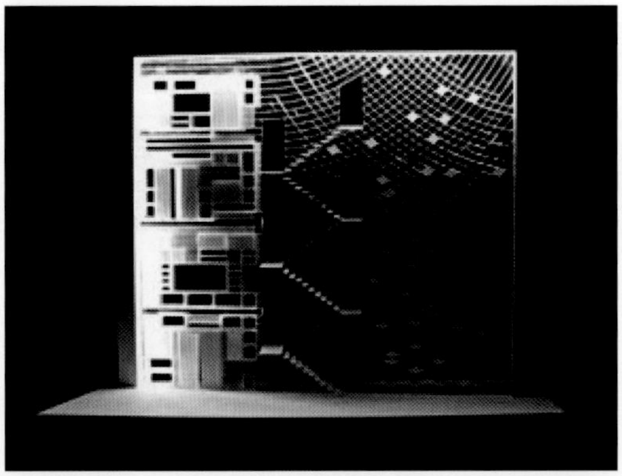

图 4.6-8 光从太阳能集热器输入光纤，然后进入采光墙与住宅内部
Light flows from the solar collectors to the cables, and then to the light wall and into the home

图 4.6-9 将采光井扩大是为了将住宅核心和空气循环整合起来。现在光线可以到达整个建筑空间，住宅的空间分配可以被重新规划
The party wall is expanded to incorporate the core of the house and the circulation. Now that daylight is available throughout the space, the organisation of the home can be reconsidered

图 4.6-10 采光墙剖面模型
Sectional model of a light wall

图 4.6-11

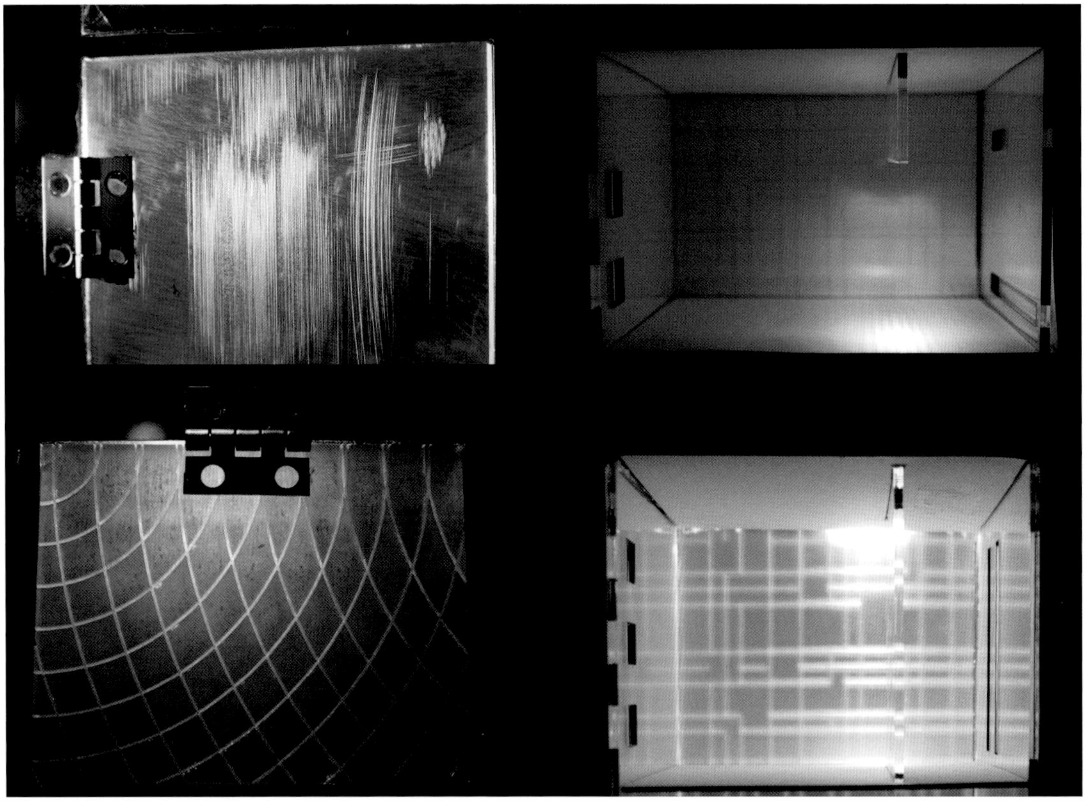

图 4.6-12

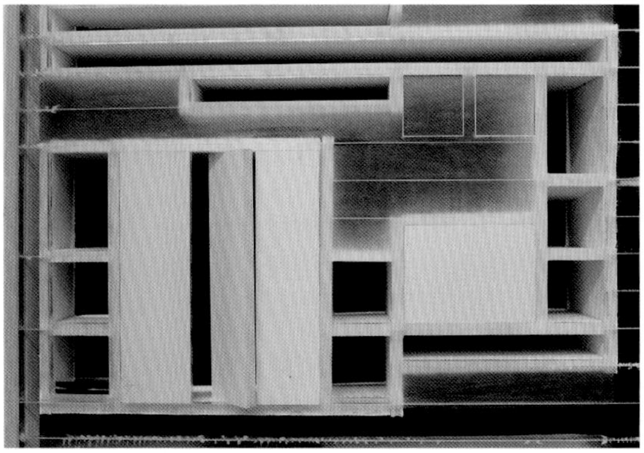

图 4.6-11 采光墙细部，光遮板。光被隐藏在遮板后面，光线的调节取决于任何既定空间的需要
Light wall detail. Light shutters. Light is always behind the shutters and can be adjusted depending on the need in any given space

图 4.6-12 采光墙细部。储藏间被嵌入采光墙内，当储藏室柜门被打开时，光线就出来了
Light wall detail. Storage units are embedded into the light wall. When the cabinet doors are opened, light comes out

图 4.6-13

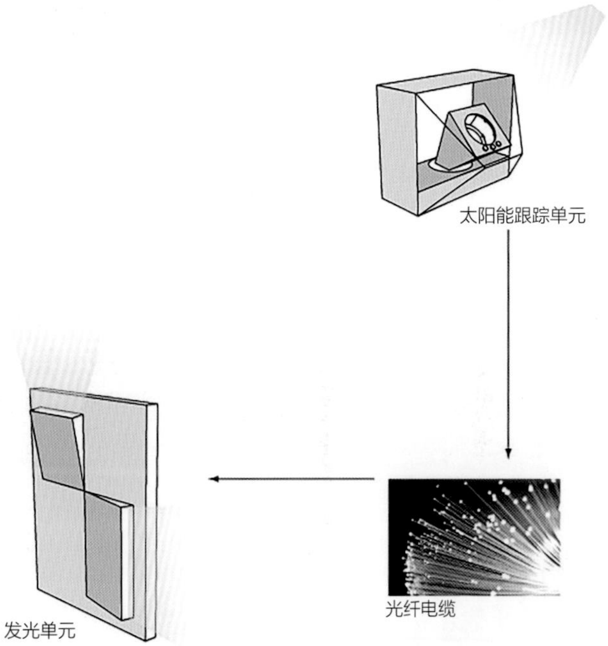

图 4.6-14

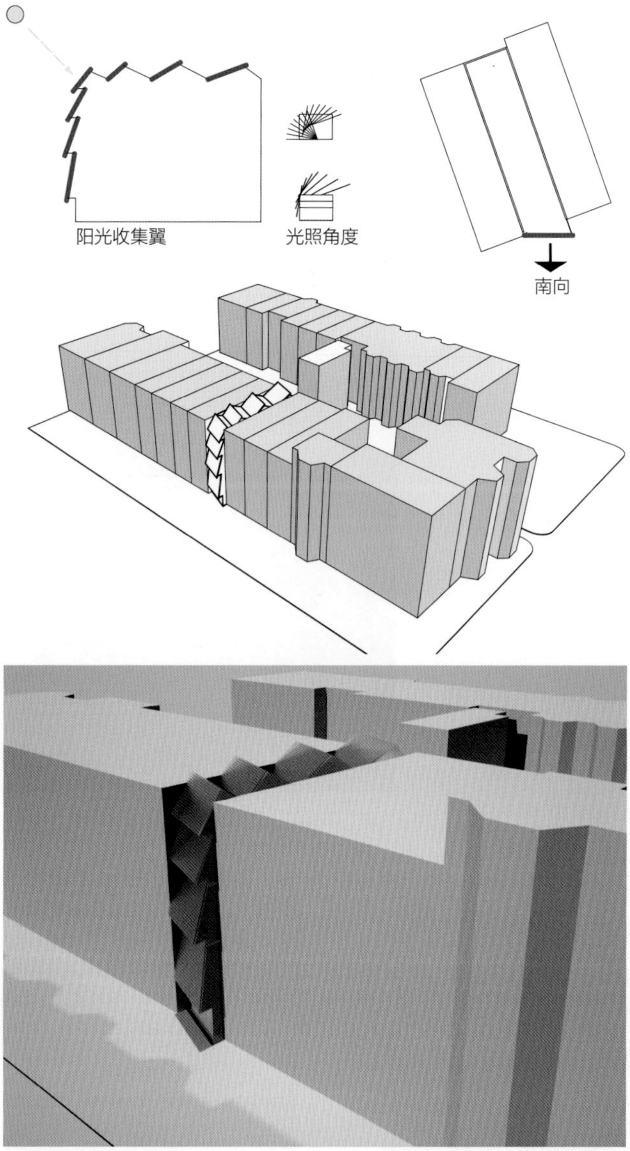

图 4.6-15

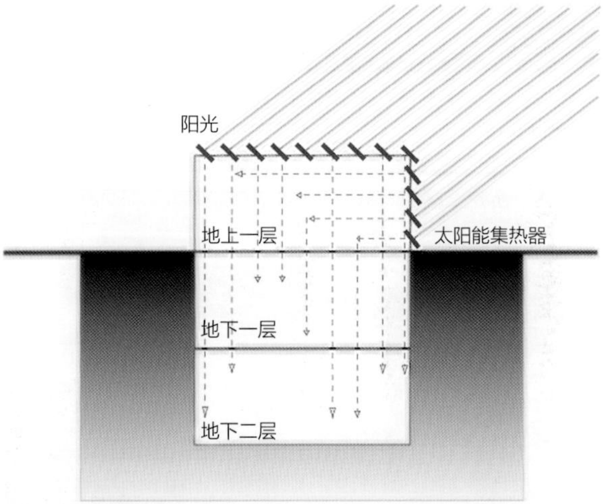

图 4.6-13 日光系统
Sunlight system

图 4.6-14 采光立面方案与效果
Facade exploration

图 4.6-15 光收集系统示意
Iceberg diagram

随着计算设备与物理世界的融合，显示器、键盘、触摸板和无数种触觉输入设备充满整个环境。城市成为一个连接虚拟和现实的接口，它的表面嵌入了读写存储器。以下项目展示了建筑在物理环境与信息流的关联行为中起到的积极作用。它们为强化信息领域—— 一种提高物理世界的大小和分辨率的虚拟影像——建立了一种门户或着力点。

As computational devices merge with the physical world, displays, keyboards, touch pads, and myriad types of haptic input devices populate the environment. The city becomes an interface that links bits and atoms, its surfaces are embedded with read / write memory. The following projects claim an active role for architecture in linking the flow of information to the physical environment. They build portals or touchdown sites for intensifying info-spheres — a virtual simulacrum of the physical world that increases in magnitude and resolution daily.

5

表面作为接口

SURFACE AS INTERFACE

5.1 IM 感应毯 IM Blanky

IM 感应毯是一种具有自建模功能的毯子，能够展示对于事物的原生认知能力。

NMinusOne 工作室，鲁道夫 · 埃尔 - 库利

IM Blanky is a self-modelling blanket exhibiting primitive cognitive capacities.

Studio NMinusOne with Rodolphe el-Khoury

A blanket that simultaneously embodies both a physical and digital presence

The (soft) Hardware

The blanket measures 7'7" x 4'2" and is composed of a distributed field of 104 soft tilt sensors. The soft sensor forms the most basic motif: the flower. The flower consists of 6 conductive petals, linked by resistors, and a conductive tassel in the centre. The flowers are grouped together into 14 larger configurations or clusters and 2 half clusters. Working as a directional marker, the tassel's contact with a petal registers a specific orientation or tilt of the blanket. The flowers are arrayed within a circular positive and negative power supply circuit, and their stems plug into a computational hub (Multiplex). The clusters are then linked together, into a larger network of clusters, each relaying the position of its flowers to a microcontroller stitched to the back of the blanket. (Arduino Lilypad)

一条同时展示现实世界与数字世界的毯子。

（软）硬件

毯子的尺寸是 7 英尺 7 英寸 ×4 英尺 2 英寸，由 104 个软倾斜传感器组成的分布式场组成。这些软传感器构成了最基本的主题——花。花朵包括 6 个与电阻相连的导电花瓣，并且在花瓣的中间引出 1 条导电的穗须。花朵组成了 14 个大型组群（或簇）和 2 个半簇。穗须作为一个定向标记，连接着记录了明确方向或毯子坡度的花瓣。花朵排列成一个环形的正负极电源电路，其茎秆插入一个计算中心（多路）。这些簇又链接在一起成为一个更大的集群网络，每个与花朵位置有关的微控制器都被缝在了毯子的背面（Arduino 板）。

软件

传感器是基于 1 个基本的六边形结构来分布的。每朵花都是 1 个六边形单体，周围环绕着其他 6 个单体。软件从 1 个单体的东西南北 4 个方向接收信息，根据六边形单体自身及周边单体的位置反映实际的坡度，在计算机中生成实质起伏的表面（Processing 语言）。

The Software

The distribution of sensors is based on an underlying hexagonal structure. Each flower occupies a hexagonal cell, surrounded by six neighbours. As the software receives diretional (N, S, E, W) input from a cell, it is able to reconstruct a slope based on the position of that cell and its immediate neighbours, generating essentially a surface of peaks and valleys. (Processing)

图 5.1-1

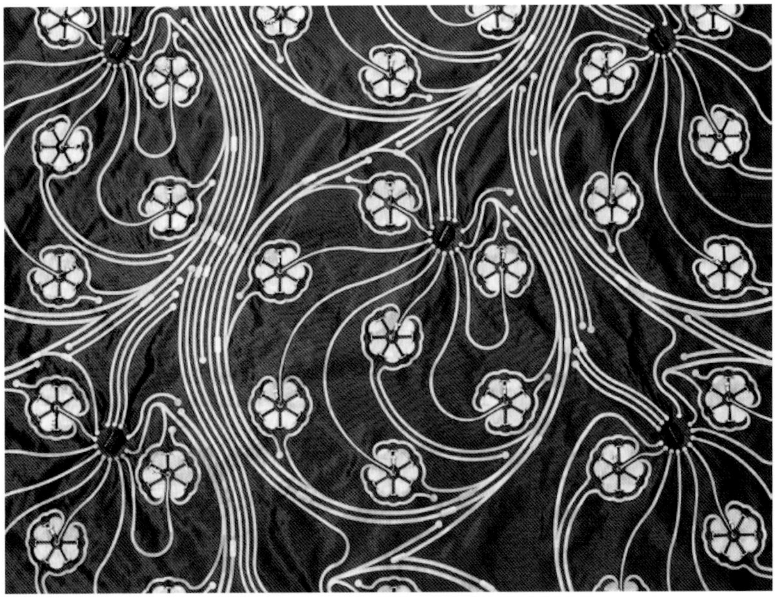

图 5.1-1 花朵组成了 14 个大型组群（或簇）和 2 个半簇。穗须作为一个定向标记，连接着记录了明确方向或毯子坡度的花瓣。材料：塔夫绸、镀镍银布、导电线、镀银珠以及由玻璃珠构成的穗须的绝缘体

The flowers are grouped together into 14 larger configurations or clusters and 2 half clusters. Working as a directional marker, the tassel's contact with a petal registers a specific orientation or tilt of the blanket. Materials: taffeta, nickel coated silver fabric, conductive thread, silver plated beads, and glass beads as tassel stem insulators

图 5.1-2

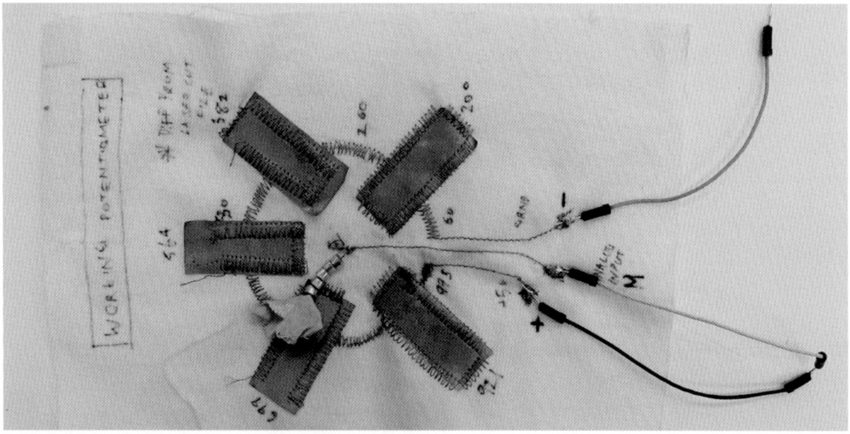

图 5.1-4

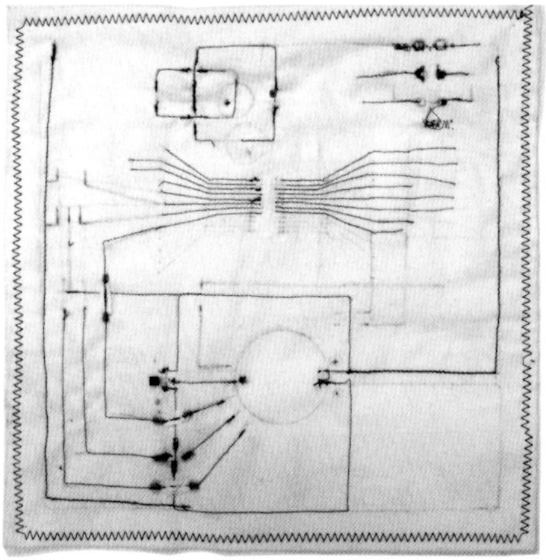

图 5.1-3

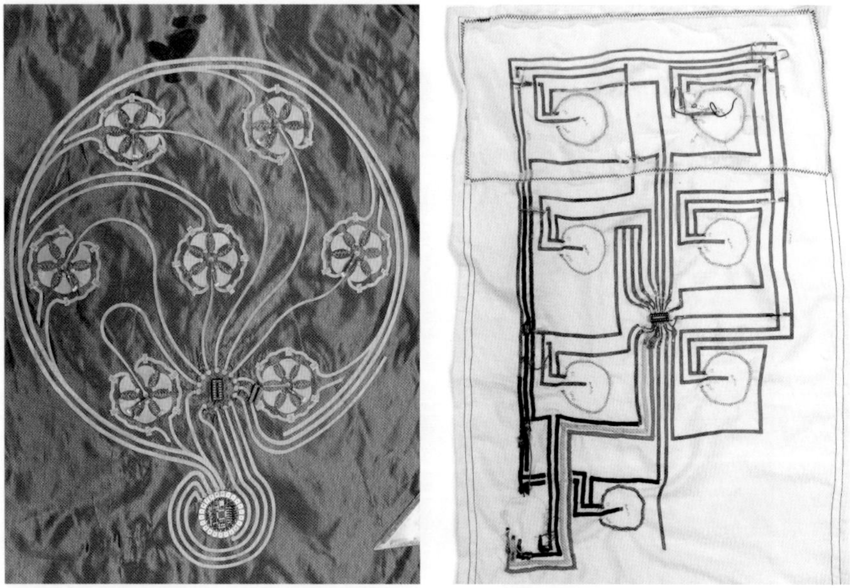

图 5.1-2 首个坡度传感器工作原型。由铜导电织物、导电线和银珠构成
The first working tilt sensor prototype. Copper fabric, conductive thread and silver beads

图 5.1-3 有机体（仿生花）与矩形构成（右图）。紧凑和均匀分布的组织比传统的笛卡尔电路更有效率
The organic (floral) vs. the orthogonal. The compact and evenly distributed organisation was ultimately more efficient than the traditional Cartesian circuit layout

图 5.1-4 测试片，由导电线连接起来的电路
Test patch, stitching circuits with conductive threads

图 5.1-5

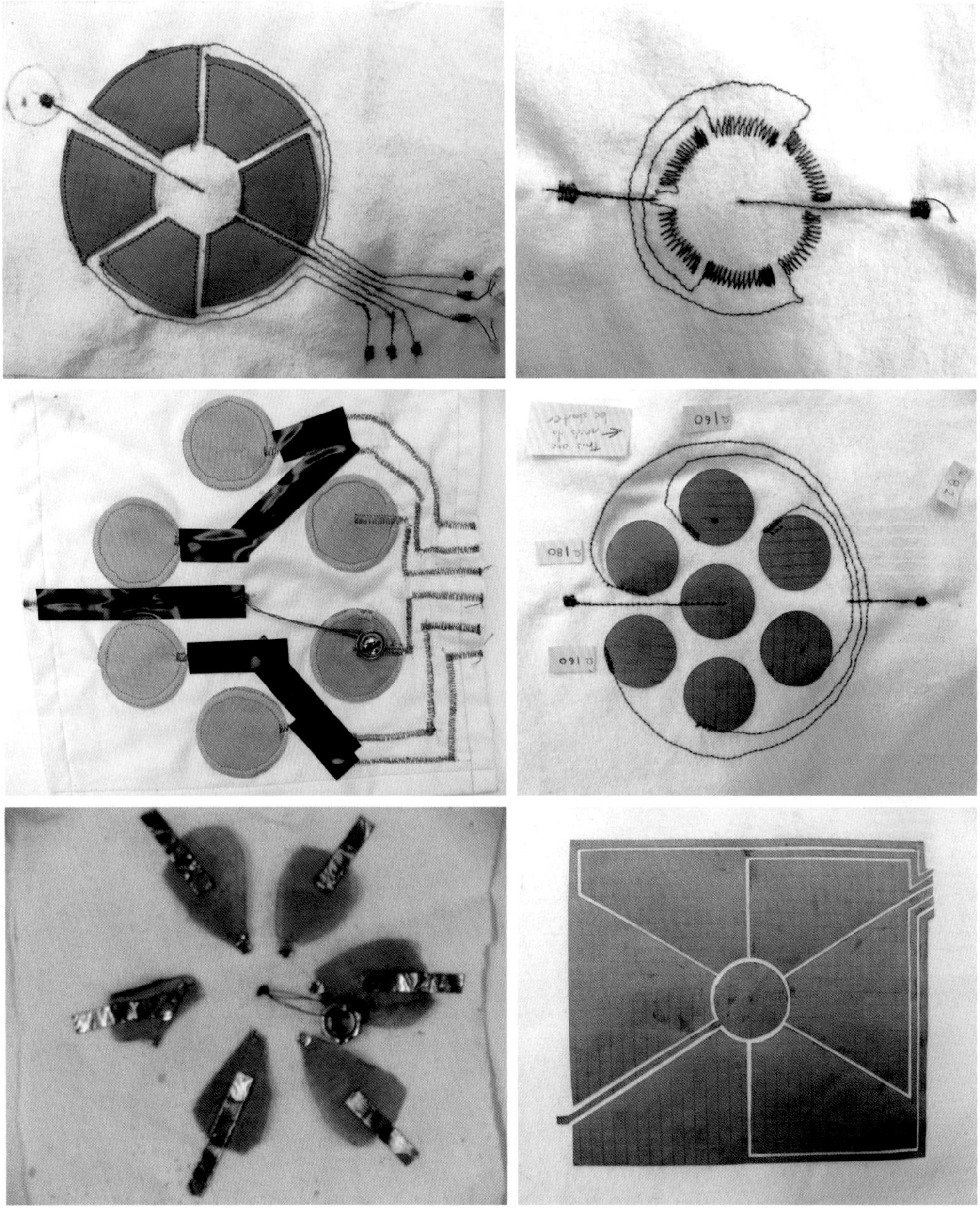

图 5.1-5 软坡度传感器的演变：探索最有效的组织结构。材料：导电织物和线、导电珠
The evolution of the soft tilt sensor: a search for the most efficient organisation. Materials: conductive fabric and thread, conductive bead

图 5.1-6

图 5.1-6 完成的感应毯与电路原理图
Completed blanket with circuit schematic

图 5.1-7

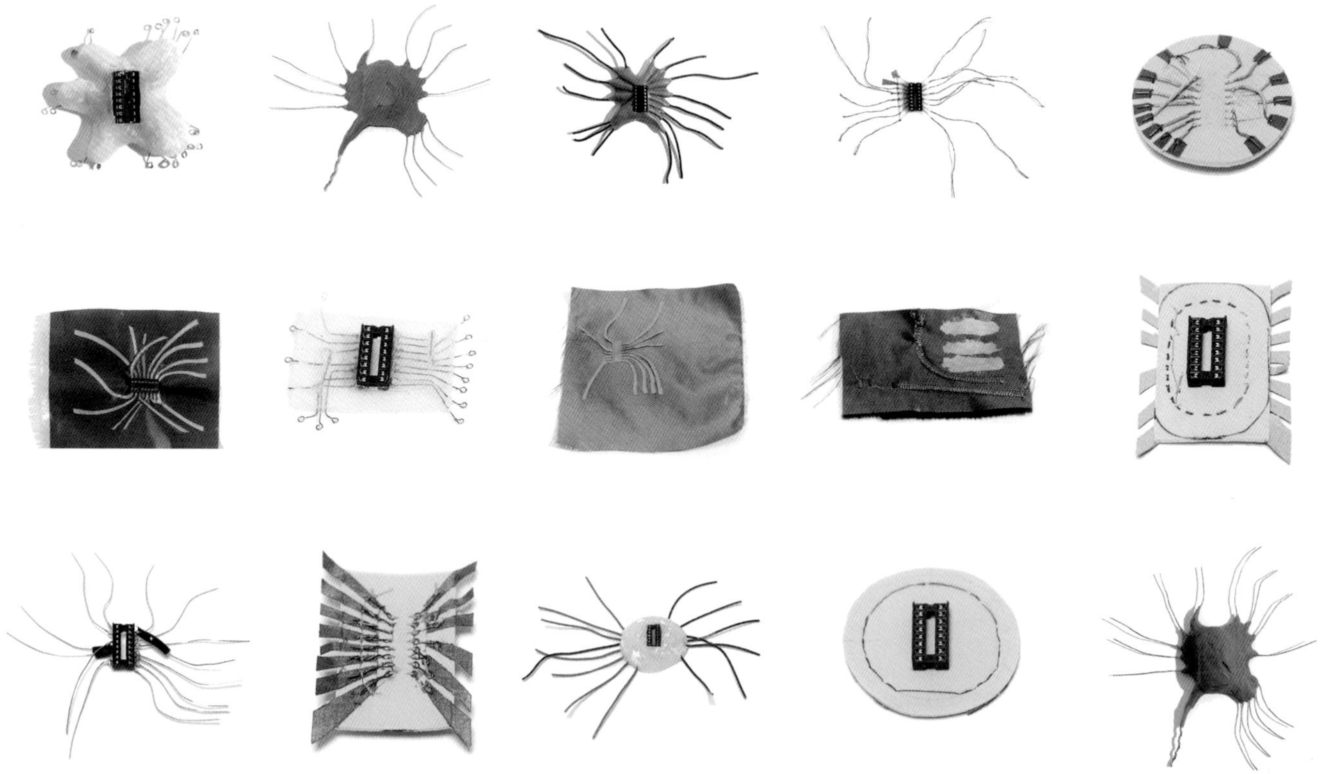

图 5.1-7 集线器 / 在形式和性能方面的多重变化
Hubs / Multiplex variations in form and performance

图 5.1-8

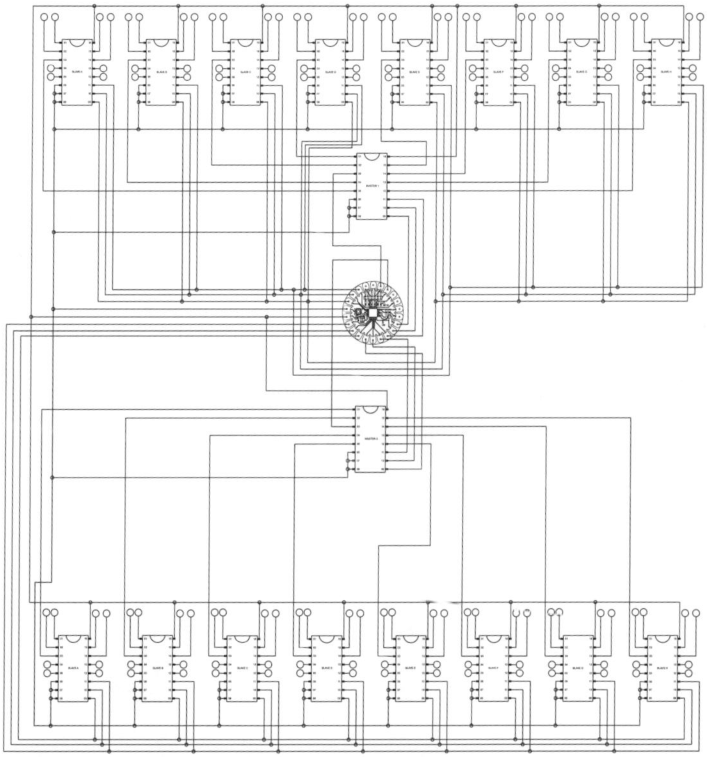

图 5.1-8 电路图
Circuit diagram

图 5.1-9 描述现实活动的实时数字复制的视频截图。感应毯使得实体和数字化可以同时存在
Stills from video depict real time digital replication of actual movement. The blanket simultaneously embodies a physical and digital presence

图 5.1-9

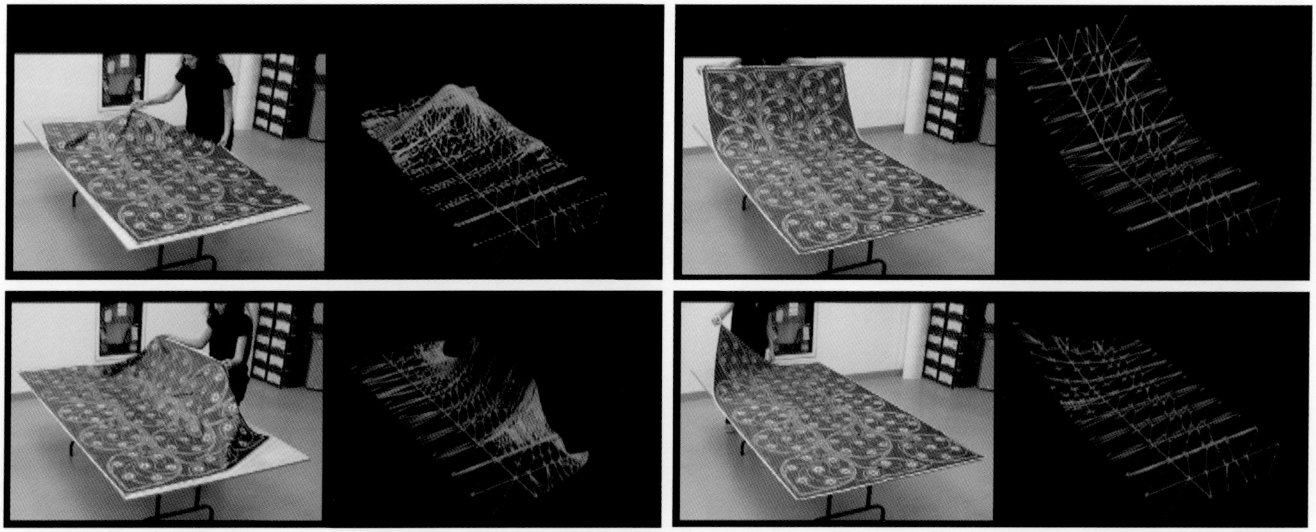

5.2 发光混凝土 Light-Emitting Concrete

利用建筑的持久性不断地发送实时信息。❶

艾达 · 尤瓦

Linking the permanence of physical built form with the temporal nature of constantly adapting information delivery.❶

Ada Juwah

发光混凝土是一种内嵌点阵 LED 的电子加强型混凝土块，可以被大量用于实体结构。在建筑的层面，发光混凝土结合用户输入识别系统，创建一种交互式低分辨率的信息门户。

“有时，我会在早餐前相信多达六个不可能的事儿。”——路易斯 · 卡罗尔，《镜子告诉爱丽丝的那些事》

发光混凝土提供了一种建成空间形式的持久性与不断适应新情况的实时信息的时间性之间的联系。通过信息的动态性与像混凝土这样的建筑材料的静态特性的结合，该构思对建立由知识的物质空间构成的信息化建筑体进行探讨，探索在

Light-Emitting Concrete is an electronically enhanced concrete block with embedded dot-matrix LEDs that can be deployed for the erection of substantial structures. In conjunction with user-input recognition mechanisms, Light-Emitting Concrete blocks create interactive, low resolution information portals at an architectural scale.

‘*Sometimes I’ve believed as many as six impossible things before breakfast.*’ — Lewis Caroll, Through the Looking Glass, and What Alice Found There

Light-Emitting Concrete provides a linkage between the permanence of the physical built form, and the temporality of real-time information that constantly adapts to new conditions. By combining the dynamic nature of information with the static characteristics of a building material such as concrete, the proposal examines the possibility of creating information architectures consisting of physical spaces

❶ 由鲁道夫 · 埃尔 - 库利与纳希德 · 纳比安指导。

❶ Guided by Rodolphe el-Khoury and Nashid Nabian.

of knowledge, exploring the potential of low-resolution display to recombine the analogue and digital aspects of architectural space that is augmented with layers of electronically mediated information.

电子介导信息增强的建筑空间如何利用低分辨显示系统，在模拟和数字方面对信息进行重组。

Light-Emitting Concrete is a building unit composed of glass fibre concrete embedded with dot-matrix LEDs. Walls built with these units provide limited levels of transparency. Meanwhile, they can seamlessly display dynamic information to engage the public. User-input recognition algorithms make this display technology interactive. Interaction is based on the concept of ‘me and my shadow’, where information is displayed and manipulated within the digital shadow of the user. Cameras capture a live image of the user, which is then displayed on the wall as the user’s shadow. This shadow is in turn controlled by the user to select and display information, becoming a window that unmasks hidden information.

发光混凝土是一种由内嵌点阵 LED 的玻璃纤维混凝土构成的建筑单元，由这种材料构成的墙体单元提供了有限的透明度。同时，它们能够通过无缝的信息显示来吸引公众的注意力，面向用户的识别算法提供了这种交互式的显示技术。这种相互作用来源于“我和我的影子”这个概念，信息的显示和操作均在用户的数码影子中完成。摄像机捕获用户的活动影像，同时把它以用户影子的形式显示在墙上。这个影子就像一个揭示隐藏信息的窗口，由与之对应的用户来选择并显示信息。

图 5.2-1

图 5.2-1 适应性强与互动信息传递
Adaptable and interactive information delivery

图 5.2-2

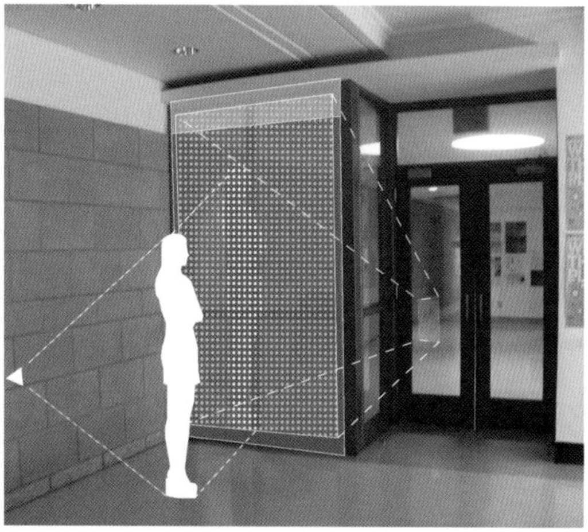

图 5.2-2 作为概念验证，利用一张打印了墙体图像的半透明薄板构成一个临时显示器，矩阵 LED 信息图案被投影到墙上形成完整的效果。通过摄像机捕获用户的运动，来解析 Processing 语言，并实时控制和显示信息图案

As a proof-of-concept, the wall was created as a temporary display, using a translucent sheet printed with a wall image. The dot-matrix LED information pattern was back-projected onto the wall to complete the effect. A camera captured the movement of the user, which the processing language interpreted to control and display the information pattern in real time

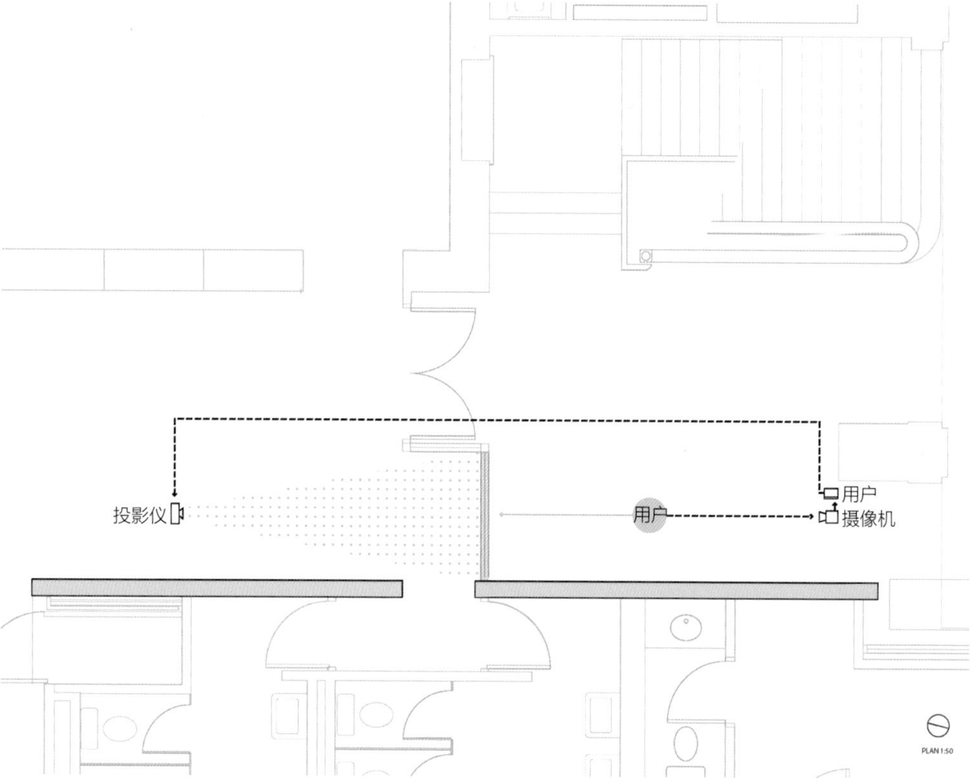

图 5.2-3

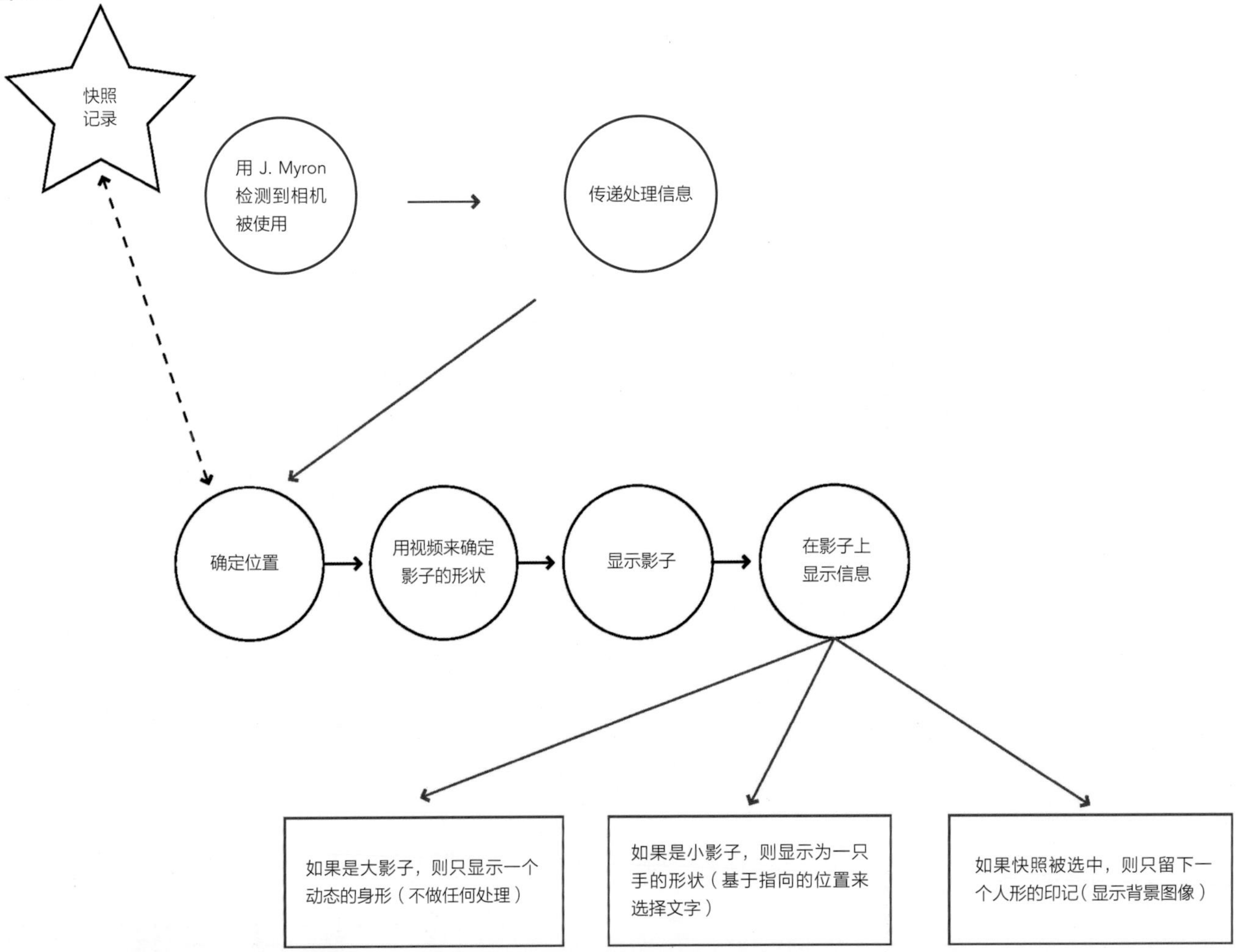

图 5.2-3 处理代码❶
Processing code❶

---

❶ 译者注：Myron 是一个跨平台、跨语言、开源、视频捕捉和计算机视觉插件，是一个核心的 C++ 对象，能够使得跨高等级语言联合编译实现封装。对 Java 和 Processing 的封装叫做 J. Myron。

图 5.2-4

图 5.2-4 用户可以通过将自己的影子对准滚动文字来决定要什么信息显示在他们的影子上面
Users could select which information to display within their shadow, by directing their shadow to touch words scrolling in a ticker to the side

5.3 环境壁纸 Ambient Wallpaper

关注房间的噪声情况，促使用户降低噪声音量。❶

阿扎德 · 穆罕默德，塔兰涅 · 米什卡尼

Bringing attention to the noise level of the room and encouraging users to decrease their sound level.❶

Azadeh Mohammadi and Taraneh Meshkani

Ambient Wallpaper is an interactive ornament. The ‘transactive intelligence’ that enhances the performance of this ornamental element allows it to respond to human activity and influence human behaviour in return. By changing visual characteristics, it provides visual hints as to the proper code of conduct within an architectural space.

Ambient Wallpaper examines the potential of technology to develop dynamic spaces which have the ability to interact with users and communicate implicit messages to them. Ambient Wallpaper is digitally enhanced with ‘transactive intelligence’ to respond to changing conditions by modifying its visual characteristics, including its colour and pattern.

For prototypical implementation, an interactive applet running in the processing environment on a computer is projected onto the architectural surfaces. The colour of the projected wallpaper changes with the sounds of the environment. With this effect, the wallpaper brings attention to the noise level of a room and encourages users to

环境壁纸是一种交互式的装饰。“智能交互”提高了这个装饰元素对于人类活动的反应性能，同时反过来也影响了人的行为。通过可变视觉化特征，在建筑空间内提供正确行为准则的视觉提示。

环境壁纸检验了发展动态空间与用户进行互动的能力，并且探讨了与用户进行隐含信息沟通的可能性。环境壁纸以数字增强提升“智能交互”，通过改变颜色和图案等视觉特性来应对不断变化的环境。

对于原型的实现，在计算机处理环境中运行交互式程序 APP，然后投射到建筑表面。被投射的壁纸颜色根据环境的声音而改变。通过这种效果，环境壁纸使得用户注意到房间的噪声等级，

❶ Guided by Rodolphe el-Khoury and Nashid Nabian.

❶ 由鲁道夫 · 埃尔 - 库利与纳希德 · 纳比安指导。

从而鼓励用户降低对话的音量。除了使用声音传感器来检测噪声级别外，这种壁纸也通过安装在投影墙上的多个距离传感器来响应用户在空间中的移动。Arduino 板收集来自距离和声音传感器的信息，并传送到处理程序中。所选波斯图案的几何性质允许使用计算算法对其视觉效果进行参数化修改，这些算法是由检测到的环境变化而触发的。

decrease the volume of their conversations. Aside from detecting noise levels with the use of sound sensors, the wallpaper also responds to the movement of users in the space by deploying multiple distance sensors installed on the projected wall. An Arduino board collects the information from distance and sound sensors and communicates it to the processing environment. Patterns of Persian tiles are used as a model for the design of the wallpaper scheme. The geometrical nature of the chosen Persian pattern allows for the parametric modification of its visual effects by using computational algorithms triggered by detected changes within the context.

图 5.3-1

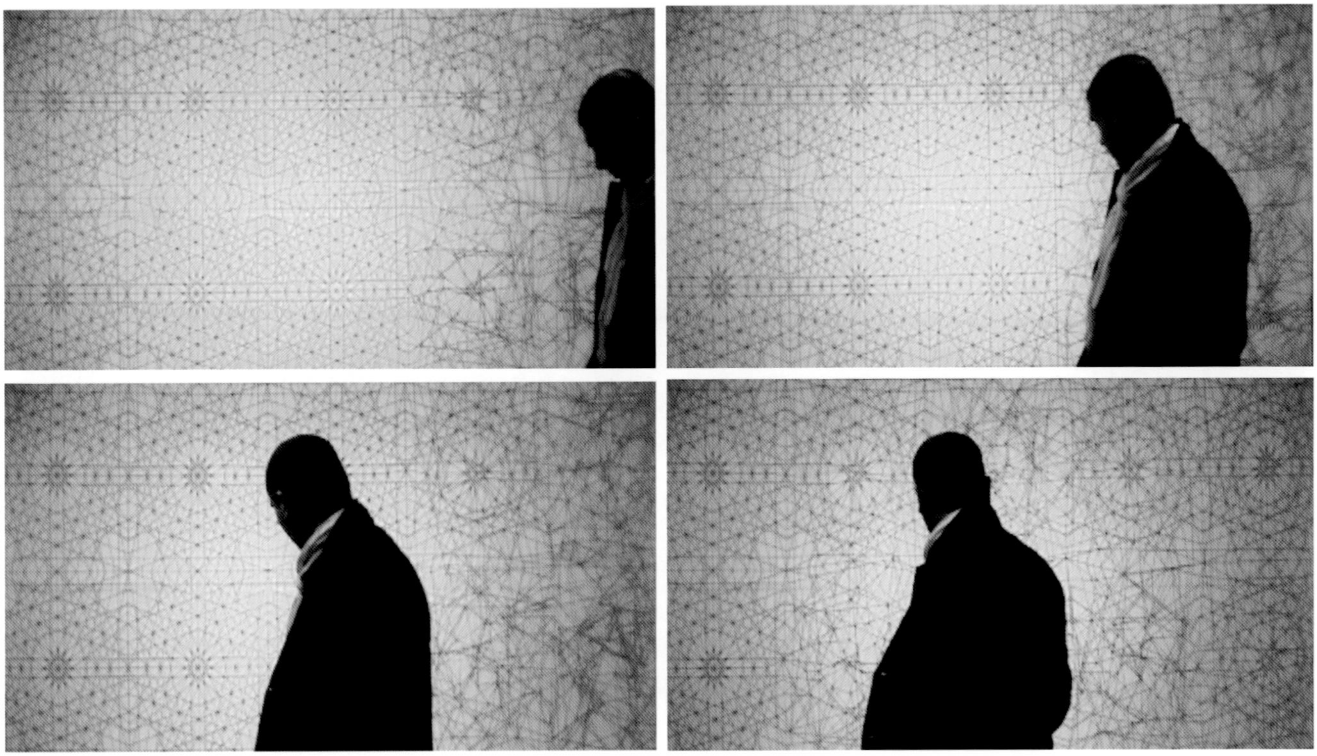

图 5.3-1 壁纸设计图案

壁纸回应了使用者在空间中的移动。我们选择波斯瓷砖的设计图案作为设计模版。这些瓷砖的图案由来自伊斯兰艺术的复杂几何图形构成，然而，它们是由圆、方形、三角形等简单的几何形状组成的。

Wallpaper design pattern

The wallpaper responses to the movement of users in the space. We chose design pattern of Persian tiles as a model of our design. The pattern of these tiles consist of complex geometries which come from the Islamic art. However, it is made from simple geometrical shapes such as circles, squares, triangles and etc.

图 5.3-2

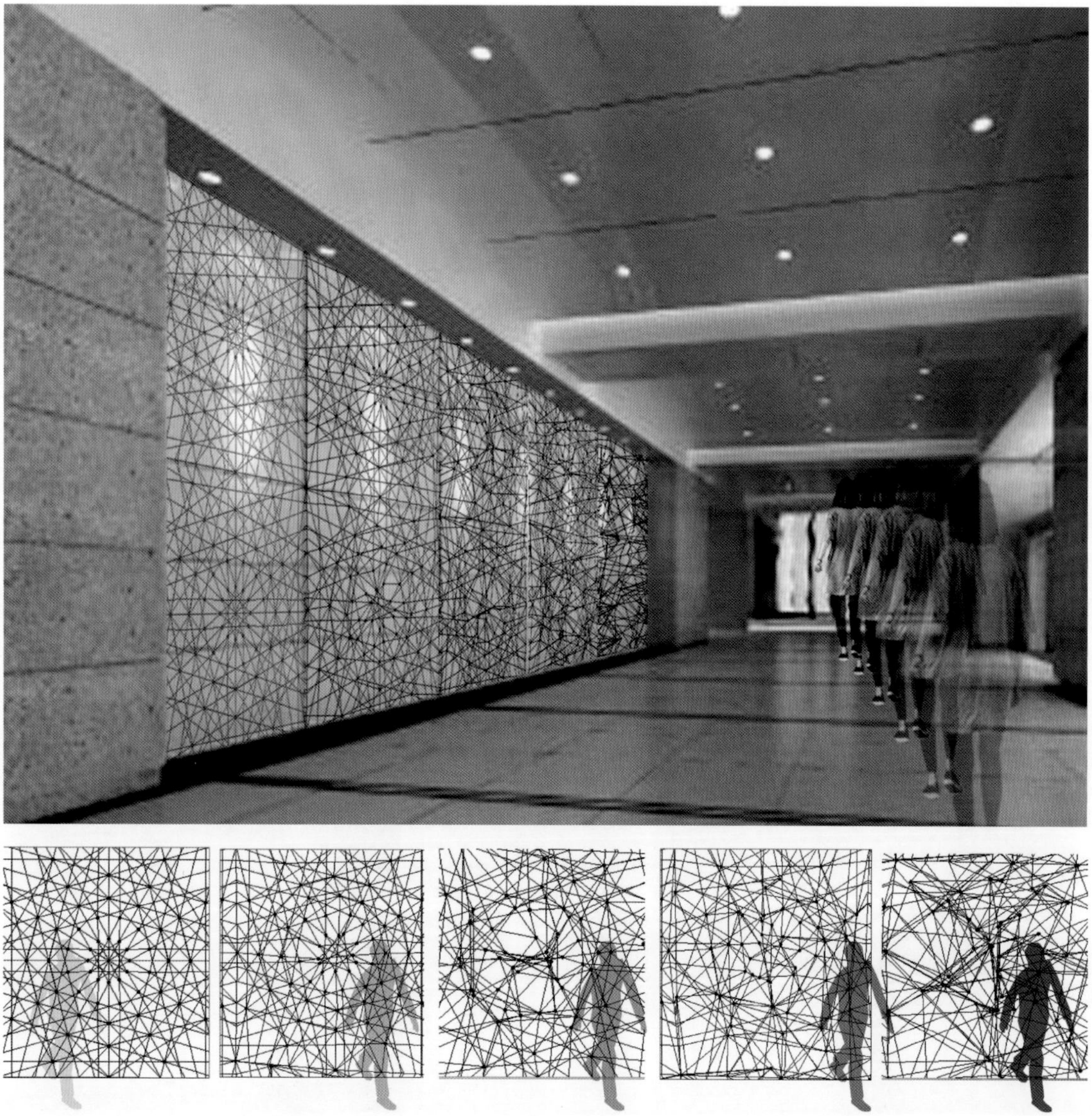

图 5.3-2 演示和示意图展示了墙纸图案对于存在和运动的响应。由伊斯兰装饰激发的灵感被分解成一种有组织的、不断变化的不规则图案
Demonstration and diagrams showing wallpapers' pattern responding to presence and movement. The Islamic ornament inspired pattern unravels into an organized chaos of increasing irregularity

图 5.3-3

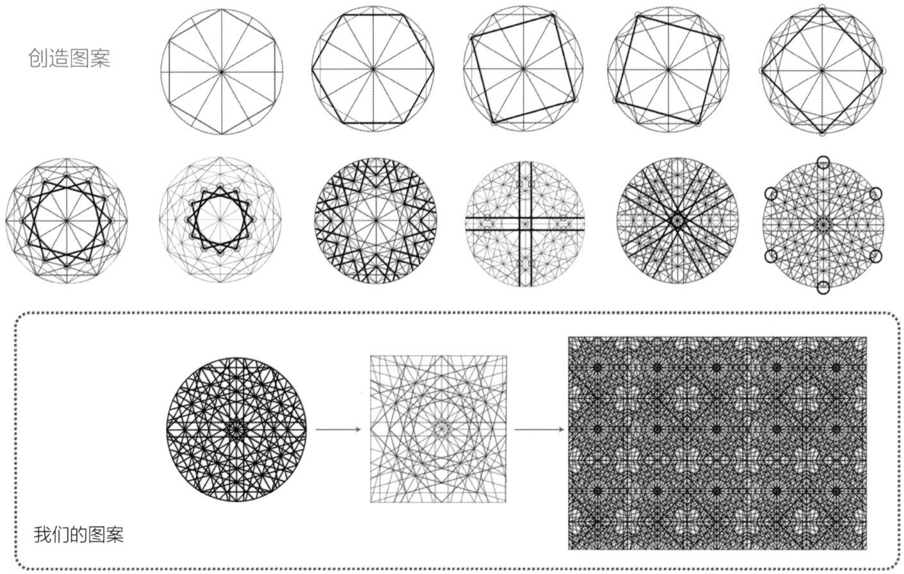

图 5.3-3 植根于伊斯兰传统的形态几何学图谱
Genealogy of pattern geometry rooted in Islamic traditions

图 5.3-4

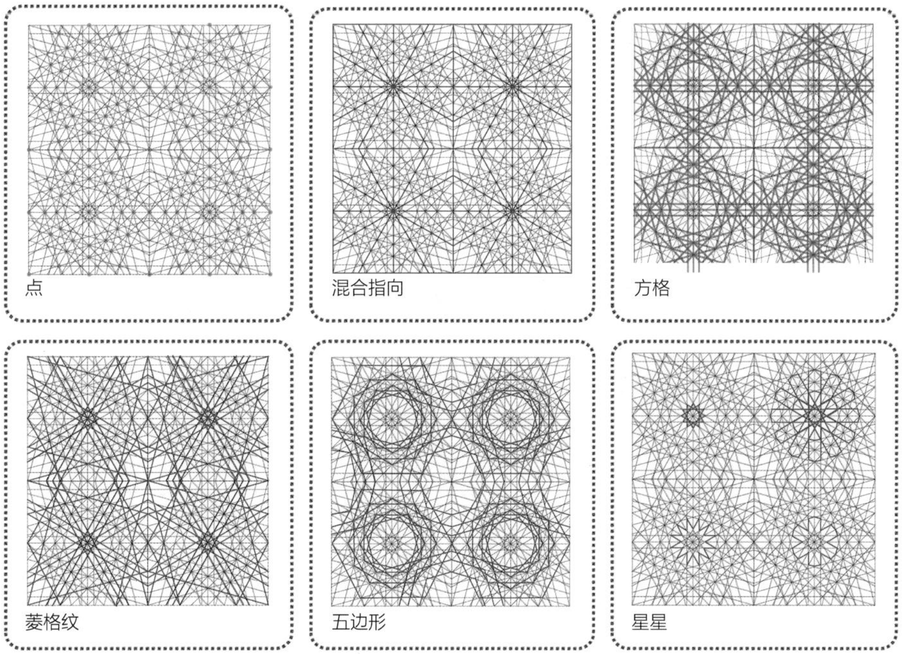

图 5.3-4 在响应模式下被突出显示的控制点
Highlighted control points within the responsive pattern

图 5.3-5

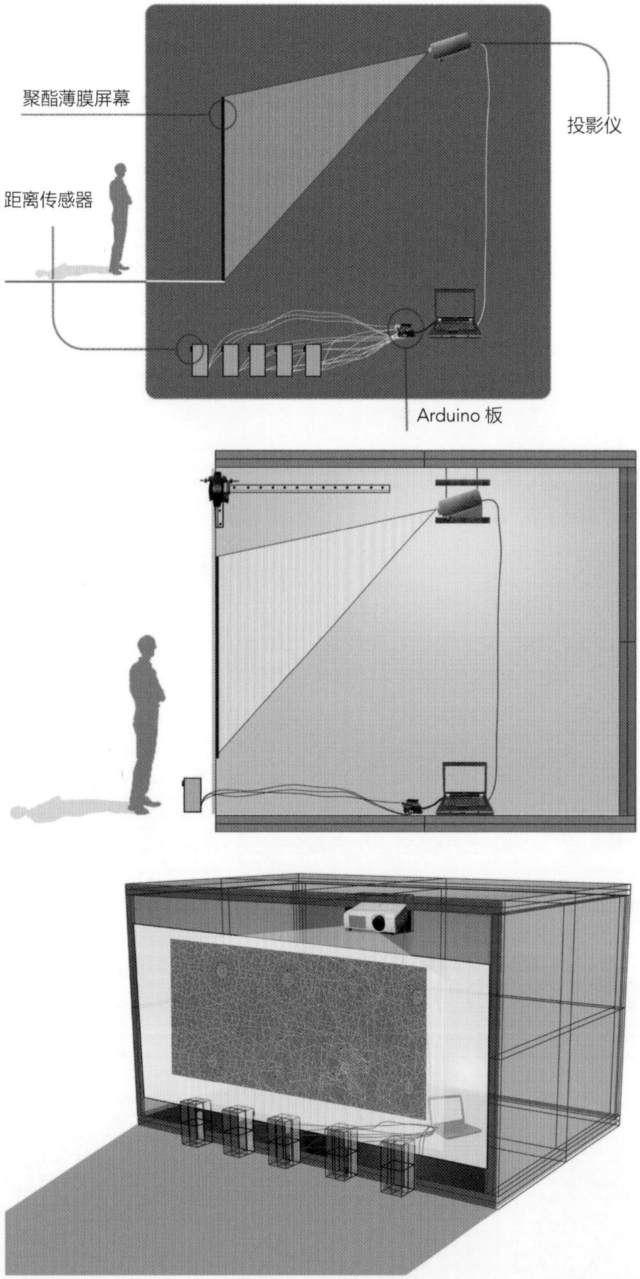

图 5.3-6

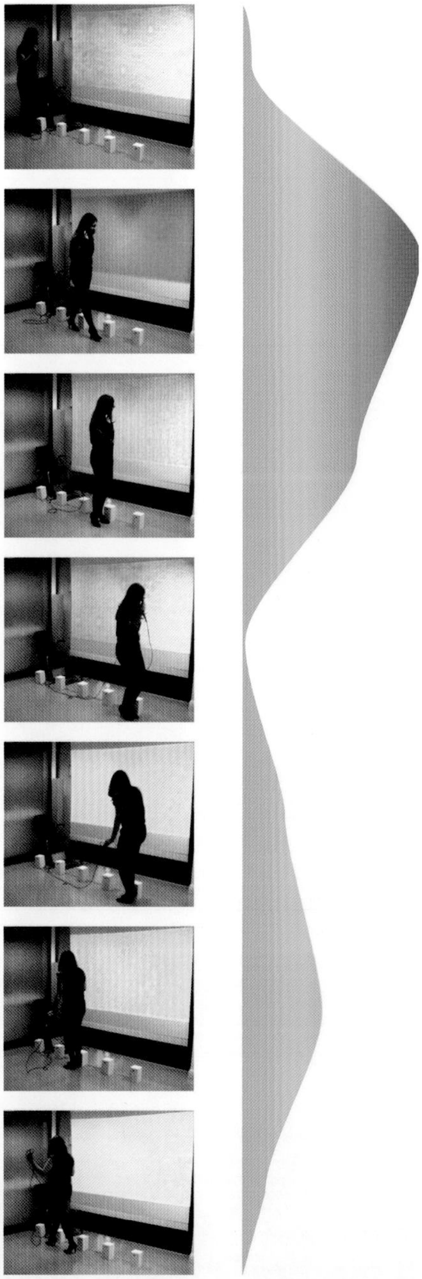

图 5.3-5 微控制传感器和执行器的安装与构造
Diagrams showing installation and configuration of micro-controlled sensors and actuators

图 5.3-6 壁纸颜色
壁纸的颜色随环境声音而变化。原始颜色是蓝色，当环境噪声增加，颜色会变得更暖直到红色。带有这种效果的壁纸引起了人们对房间噪声等级的注意，并鼓励使用者降低谈话的音量。颜色的变化向我们每个人展示了建筑如何在不同地方作出反应的能力。
Wallpaper color
The color of the wallpaper changes with the sound of the environment. The original color is blue, as the noise of the environment increases, the color changes to warmer color to the extend of red. With this effect the wallpaper brings attention to the noise level of the room and encourages users to decreases the sound level of their conversations. The transformation in the color shows the ability of Architecture to each us how to react in different places.

图 5.3-7 波斯瓷砖的历史（釉面砖）
History of Persian tile (glazed brick)

图 5.3-7

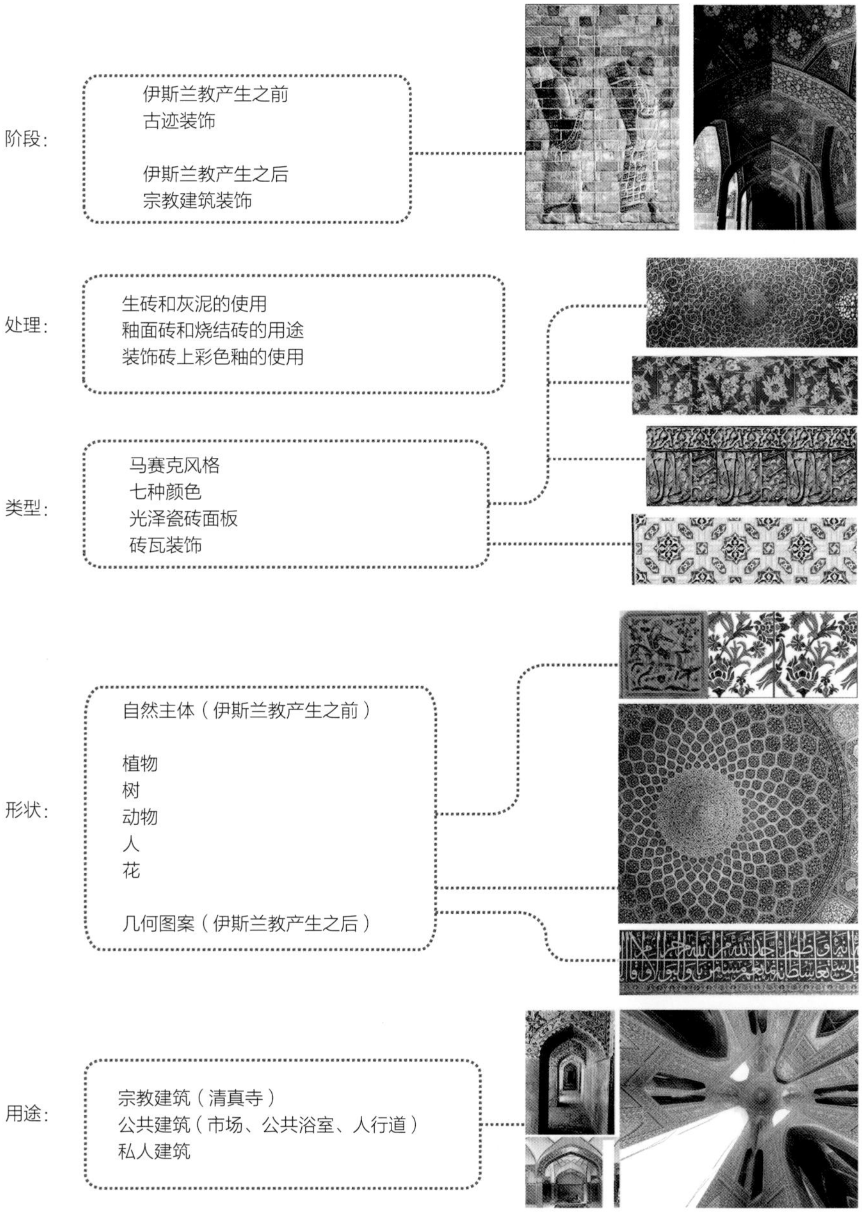

5.4 数字化窗户 Digital Window

数字化显示窗户的“另一面”。

NMinus One 工作室

A window which digitally displays the ‘other side’.

Studio NMinusOne

The ‘Digital Window’ produces a digitally mediated transparency. The Digital Window appropriates a common hardware application found in popular image capturing devices.

The effect accidentally produced by these devices is a digitally mediated transparency, or invisibility. It is this ‘side effect’ that the digital window opportunistically assimilates. The digital window is a window which digitally displays the ‘other side’. Besides relaying the ‘other side’, the window can display a variety of other things: TV, the Internet, or the ultimate reality TV. Similar to the one-way mirror, one can only look out from the window, but not in.

“数字化窗户”可以利用普通图像捕捉设备中常见的硬件，产生一种数字化介导的透明度。

由这些设备产生的数字化介导透明度或隐性化对周边环境无意中产生了影响，这是数字化窗户有时会呈现出来的“负面影响”。数字化窗户能够数字化地显示另一边的影像，除此之外，还能显示各种东西，如电视节目、互联网，或者是终极的实境电视。类似于只能从窗户向外看而不能往里看的单向镜子。

图 5.4-1

图 5.4-2

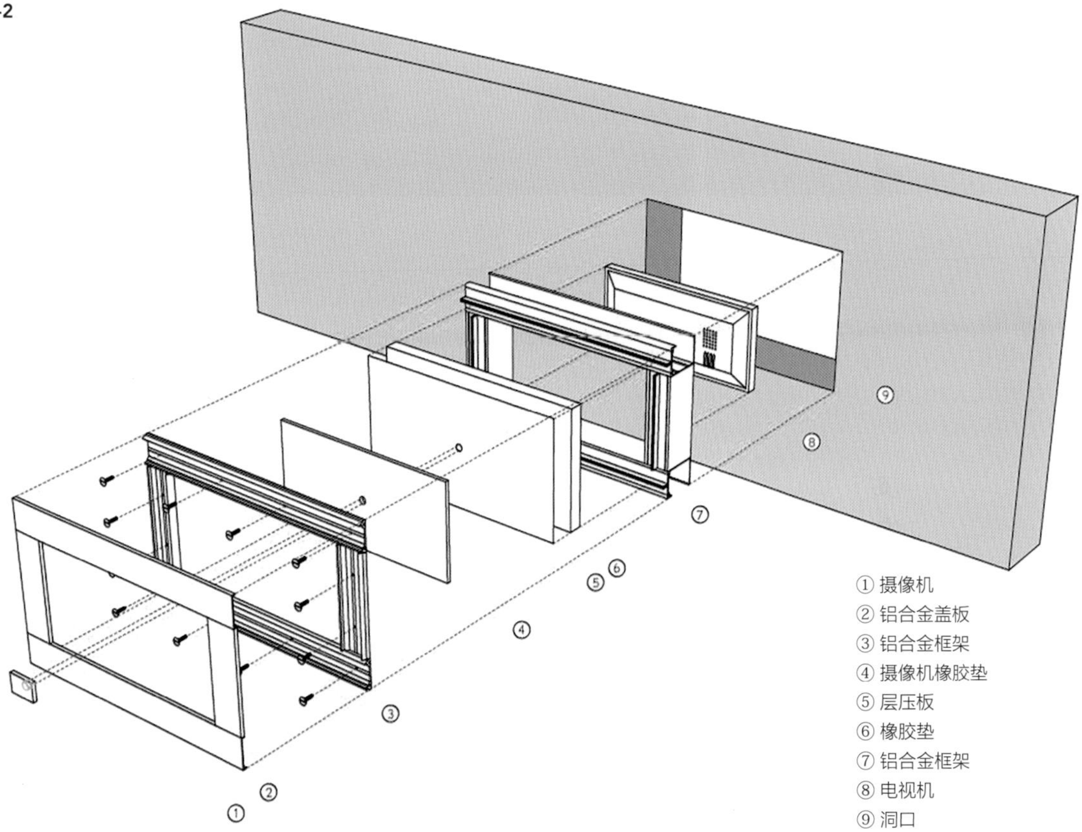

图 5.4-1 作为摄像机或屏幕的窗户
Window as camera and screen

图 5.4-2 数字化窗户装配图
Digital window assembly

图 5.4-3

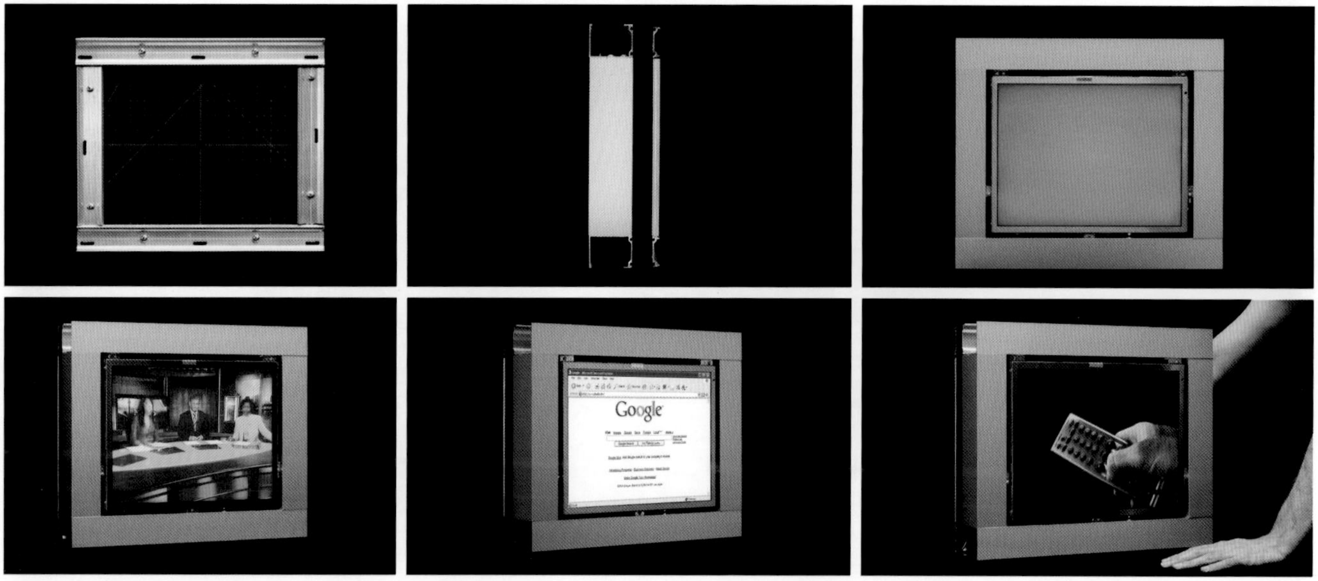

图 5.4-4

图 5.4-3 外部、侧面和内表面
Exterior, side and interior surface

图 5.4-4 不同的显示场景
Different display scenarios

5.5 窗户 2.0 Window 2.0

根据观看者的位置来动态调整虚拟空间视角的响应系统。

KLF 事务所

A responsive system dynamically adjusts the perspective of the virtual space according to the position of the viewer.

Khoury Levit Fong

“窗户 2.0”追踪观察者的位置并据此调整虚拟空间，利用反应灵敏的视频影像处理系统来对现实空间进行数字化增强。

“窗户 2.0”是为了虚拟、增强或者构建全新的物理窗口的光学和视觉现象而设计的，应用范围从电话会议、思科网真[1]到虚拟空间及透明度的拓展。

“窗户 2.0”通过两个途径来生成透视效果：

“窗户 2.01”利用网络摄像头的视频轨迹来对观察者的位置进行准确定位，并可动态控制安

[1] 译者注：telepresence，思科网真是一种全新的系统技术，它结合思科在智能化 IP 网络、统一通信、超高清 IP 视频、空间 IP 音频、交互式协同组件、数字电影、灯光环境、人体工程等领域一系列的技术突破及集成技术，成功进行了系统性创新，实现跨越空间和技术障碍的真实体验。

‘Window 2.0’ uses responsive video imaging to digitally augment physical space by tracking the position of the viewer and adjusting the virtual space accordingly.

‘Window 2.0’ is designed to simulate, enhance or entirely reconfigure the optics and the visual phenomenon of a physical window. Applications range from teleconferencing and telepresence to virtual spatial extension and transparency.

‘Window 2.0’ produces responsive perspective effects by means of two methods:

‘Window 2.01’. The position of the viewer is pinpointed by means of video tracking with a webcam. It dynamically controls the orientation of the viewing camera mounted on a servo motor. The responsiveness of the camera can be finely calibrated to react very sensitively to minute signals in body language, shifting the view in response to a slight shoulder rotation or to more closely imitate the interaction of viewers with physical windows in real space.

'Window 2.02'. The position of the viewer, tracked with a webcam, controls the virtual perspectival orientation of the live video feed by means of digital processing.

装在伺服电机上的观察摄像机的方向。摄像机的敏感度可以精细校准由身体语言的瞬间动作所引发的反应，可以因一个轻微的肩膀转动而移动视图，或者通过一个现实空间中真实的窗口来更准确地模拟与观察者的互动。

“窗户 2.02”：由网络摄像头来追踪观察者的位置，通过数字处理装置来控制实时视频的虚拟透视方向。

图 5.5-1

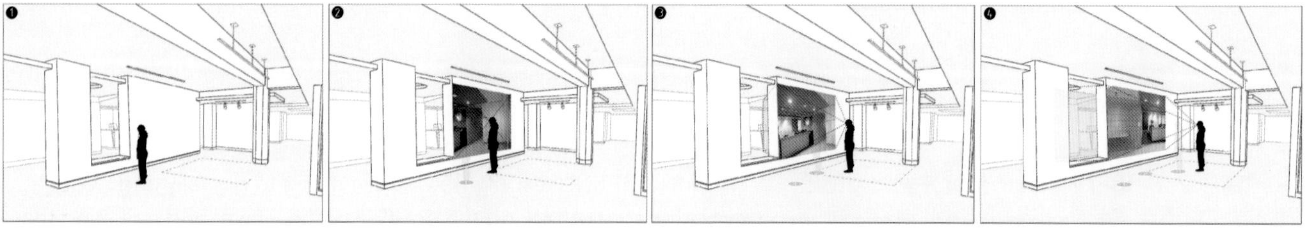

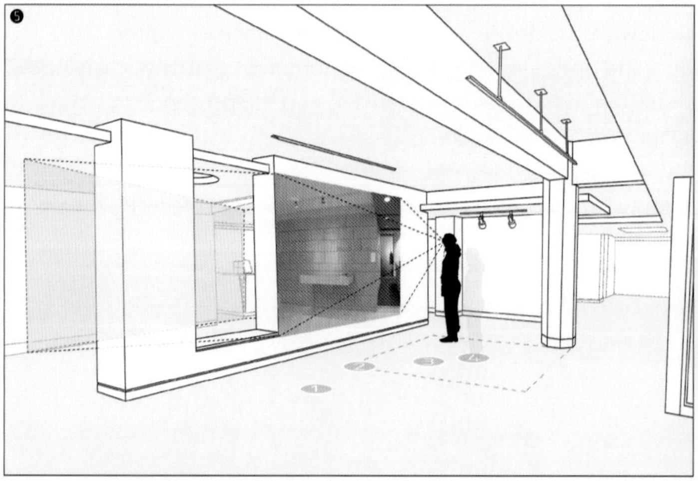

图 5.5-2

图 5.5-1 行为响应图
Diagram of responsive behaviour

图 5.5-2 视频演示截图
Still from video demonstration

图 5.5-3

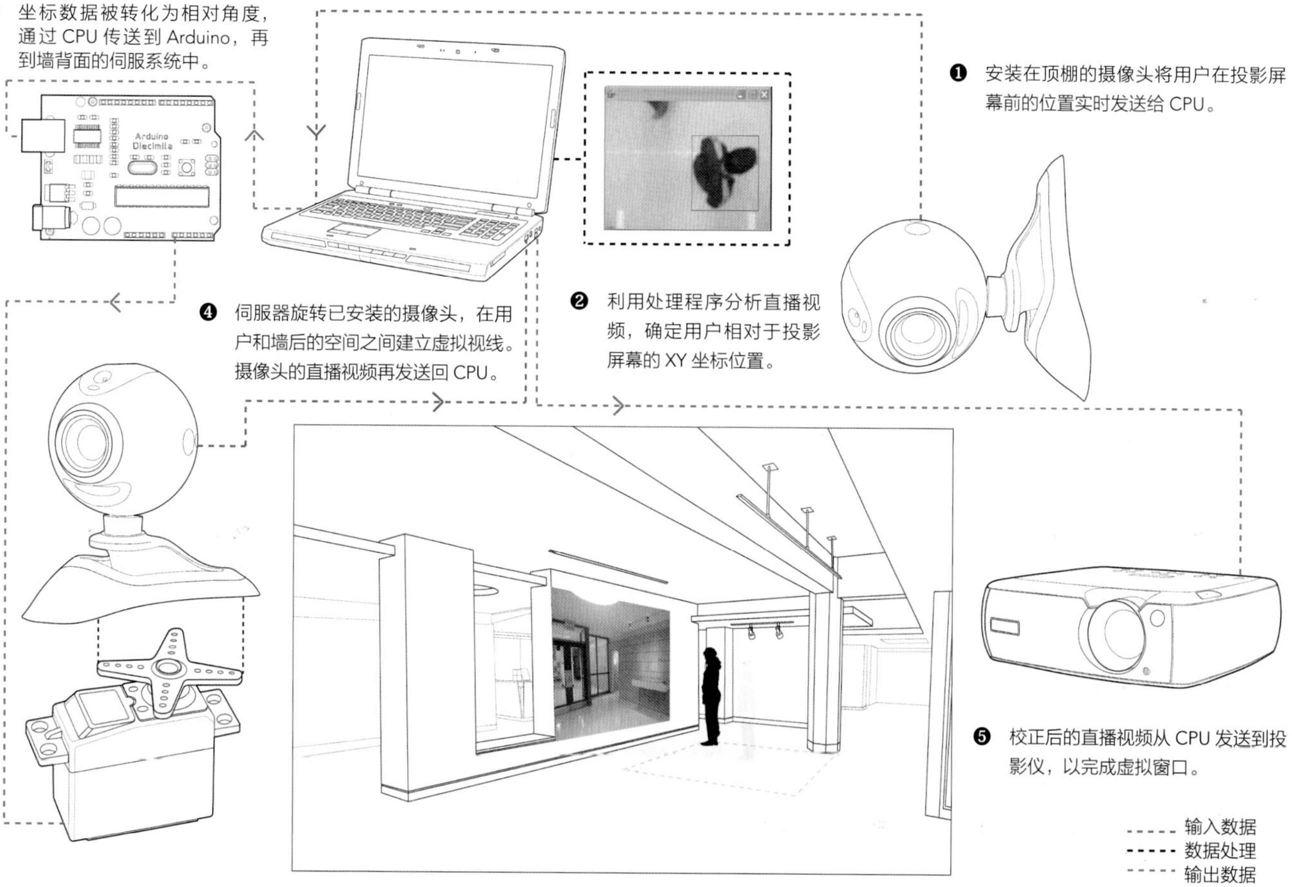
❸ 坐标数据被转化为相对角度，通过 CPU 传送到 Arduino，再到墙背面的伺服系统中。
Arduino Diecimila
❶ 安装在顶棚的摄像头将用户在投影屏幕前的位置实时发送给 CPU。
❷ 利用处理程序分析直播视频，确定用户相对于投影屏幕的 XY 坐标位置。
❹ 伺服器旋转已安装的摄像头，在用户和墙后的空间之间建立虚拟视线。摄像头的直播视频再发送回 CPU。
❺ 校正后的直播视频从 CPU 发送到投影仪，以完成虚拟窗口。
输入数据
数据处理
输出数据

图 5.5-3 视频跟踪、摄像头 / 伺服系统控制图
Diagram of video tracking, and camera / servo control

图 5.5-4

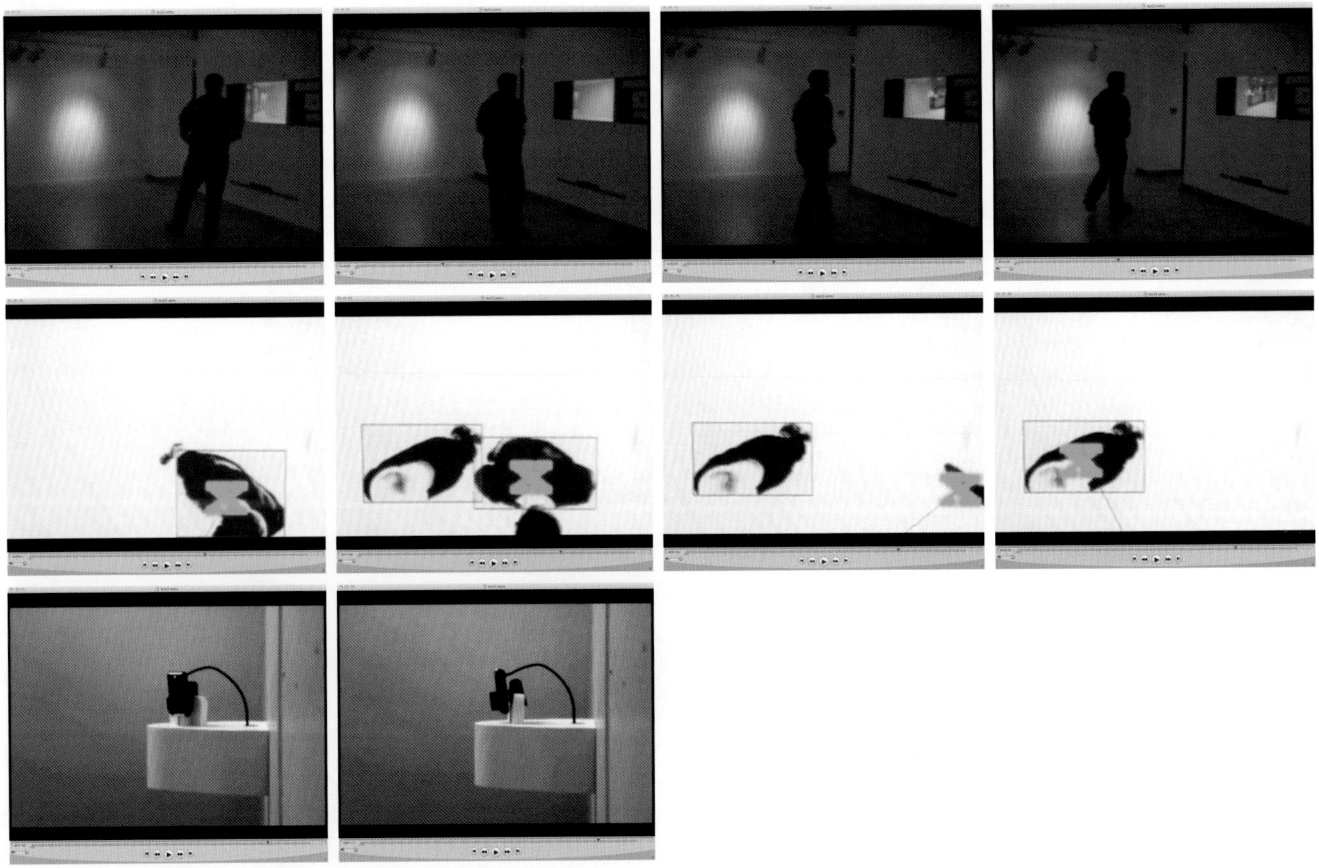

图 5.5-4 捕捉装置运行的三要素的视频截图：用户与窗口（屏幕）的互动、视频跟踪和锁定过程，以及安装伺服系统摄像头
Video stills capturing the three main elements of the device in operation: the interaction of the user with the window / screen, the video tracking and locking process and the servo-mounted webcam

5.6 我的城市 MyCity

“我的城市”通过建筑与信息技术的结合，赋予建筑可感知的体块与时空表皮，使其成为与社会连接的紧密通道。

KLF 事务所，马克 · 琼斯（新兴软件）

‘MyCity’ aligns architecture with information technology to empower the sensate body and its spatio-temporal envelope as the intimate locus of social connection.

Khoury Levit Fong with
Mark Jones (Emergent Software)

“我的城市”探索环境对于主体行为和特性的针对性回应能力。

个性化连接意味着扩大对象的影响范围。在这种情况下，任何有意愿的参与者个人的网络相簿画廊将会展示到公共空间中。通过手机就能方便实现，可以允许个人识别他或她的网络相簿账户，随后选择公开展示那些被选出来的个人照片。

“我的城市”鼓励个人将他们的网络照片转化到现实空间中，让公共空间去契合他们的个人使用和情感表达。

‘MyCity’ explores the capacity for environments to respond pointedly to a subject’s actions and characteristics.

The individuated connectedness is meant to expand the subject’s sphere of influence, projecting in this case, the personal Flickr gallery of any willing participant onto public space. This is achieved simply with the help of a cell phone that allows an individual to identify his or her Flickr account and trigger the public display of selected personal images.

‘MyCity’ encourages individuals to translate their web presence into a physical reality allowing them to appropriate public space for their own use and expression.

图 5.6-1

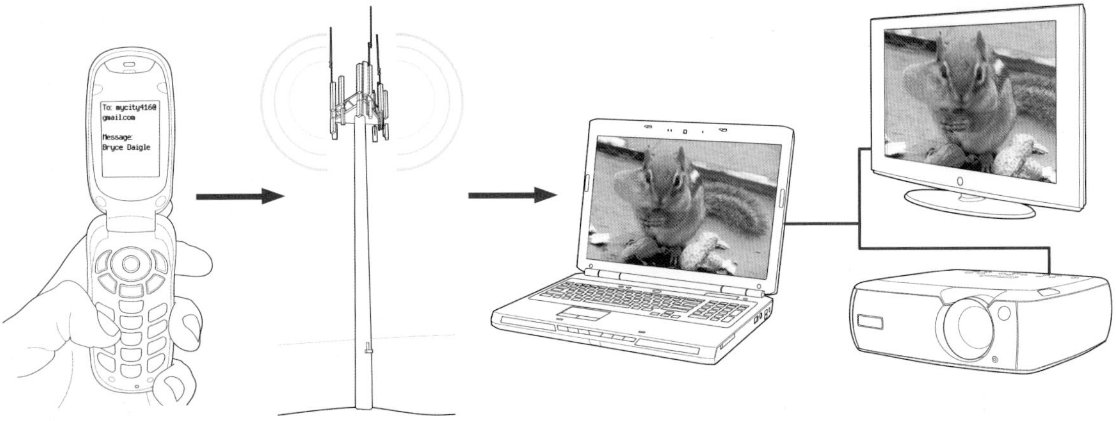

发送您的网络相簿用户名到 mycity416@gmail.com

消息通过无线网络传递

程序加载用户的网络相簿照片流

图片被显示在电脑屏幕和投影屏幕上

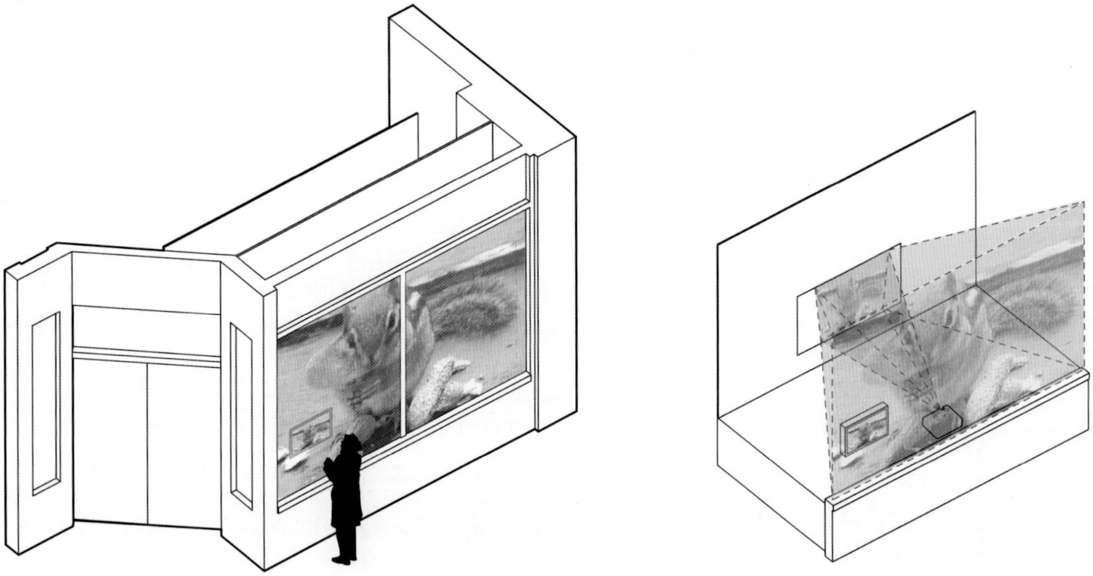

网络相簿的照片流通过带有内置 LCD 监视器的大窗口投影仪展示出来，以获得白天和夜晚的最佳观看效果

图 5.6-1 用户参与模式图解
Diagram explaining the mode of user participation

图 5.6-2

图 5.6-2 装有背投影的工作视图
View of installation with rear projection in operation

5.7 无穷集合 Aleph TO

街头实况的镜像投影。

KLF 事务所，纳希德·纳比安

A projected image mirrors the scene on the street.

Khoury Levit Fong with Nashid Nabian

'Aleph TO' uses digital projection to create a responsive image overlooking the street. Triggered by the proximity of the viewer, it is constantly changing to reflect the scene on the street.

Aleph TO's resolution varies with pixels diminishing or augmenting in size depending on the proximity of the viewer. Each pixel is a photograph that is grabbed in real time from Flickr galleries on the web. The continuously updated photographs, identified and selected by keywords in their tags, capture concurrent scenes throughout the city.

'When I opened my eyes, I saw the Aleph.' 'The Aleph?' I repeated. 'Yes, the only place on earth where all places are seen from every angle, each standing clear, without any confusion or blending.' — Jorge Luis Borges, The Aleph.

"无穷集合"利用数码投影创建一个俯瞰街道的响应图像。它由观看者的接近而触发，并且不断地变化，以反映街上的场景。

"无穷集合"根据接近的观察者来增减像素分辨率。每个像素都是一张从网络相簿图库实时抓取的照片。不断更新的图片是根据标签的关键词从整个城市的实时场景中确认和捕捉的。

"当我睁开眼睛，我看见了阿莱夫。""阿莱夫？"我重复到。"是的，一个能从各种角度观察的地方，每一个停留都很明确，没有任何困惑或混乱"——豪尔赫·路易斯·博尔赫斯，《阿莱夫》

图 5.7-1

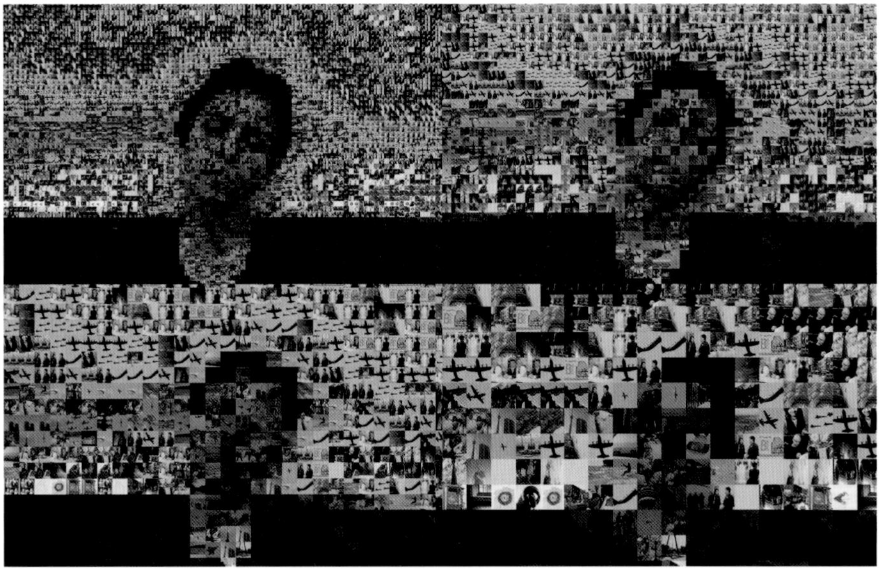

图 5.7-1 运动和接近度传感器感知观察者相对于镜子 / 屏幕的距离，并成比例地改变图像的分辨率。
The resolution of the image changes in proportion to the distance of the viewer from the mirror / screen measured by motion and proximity sensors

图 5.7-2

图 5.7-2 在多伦多的“2009 白色夜晚”活动中运行的装置
View of the installation in operation during Toronto's ‘2009 Nuit Blanche’ event

5.8 可连接的遮阳 Linking Shade

整个城市由串起来的陶瓷珠提供遮阳和动态照明。

KLF 事务所，KR 建筑事务所

Strung ceramic beads offer shading and dynamic lighting citywide.

Khoury Levit Fong with
Kuth Ranieri Architects

Football-shaped tiles, designed to be strung like beads on string are suspended over public space in a variety of structural configurations and shading scenarios in Phoenix, Arizona.

The strung ball-tiles can hang by cables from concrete columns along sidewalks as well as in park settings. They can also shade public alleys or small urban squares by attaching directly to the sides of buildings. The perforated ball-tiles come in a variety of designs and colours to achieve a range of compositional and environmental ambitions. Variations include embedded photovoltaic cells, motion sensors, and LEDs allowing tiles to function as autonomous and responsive light sources. At night, the tiles glow in patterns of colour controlled interactively by sensors responsive to movement and any number of programmable inventions.

The materiality of the ceramic ball — which includes recycled ingredients — and their simple moulded production coincides with the economy and simplicity of vernacular traditions of moulded roof tiles. It is able to draw upon experiences of common familiarity while doing so in a manner that refashions a familiar material according to up-to-date and emerging technologies — technologies which are themselves reshaping our experience as an urban public.

在美国亚利桑那州的凤凰城，设计成串珠状的橄榄球形瓷砖，以各种各样的结构形式悬挂在公共空间中提供遮阳。

这些球状瓦片可以用悬索悬挂在人行道和公园里的混凝土柱子上。它们还可以直接附着在建筑物的侧面，为公共巷道或城市小广场遮阳。穿孔陶瓷球具有多样的设计和颜色，以实现一系列的创作和环境设计目标。这些变化包括嵌入式光伏电池、运动传感器和发光二极管，允许陶瓷球能够自发光或作为感应光源。晚上，陶瓷球可发出彩光，这种光是由响应运动的传感器和任意数量的可编程发明交互控制的。

陶瓷球的材料（其中包括回收材料）及其简单的铸模生产方式与当地传统的模压屋顶瓦的经济简便相吻合。它能够利用熟悉的经验，同时根据最新和新兴技术重新设计熟悉的材料。这些新兴技术本身正在重塑我们作为城市公众的体验。

图 5.8-1

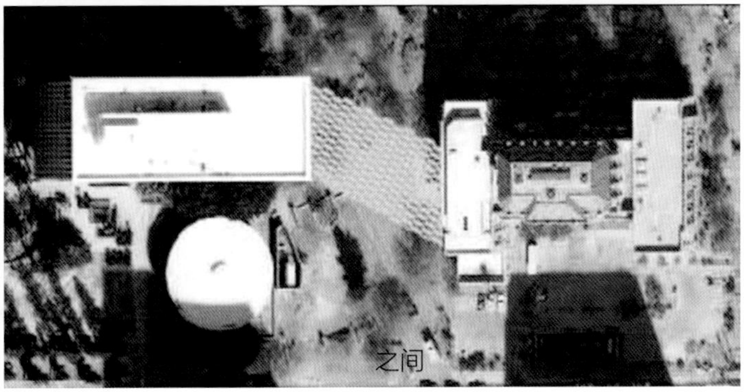

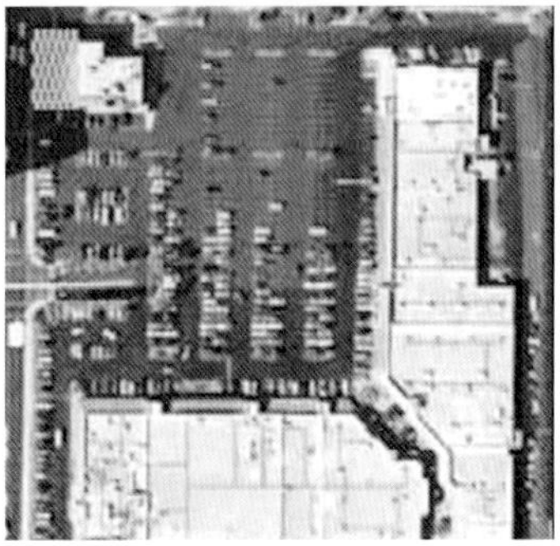

图 5.8-2

图 5.8-1 针对不同城市环境的三种布置场景
Three deployment scenarios for varying urban conditions

图 5.8-2 瓷球原型
Prototype of ceramic tile

图 5.8-3

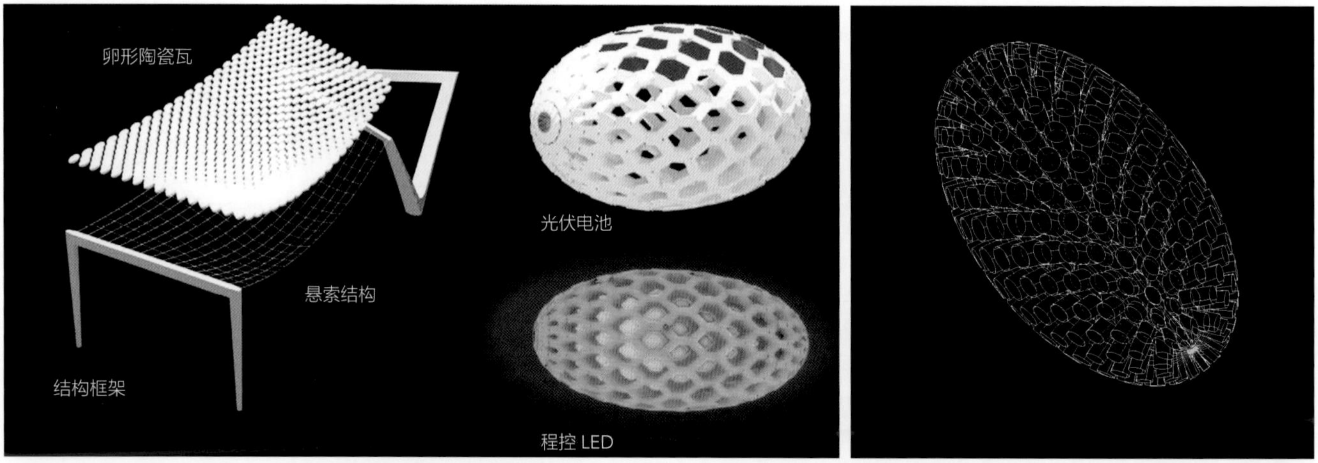

图 5.8-4

卵形陶瓷瓦的材料特性及其铸模生产方式与当地传统铸模制的经济性与简便性相吻合，并根据新兴技术重新设计熟悉的材料。

图 5.8-3 嵌入 LED 光源和光伏电池的瓷球及其结构框架
Structural frame and tiles showing embedded LEDs and PV cells

图 5.8-4 遮阳棚下的公共空间视图
View of public space under the canopy

5.9 斯特拉特福德广场 Stratford Square

在公共广场上悬挂一种可编程的网状虚拟屋顶。

KLF 事务所

A suspended programmable net forms a virtual roof for a public square.

Khoury Levit Fong

美国康涅狄格州斯特拉特福德的市集广场拟建的顶篷是一个突出的建筑元素，它遮盖在公共汽车站上面，充当一个数字屋顶（屏幕），并能调节声音。

由木材制成的顶篷是张拉整体结构的一部分，该结构布置在广场上方，形成一个虚拟的屋顶。钢索网的首要作用是作为受压构件来悬挂木质顶篷，反过来，它又有助于锚定和稳定钢网。

支撑顶棚的钢网也支撑着一排光纤电缆。这个顶篷与辅助装置通过由程序编写的各种各样的图案照亮整个广场。可能的照明场景包括基于网络的交互式编程和由传感器触发的实时响应。

通过嵌入式的传感器，木质顶篷就像一个超大的扬声器，它能够根据周围环境的城市噪声成比例地放大水景的声音。

The proposed canopy for Market Square in Stratford, Connecticut is a prominent architectural element that shelters a bus-stop, acts as a digital roof / screen, and mediates sound.

The canopy, made of wood, is integral to a tensegrity structure that is deployed above the square to form a virtual roof: the primary purpose of the steel-cable net is to suspend the wood canopy while the canopy's compressive members, conversely, serve to anchor and stabilise the steel net.

The steel net holding the canopy also supports an array of fibre-optic cables. It is designed, along with auxiliary fixtures, to light the square in a variety of programmable patterns. Possible lighting scenarios include interactive web-based programming and sensor triggered real-time responsiveness.

By means of embedded transducers, the wood canopy performs like an oversized loudspeaker. It amplifies the sound of the cascading water in proportional response to the ambient city noise.

图 5.9-1

图 5.9-1　面向市集广场的顶篷视图
View towards Market Square

图 5.9-2

图 5.9-2　顶篷下方或周边空间可以提供各类活动，增强公众的参与感
Various programmatic activities are accommodated under and around the net, enhancing public use

图 5.9-3

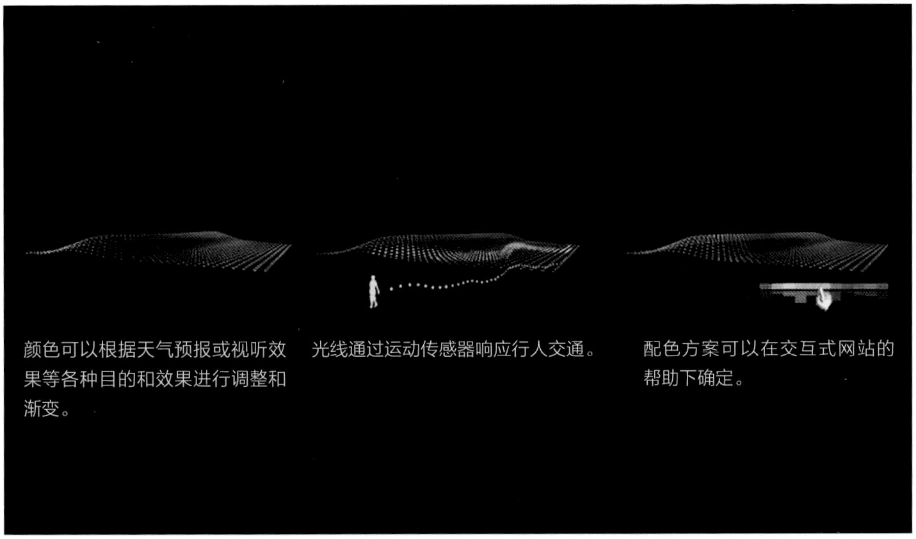
颜色可以根据天气预报或视听效果等各种目的和效果进行调整和渐变。
光线通过运动传感器响应行人交通。
配色方案可以在交互式网站的帮助下确定。

图 5.9-3 照明场景
Lighting scenarios

图 5.9-4 顶篷色彩会根据天气预报或视听效果等各种不同的效果和目的进行调整和渐变
The colour adapts and gradually shifts for a variety of effects and purposes such as weather forecasting or audiovisual spectacle

图 5.9-5 光线通过运动传感器强烈响应步行交通
The light intensely responds to pedestrian traffic by means of motion sensors

图 5.9-4

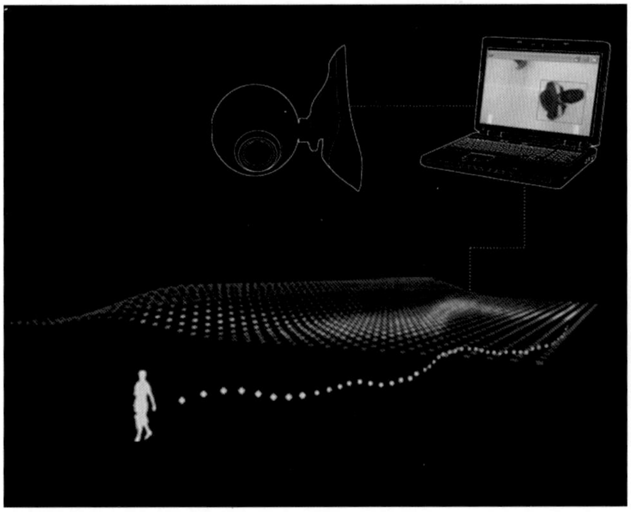

图 5.9-5

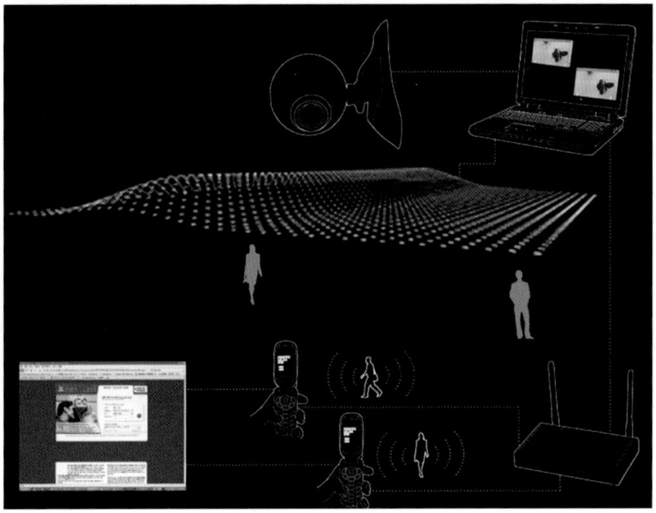

图 5.9-6

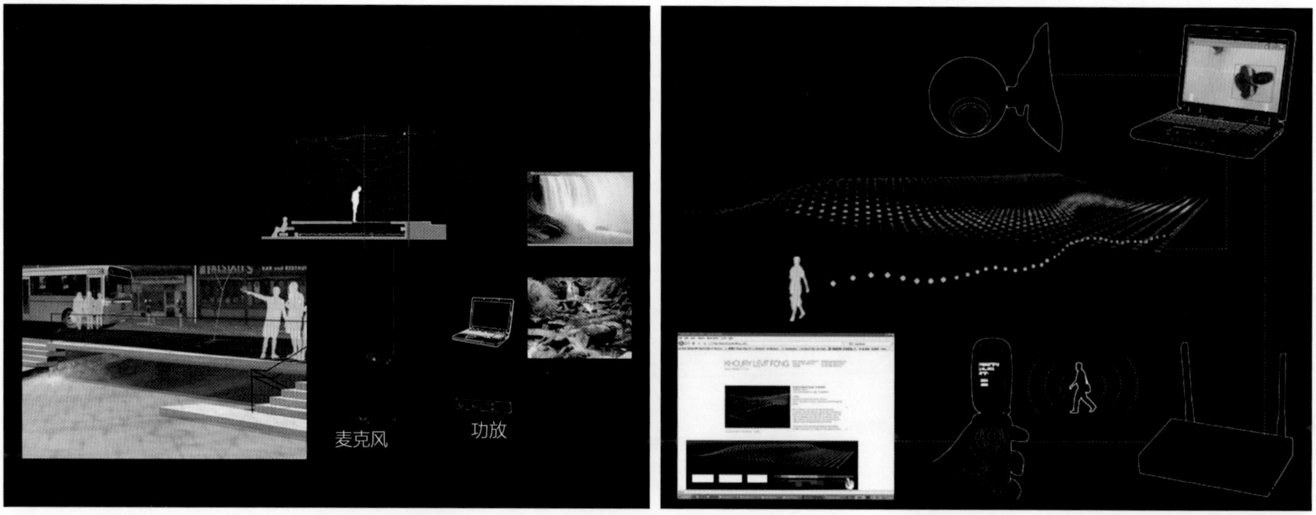

图 5.9-6 配色方案可通过一个交互式网站来设定
The colour scheme can be decided through an interactive website

图 5.9-7

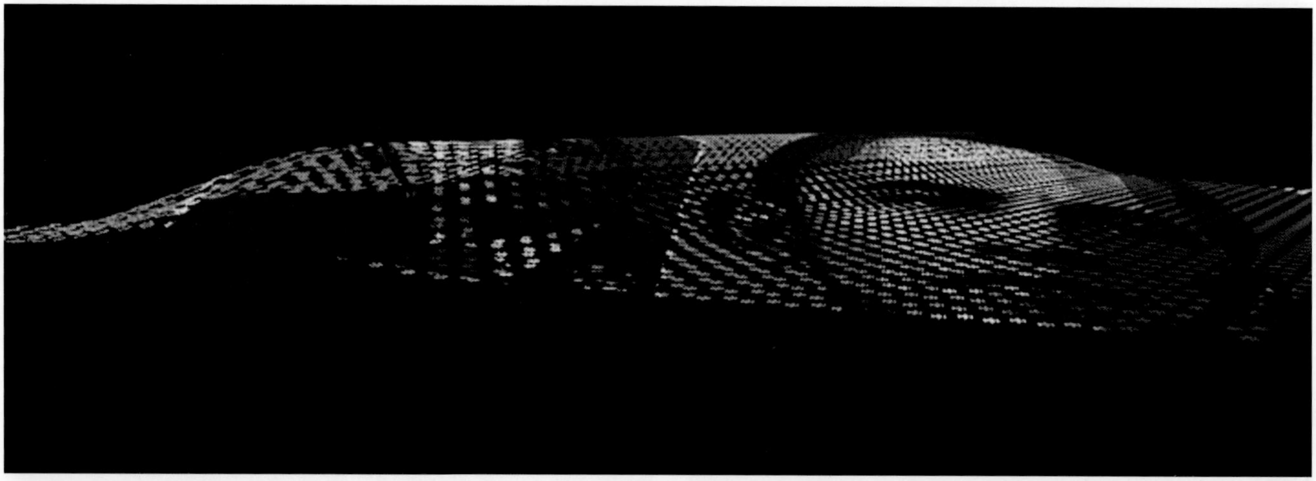

图 5.9-7 数字顶篷
A digital roof

图 5.9-8

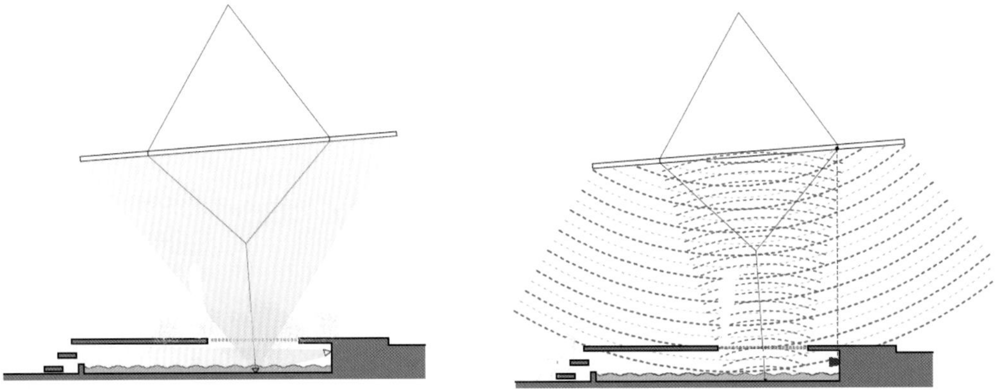

图 5.9-9

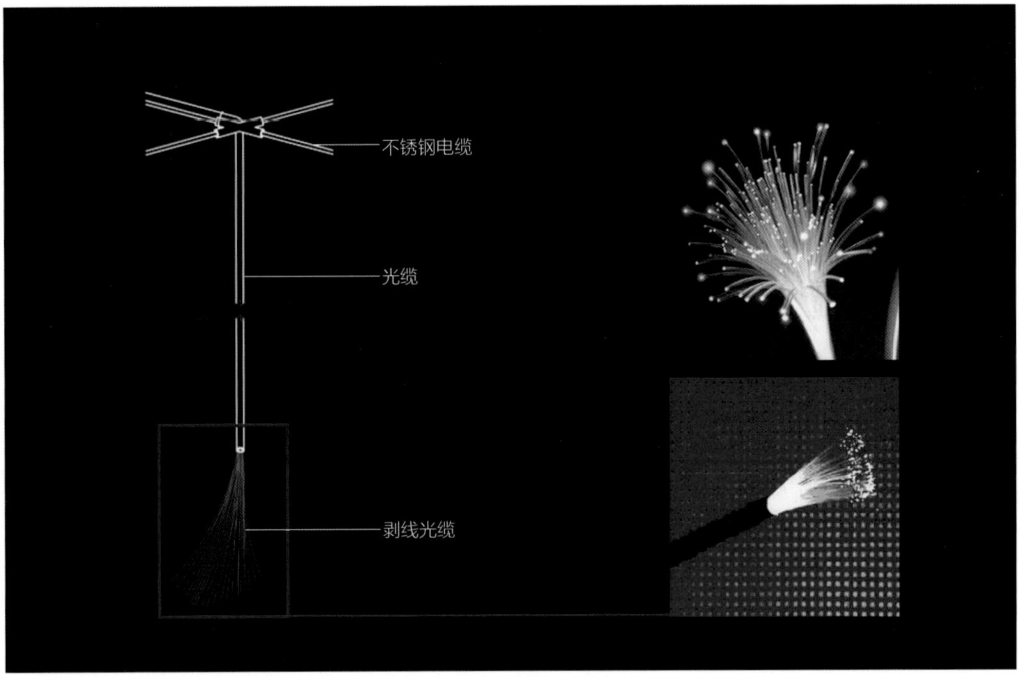

图 5.9-8 声音与光
Sound and Light

图 5.9-9 支撑顶篷的辅助装置
Auxiliary fixtures to holding the canopy

5.10 超级块 Superblock

嵌入式 LED 和无线通信将玻璃块结构体转变成一个可变分辨率的显示器。

KLF 事务所，纳希德 · 纳比安

Embedded LEDs and wireless communication transform structural glass blocks into a variable resolution display.

Khoury Levit Fong with Nashid Nabian

A standard 8 x 8 LED array is embedded into a generic glass block to form a digital display. The image displayed is received wirelessly by means of an embedded ZigBee device.

The LED pixels are extended with an egg crate-like diffuser to cover the entire face of the brick. The Superblock can function alone as a low resolution 64-pixel display or can be aggregated with other blocks in a variety of configurations and scales. When Superblocks wirelessly detect and situate other adjacent blocks they re-map the image over the entire field that is constituted by the aggregated blocks. The design and wireless connectivity are meant to enable adaptation to different conditions and facilitate installation. This is a plug-and-play product that can be used for building wall partitions at different scales and in a variety of spatial and structural applications.

普通玻璃块被嵌入一个标准的 8×8 LED 阵列中，形成一个数字显示器。其显示的图像是由嵌入的 ZigBee 设备接收无线信号来生成的。

LED 像素利用方格形扩散器覆盖整个玻璃砖表面。超级块可以作为一个 64 像素的低分辨率显示器单独发挥作用，或者与其他块体通过各种配置与比例聚合起来共同工作。当“超级块”通过无线信号检测和定位其他相邻块体时，它们将图像重新映射到由聚合块组成的整个区域。这样的设计和无线连接方式意味着能够适应不同的条件且安装方便。这是一个即插即用的产品，可以在不同尺度，以及各种空间和结构应用中用于建造墙体隔断。

图 5.10-1

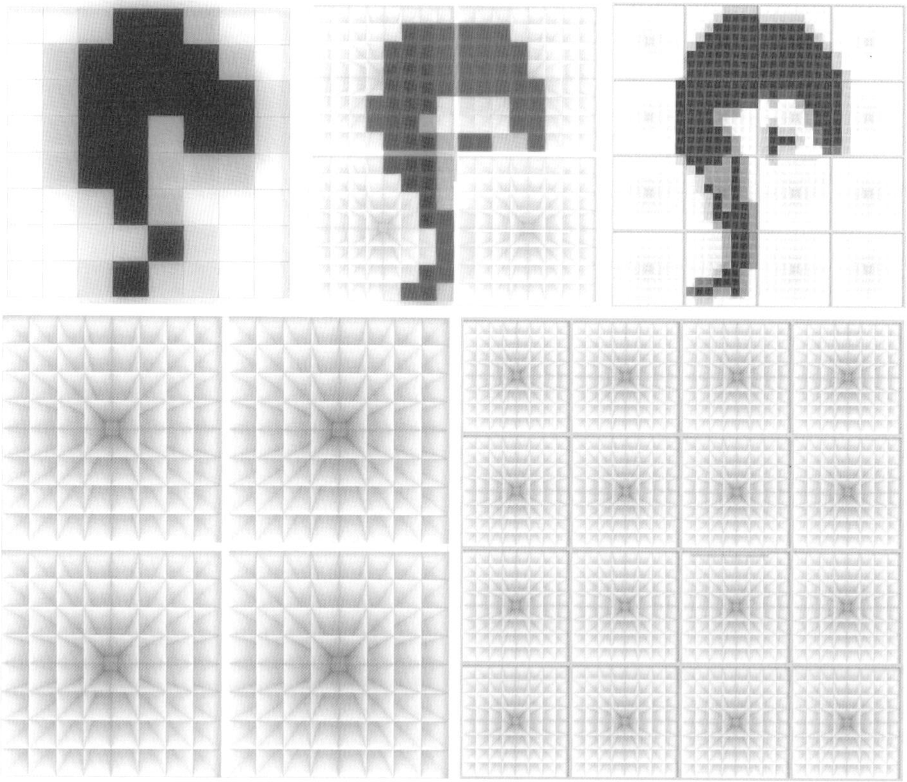

图 5.10-2

图 5.10-1 图像分辨率和尺寸通过无线信号自动适应块体配置
Image resolution and dimensions wirelessly and automatically adapt to block configuration

图 5.10-2 家庭室内空间中由“超级块”构成的墙
A wall of ‘Superblocks’ in a domestic interior

图 5.10-3

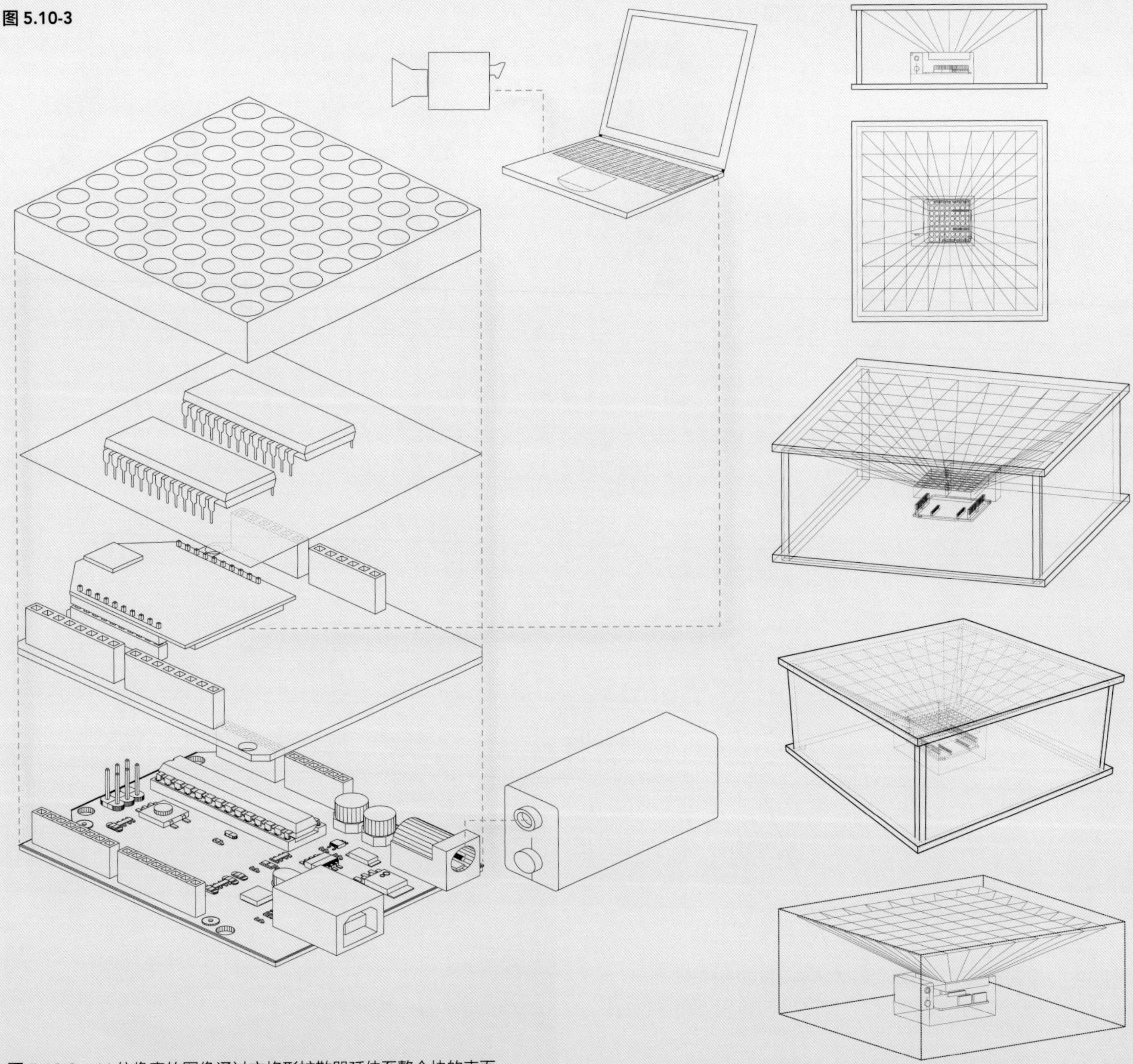

图 5.10-3 64 位像素的图像通过方格形扩散器延伸至整个块的表面
A 64-pixel image is extended over the entire block face by means of a carton-like diffuser

图 5.10-4

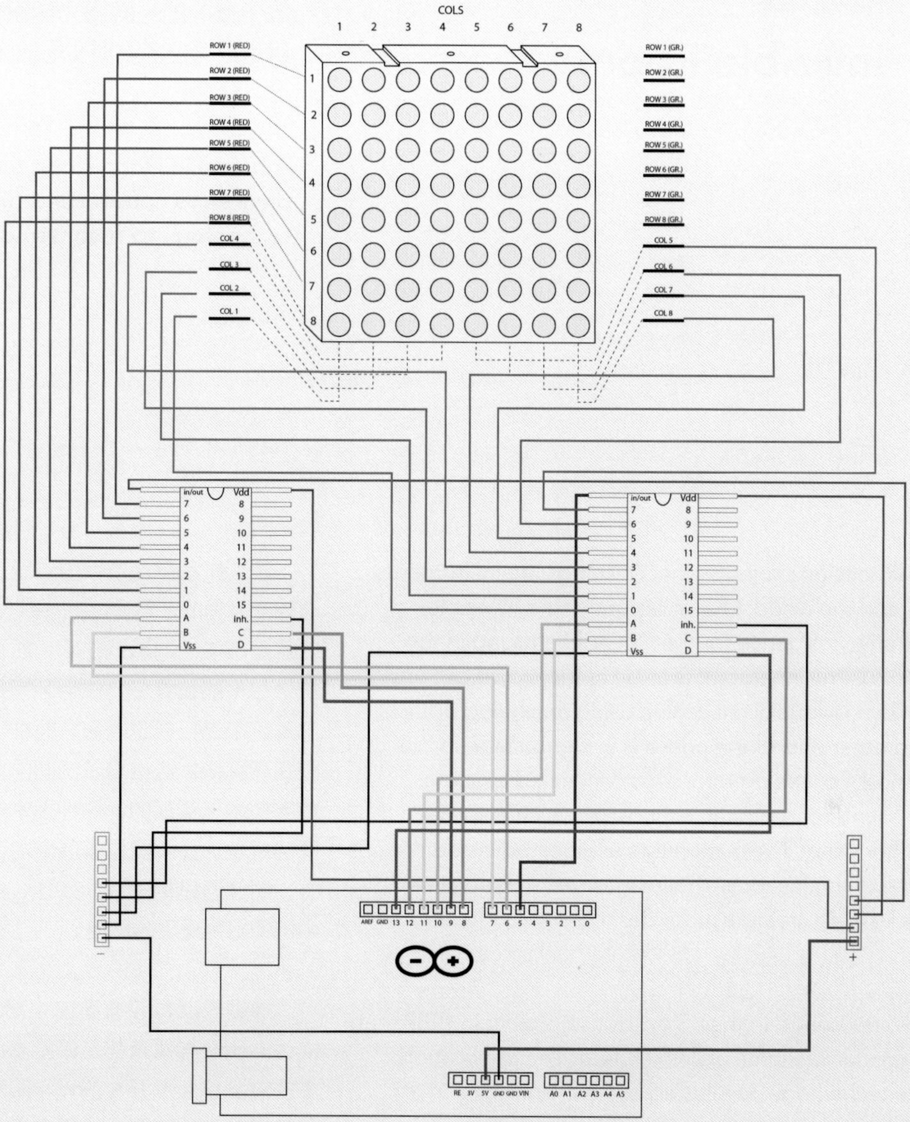

图 5.10-4 Arduino 驱动原型电路图
Circuit diagram of Arduino powered prototype

5.11 隐形屋 Invisible Room

位于香港的一间用光学设备隐藏自身并且用数码技术重现实景的房子。❶

塞缪尔 · 高

A house in Hong Kong uses optical devices to conceal and digital technology to reconstruct the real.❶

Samuel Ko

The thesis of this studio design project is that buildings are instruments to mediate ourselves and the world. I have designed this 'invisible house' as a demonstrative residence for the CEO of Samsung in the city centre of Hong Kong. The inhabitant is to be in the crowd but not to be seen. The bedroom is visually concealed using an analog optical system of mirrors, and other part of the house is in camouflage in the surroundings. The concept of seeing through a concealed space is also achieved by the counterpart digital technology which displays live images of the 'other side'. These monitors are located inside the bedroom and on some of the facades. The images may be processed by relevant programs so as to create a more desired environment in a reconstructed reality.

Diller and Scofidio described in their Slow House project in 1989 that bricks and pixels are irreducible units of architecture. The application of optical camouflage technology developed by Keio University in Japan also challenged the idea of transparency. My research was later

❶ Guided by Rodolphe el-Khoury.

该工作室设计项目的主题是建筑是调节我们与外部世界的工具，“隐形屋”是三星首席执行官在香港市中心的示范住宅，居住者身处拥挤的人群中却不能被看见。住宅卧室是通过模拟镜面光学系统来进行视觉隐蔽，而住宅的其他部分则隐蔽在环境中。透过隐藏空间的概念也是由对应的数字技术来实现的，这种技术能够展示另一面的鲜活景象。显示器位于卧室内部与一些建筑立面上，图像通过相关程序处理，从而在重建的现实中创造出更理想的环境。

迪勒与斯科菲迪奥在 1989 年的慢屋项目中描述：砖块与像素都是最基本的建筑单元。由日本庆应义塾大学开发的光学隐蔽技术的应用也挑战了透明性的概念，该研究后来将此浓缩为一个

❶ 由鲁道夫 · 埃尔 - 库利指导。

能比较模拟和数字化的潜力和性能的矩阵。研究通过对视觉和听觉数据的提取、叠加、转变或处理，用模拟或数字化的方法深入研究重建现实的时空性质变化。

例如，为了看到你的背面，我可以在普拉达商店的更衣室里放置一组镜面，或者用延迟 12 月 5 日的显示器。数字技术能产生的另一个时间特性就是 1~12 小时的时间延迟，以缓解旅客或夜班工人的时差反应。这个矩阵的结论是数字化更有效，尽管如此我也欣赏模拟。所以数字化与模拟既要结合又要形成对比，以最大限度地发挥潜力。

与此同时，我也用这些模型探讨了隐藏空间不可见性的概念，另一方面，数字化对应模型和应用程序可以帮助处理实时数据，后来发展成这个住宅作为最终的设计。

隐形屋设置在香港一个人口稠密地区的高层建筑的屋顶，它的目的是要尽可能地减少对于相邻建筑的视线遮挡。但是这个玻璃建筑不是完全透明看不见的，由于镜面系统形成的隐形视觉幻象而消失的只是二层的主卧室。在隐形卧室的内部用四块数字显示屏替代窗户，通过程序处理对现实世界重组，并在屏幕上调节对于外界的感知。例如，可以删除或重组窗外的景色来实现预期的效果。时间也可以通过延迟和视频直播来重塑，或许也可以从重复模式中预见未来数字窗口的景象。

condensed into a matrix which compares the potentials and performance of the analog and the digital. By means of extraction, superimposition, mutation or manipulation of visual and acoustic data, I looked into the changes of spatial and temporal qualities of a reconstructed reality using either the analog or the digital method.

For instance, in order to see your back, I either use this set of mirrors or a monitor that displays with 5 Dec delay in the changing room of Prada shop. Another temporal quality which the digital can produce is the 1-12 hours delay for relieving jet lag for traveler or night shifters. The conclusion of this matrix is that the digital does more effectively. Nevertheless I also appreciate the analog so the digital is to be incorporated in conjunction and in contrast with the analog to maximize the potential.

Simultaneously I have explored the idea of invisibility of a concealed space with these maquettes and on the other hand, a digital counterpart model with an apps that can help processing live data. It then developed into this house as a final design.

The invisible house occupies the roof of a mid-rise building in a very densely populated area of Hong Kong. It is meant to be as transparent as possible to minimise the obstruction of the view from adjacent buildings. But glass buildings are not entirely transparent and hardly invisible. What is invisible here is the master bedroom on the upper floor that is designed to virtually disappear thanks to a system of mirrors that creates an optical illusion of invisibility. Inside the stealth bedroom, four digital displays designed to substitute windows mediate the perception of the exterior by overlaying a programmed and wilfully reconstructed version of the real. For instance, elements of the view can be deleted or reconstituted to achieve desired effects. Time also could be refashioned with delayed and manipulated video feed, allowing perhaps a view from the digital window into the future by extrapolating from repeating patterns.

图 5.11-1

图 5.11-1　从对面建筑看隐形屋
View of Invisible House
from opposite building

图 5.11-2

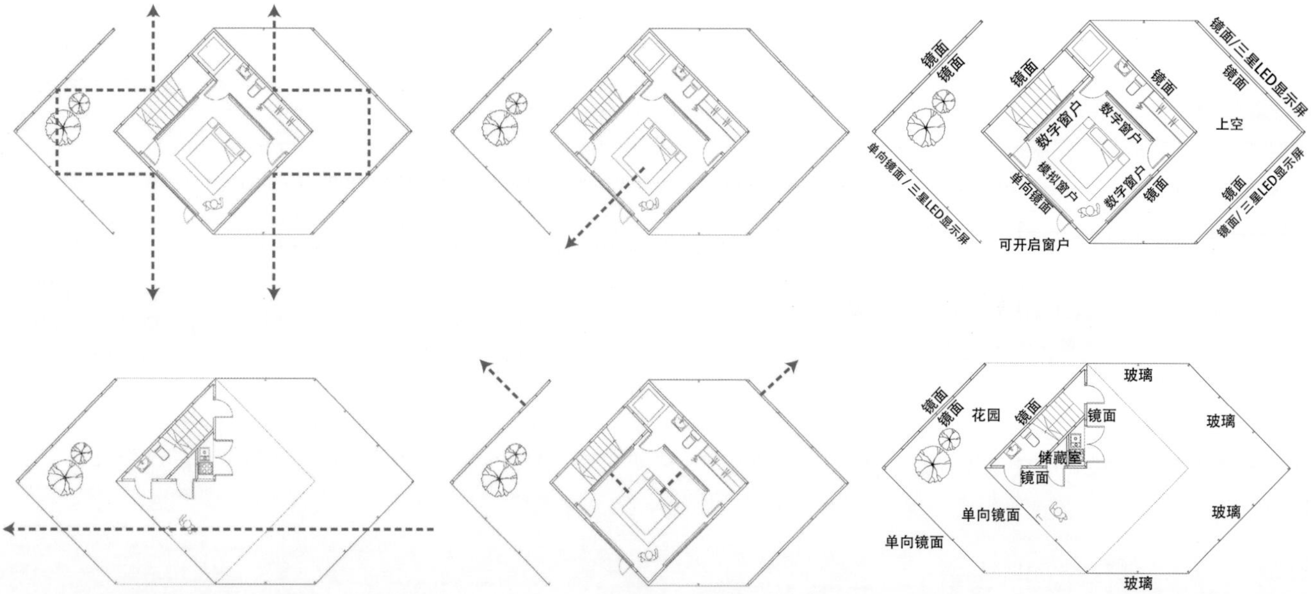

图 5.11-2　建筑光学平面
Plan view optical diagrams

图 5.11-3

图 5.11-3 隐形屋二层消失的模型展示
Model demonstration of the disappearance of the upper floor room

图 5.11-4

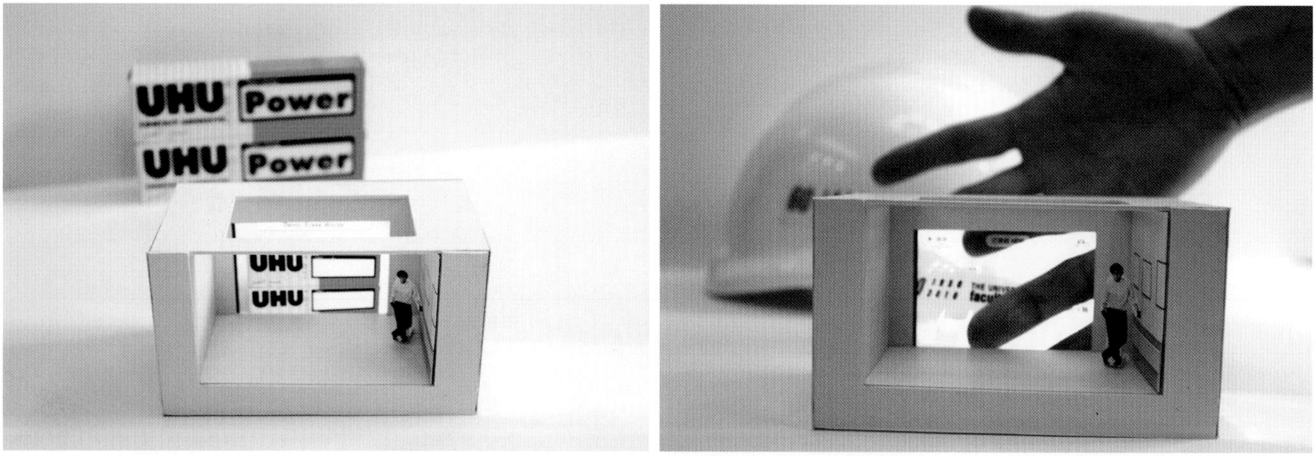

图 5.11-4 数字窗口重塑现实世界——在这个演示中，一个 iphone 应用程序无缝擦除实时视频图像中的文本
Digital window reconstructs the real — in this demonstration an iphone application seamlessly erases text from a live video image

图 5.11-5

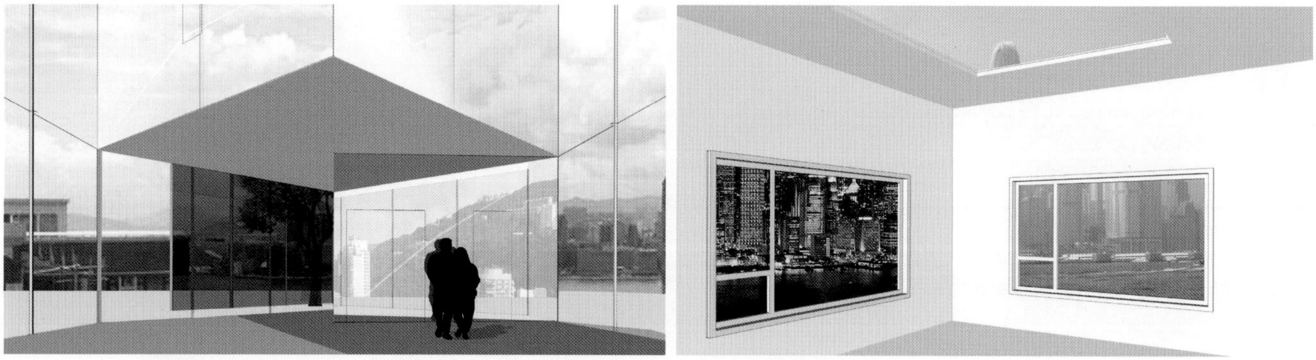

图 5.11-5 隐形屋里的数字窗户通过一个双向镜子提供电子媒介的视野
Digital windows offer electronically mediated views through a two way mirror in the camouflaged room

图 5.11-6 建筑剖面
Section

图 5.11-7 声学矩阵
Acoustic matrix

图 5.11-6

图 5.11-7

声学				
	空间		时间	
	模拟	数字化	模拟	数字化
提取 建议不限于缩小现实	实时 技术：吸收 效果：体积缩小	实时 技术：声音删除 效果：留空的私密性		1 秒的延迟 技术：提取 效果：顺滑
叠加 建议不限于放大现实		实时 技术：白噪声注入 效果：来自幻觉的私密性		实时 + 录音 技术：声音破碎化 效果：来自混合的私密性
转变 建议不限于增强现实	实时 技术：排列 效果：集中	实时 技术：指定压力 效果：强化理解		及时回音 技术：记录 效果：回忆
操纵 建议不限于重构现实		实时 技术：回声 效果：空间幻觉	实时 技术：私语美术馆 效果：回声	实时 技术：转化 效果：理解

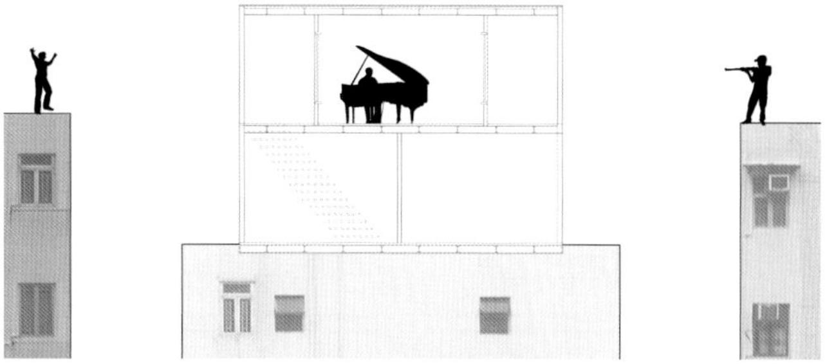

图 5.11-8

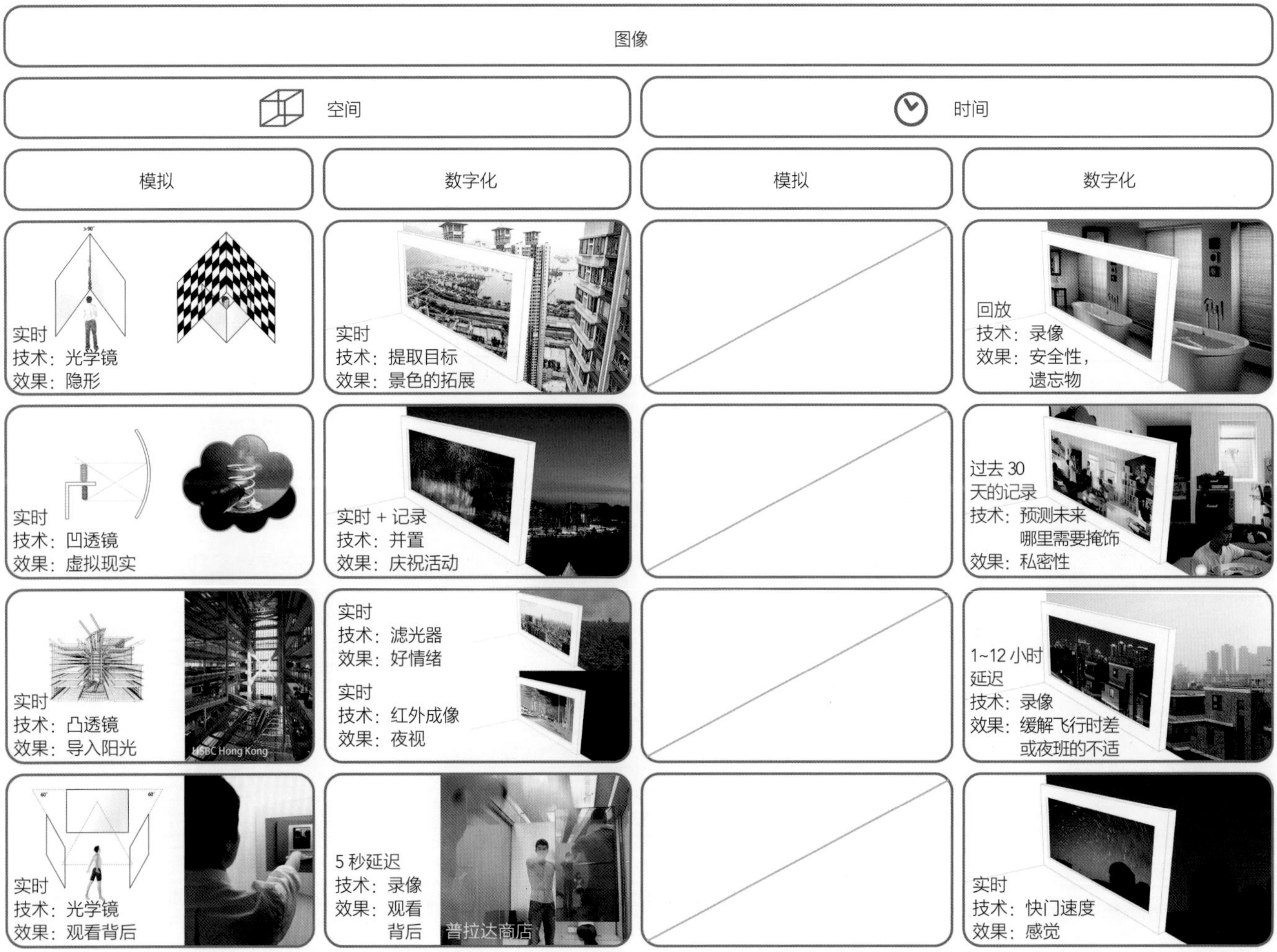

图 5.11-8 通过电子媒介重构现实景色可能性的视觉矩阵
Matrix detailing the possibilities for the reconstitution of the real through electronic mediation

5.12 微环境装置 Micro-environment Gear

微环境装置结合了医疗设备的性能，用时尚为自我提升和个性表达提供了空间。❶

斯泰西 · 沃斯

Micro-environment Gear merges the performance of medical gear with fashion's affordance for self-augmentation and personal expression.❶

Stacie Vos

随着全球流行病发生的持续上升，我们在 21 世纪的全民健康将越来越依赖于在城市范围内追踪和识别传染病所使用的有形和虚拟设备。该项目提出了在高密度城市环境中将衣服、一幢亲密的建筑作为场所来减少空气中病菌的传播，同时提供个性化的心理安慰和个性表达。

以运动的个人（即身体）为尺度的干预，而不是包裹他的外壳，是使疾病传播最小化的最佳时机：个人设备可以预测在什么情况下一个人最有可能被感染，并采取保护措施。该项目提出了一系列的衬衫和个性化的口罩。这些衬衫配备了温度和电流响应传感器，一旦检测到人身体状态的变化，就会松开用形状记忆线缝制的衬衫领口，以覆盖穿着者的面部。这种衬衫会随着生物传感

As the frequency of global pandemics continues to rise, our collective health in the twenty-first century will increasingly depend on the physical and virtual apparatuses used for tracking and identifying infectious diseases within the urban context. This project proposes clothing, an intimate architecture, as a site capable of mitigating the spread of airborne germs while providing customized psychological comfort and personal expression in dense urban environments.

An intervention at the scale of the mobile individual, the body, rather than one's collective enclosure, is opportune for minimizing the spread of diseases: personal devices can anticipate the conditions when one is most likely to become infected and respond with protective measures. The project proposes a series of shirts and personalized face masks. The shirts are equipped with temperature and galvanic response sensors. Upon detecting a change in the state of the body, the shirt's collar – sewn with shape memory wire – is released to cover the wearer's face. The shirt mutates with the subconscious physical signals detected

❶ 由克里斯托 · 马尔科普洛斯指导。

❶ Guided by Carol Moukheiber.

by biosensors. The body's internal nervous system is extended to the exterior, manifesting itself in the electric circuit that is incorporated into the textile.

The design of the mask takes current advanced medical mask technology — such as silver nanoparticle impregnated filter textiles for germ deactivation — and adds performance through form. By multiplying the mask's surface area through the folding of the fabric, an increased level of protection is gained. The prototypical medical gear then transforms towards a more complex form, one that converges with the presentation and augmentation of the self through fashion.

The project builds on current trends in wearable computing, along with the increasing proliferation of self-monitoring health initiatives, biosensors and electronic hobbyist tools in customizing everyday life. As ubiquitous technologies become more adept at sensing elements invisible to the human senses and acting on this function invisibly, to what extent do we desire a visible representation and form for these operations?

This project emerged out of an interest in airports as the frontlines of disease control. The prevalence of biometric scanners for temperature detection, airport quarantine rooms and on-board health checks all collapse here onto the surface of the body. The clothes argue for an integration of systems designed to enhance the perception of our milieu.

器探测到的潜意识物理信号发生改变，就像身体内部的神经系统延伸到外部，表现在融入织物的电路中。

口罩的设计采用了当前先进的医用口罩技术，如用于细菌灭活的银纳米浸渍滤布，并通过外形增加了性能。通过折叠织物增加口罩的表面积，提高了防护安全等级。最初的医疗设备转换成更复杂的形式，通过时尚将自我呈现和提升相融合。

该项目是基于当前可穿戴计算的趋势，以及在日常生活中定制自我监测健康计划、生物传感器和电子设备爱好者的日益广泛。随着泛在技术变得更适用于感知人类不可感知的元素，并在无形中充当了这种功能，在多大程度上需要为这些无形的功能提供一种有形的表现和形式呢?

该项目成为机场疾病控制的第一道防线，用于温度检测、机场检疫室和机上健康检查的生物识别扫描仪在人体表面随处可见。这些衣服主张整合各种系统以提高人们对环境的感知。

图 5.12-1

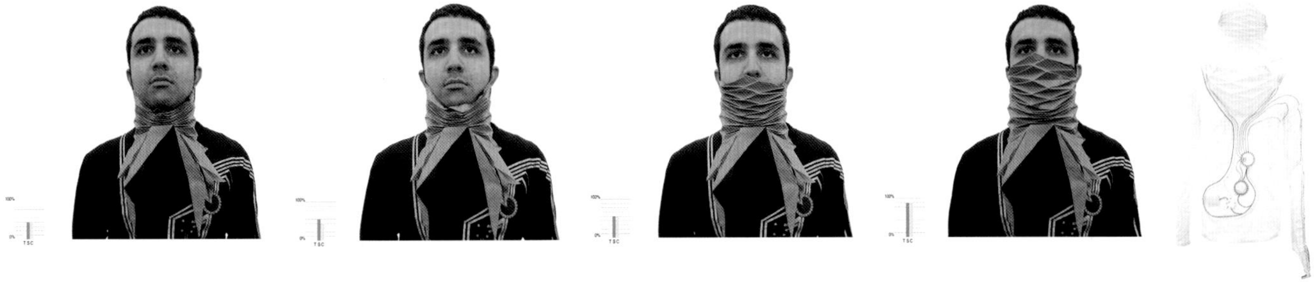

图 5.12-1 一旦察觉到身体状态的变化，衬衫领子就会松开，遮住穿戴者的脸。成分：荷叶状开源硬件板、导电面料、形状记忆合金（SMA）线
Upon detecting a change in the state of the body, the shirt's collar is released to cover the wearer's face. Components: LilyPad Arduino, conductive fabric, shape memory alloy (SMA) wire

图 5.12-2

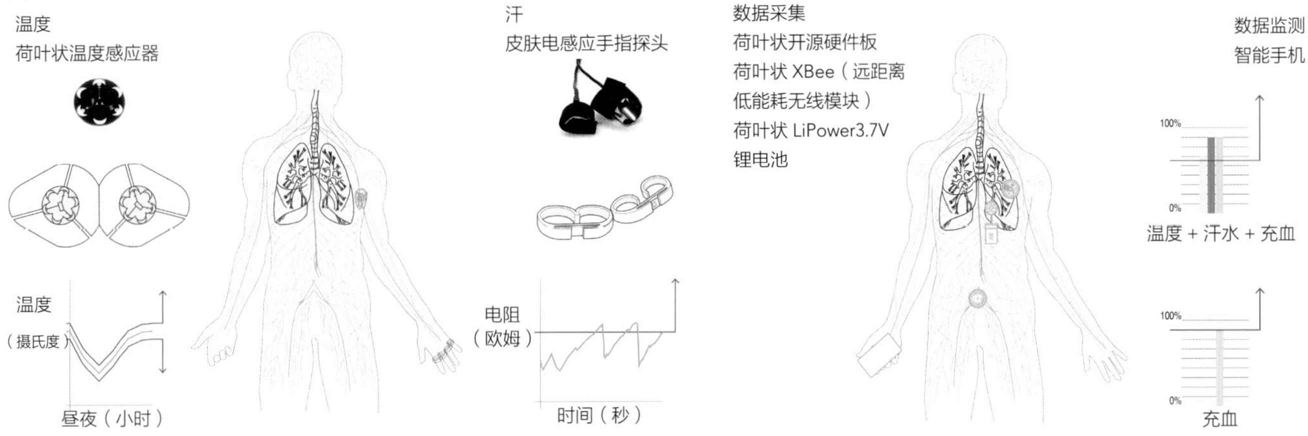

图 5.12-2 该衬衫集成了温度和皮肤电反应（GSR）传感器。戴在手指上的戒指形状的电极可以检测到由于佩戴者的情绪状态或压力引发的汗液增加所导致的电流活动变化
The shirt integrates temperature and galvanic skin response (GSR) sensors. Electrodes in the shape of a ring around the finger detect changes in the electrical activity due to increased sweat triggered by a wearer's emotional state, or stress

图 5.12-3

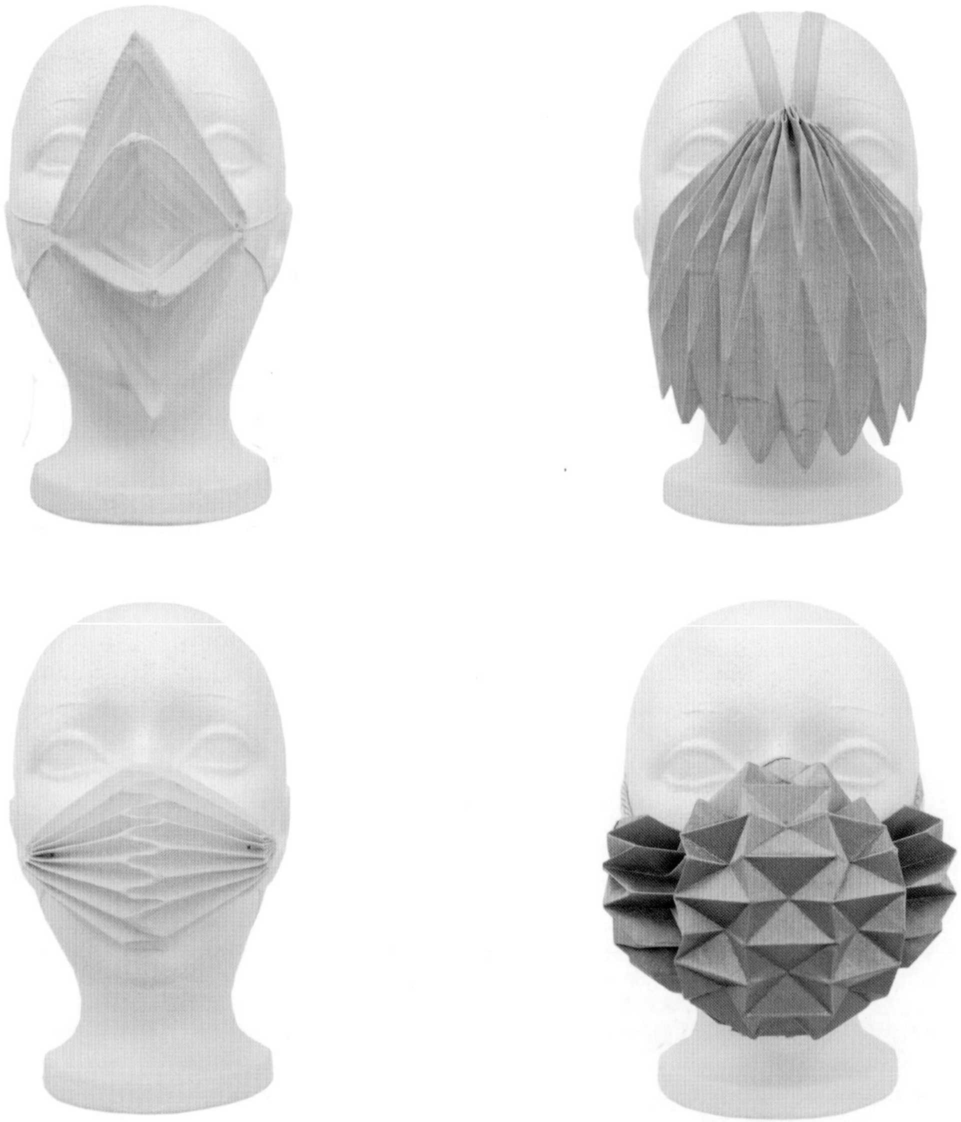

图 5.12-3 口罩作为自我表达的工具，同时保护佩戴者。通过以织物折叠的方式增加口罩的表面积，增加了防护安全等级。材料：丝塔夫绸银纳米过滤器
The mask acts as a vehicle for self-expression while protecting the wearer. By multiplying the mask's surface area — through the folding of the fabric — an increased level of protection is gained. Material: Silver nanoparticle filter with silk taffeta

5.13 数字时代的涂鸦和建筑 Graffiti and Architecture in the Digital Age

数字技术拓展了涂鸦的定义，并延伸其作为参与性艺术形式的范畴。

贾维德 · JAH，指导：鲁道夫 · 埃尔 - 库利

Digital technology expands the definition of graffiti and extends its reach as a participatory art form.

Javid JAH,
Advisor Rodolphe-el-Khoury

本文旨在探讨用数字技术文化拓展涂鸦的定义，并延伸其作为参与性艺术形式的范畴。它由一系列的实验组成，可以将艺术或实践延伸向不同的方向，寻求以令人兴奋的方式帮助每个人塑造建成环境的形象。

This thesis is about exploring the ways in which the culture of digital technology expands the definition of graffiti and extends its reach as a participatory art form. It consists of a series of experiments which take the art / practice in different directions, seeking to facilitate exciting ways in which individuals can shape the appearance of the built environment.

标签

我的第一个实验是基于这个语言事件——标签。作为威尼斯双年展加拿大展区设计方案的一部分，我构想被标记的标签如何能让人们在画廊空间内体验涂鸦和它的空间品质。

Tagging

One of my first test was based on this linguistic accident — #TAGGING. As part of a design proposal for the Canadian Exhibition at the Biennale in Venice, I imagined how the tagged tag could enable individuals to experience graffiti and its spatial qualities while being in the confines of the gallery.

The Digital Can & E-Facade

This investigation imagines the logistics of a city in which graffiti were no longer transgressive — ie artists painted with digital ink. It proposes a scenario where property owners could outfit their buildings in responsive screens using e-ink technology — a low-energy, solar-powered, infra-red sensoring system of tiles forming a facade. The E-Facade works in conjunction with the Digital Spray Can, enabling artists to write graffiti with infra-red light, digitizing artworks that now have the potential to be shared, downloaded and archived.

数字盒和数字表皮

这项调查构想了在一个城市中涂鸦不再是违法的——艺术家可用数字墨水作画。该方案提出了一种设想，即业主可以使用数字墨水技术为他们的建筑物安装响应式屏幕—— 一种由具有低能耗、太阳能供电、红外线感应系统的瓷砖构成的建筑表皮。表皮与数字喷射器协同工作，使艺术家能够利用红外线进行涂鸦，数字化的艺术品具有共享、下载和存档的潜在优势。

图 5.13-1

图 5.13-1 JAHREALITY 是一个应用程序，可以利用街头艺术来增强现实。它利用涂鸦和资源开放技术在公共区域举行分享活动。通过这种移动应用，涂鸦艺术品变成了一个窗口，用户可以通过它进入交互式的增强现实：当虚拟和实体在公共领域发生碰撞时，这种以视频、动画、3D 对象和聊天论坛形式构成的信息就可以被接收、分享与改变

JAHREALITY is an app that utilizes street art to augmented realities. It invites participatory activity in the public realm fueled by graffiti and open-source technologies. Through this mobile application, graffiti artworks become portholes through which users can access an interactive augmented reality: information in the form of video, animation, 3-D objects and chat forums can be accessed, shared and altered as the virtual and physical collide on the surfaces of our public realm

图 5.13-2

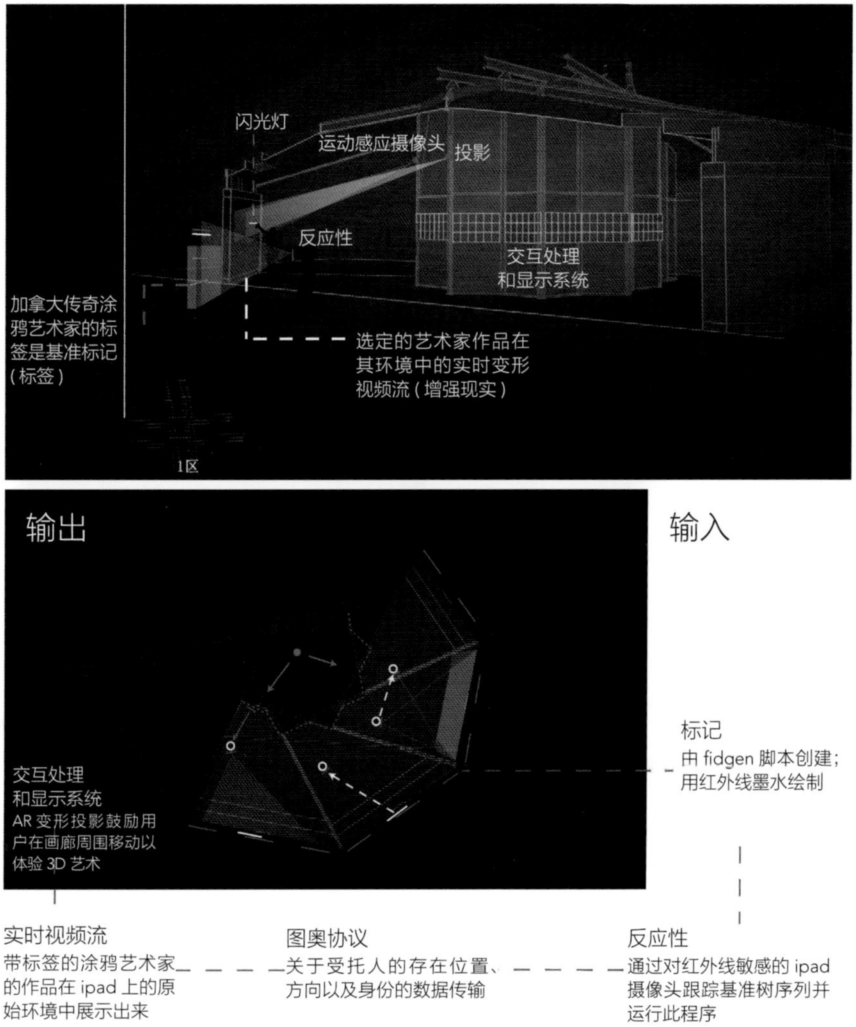

图 5.13-2 标签化是一个数字化与涂鸦交集的重要时刻。本质上，我们所说的涂鸦文化的基本行为——标签——也是新一代 ICT 系统的支柱：从脸书到谷歌——拼凑式标签。在这些试验中，这些标签充当了拼凑式标签（基准标签）的作用，它触发一个增强现实（3D 对象）。增强现实的有趣之处在于，它的虚拟呈现具有精准的物理位置，运行增强现实的程序会对图像进行校准，使其看起来好像占据了一个物理空间

Tagging is an important moment of intersection between the digital and graffiti. Essentially what we refer to as a fundamental act of graffiti culture — the tag — is also the anchor a generation of ICT systems: from Facebook and Goggle — the hashtag. In this experiment, the tag operates as a hashtag (a fiducial marker) which triggers an augmented reality (3D object). The interesting condition about augmented reality is that its virtual representation has a precise physical location, and a program running augmented reality will anamorphically adjust the image to appear as though it occupies a physical space

图 5.13-3

图 5.13-3 JAHREALITY 的工作原理图，突出显示用户如何通过该应用程序的关键功能进行路径查找、体验和参与
Schematic diagram of how the app works, highlighting how users can way-find, experience and participate through the key functions of JAHREALITY

图 5.13-4 这种数字表皮与数字喷射罐协同工作，使艺术家可以用红外线光绘制涂鸦，数字化的艺术品具有共享、下载和存档的潜在优势
The E-Facade works in conjunction with the Digital Spray Can, enabling artists to write graffiti with infra-red light, digitizing artworks that now have the potential to be shared, downloaded and archived

图 5.13-3（续）

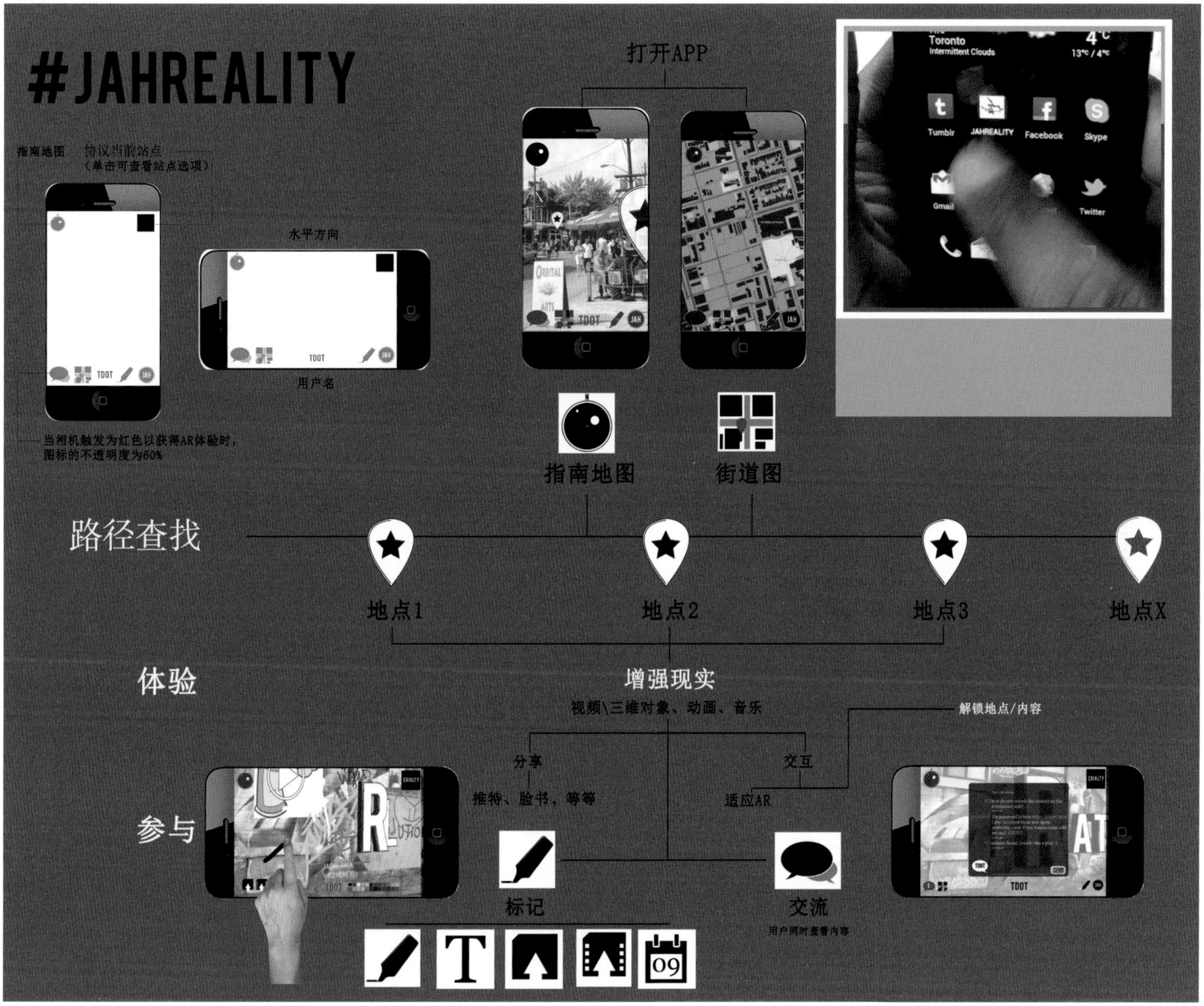

#JAHREALITY
指南地图
协议当前站点
（单击可查看站点选项）
水平方向
用户名
当相机触发为红色以获得AR体验时，
图标的不透明度为60%
打开APP
指南地图
街道图
路径查找
地点1
地点2
地点3
地点X
体验
增强现实
视频\三维对象、动画、音乐
解锁地点/内容
分享
推特、脸书，等等
交互
适应AR
参与
标记
交流
用户同时查看内容

图 5.13-4

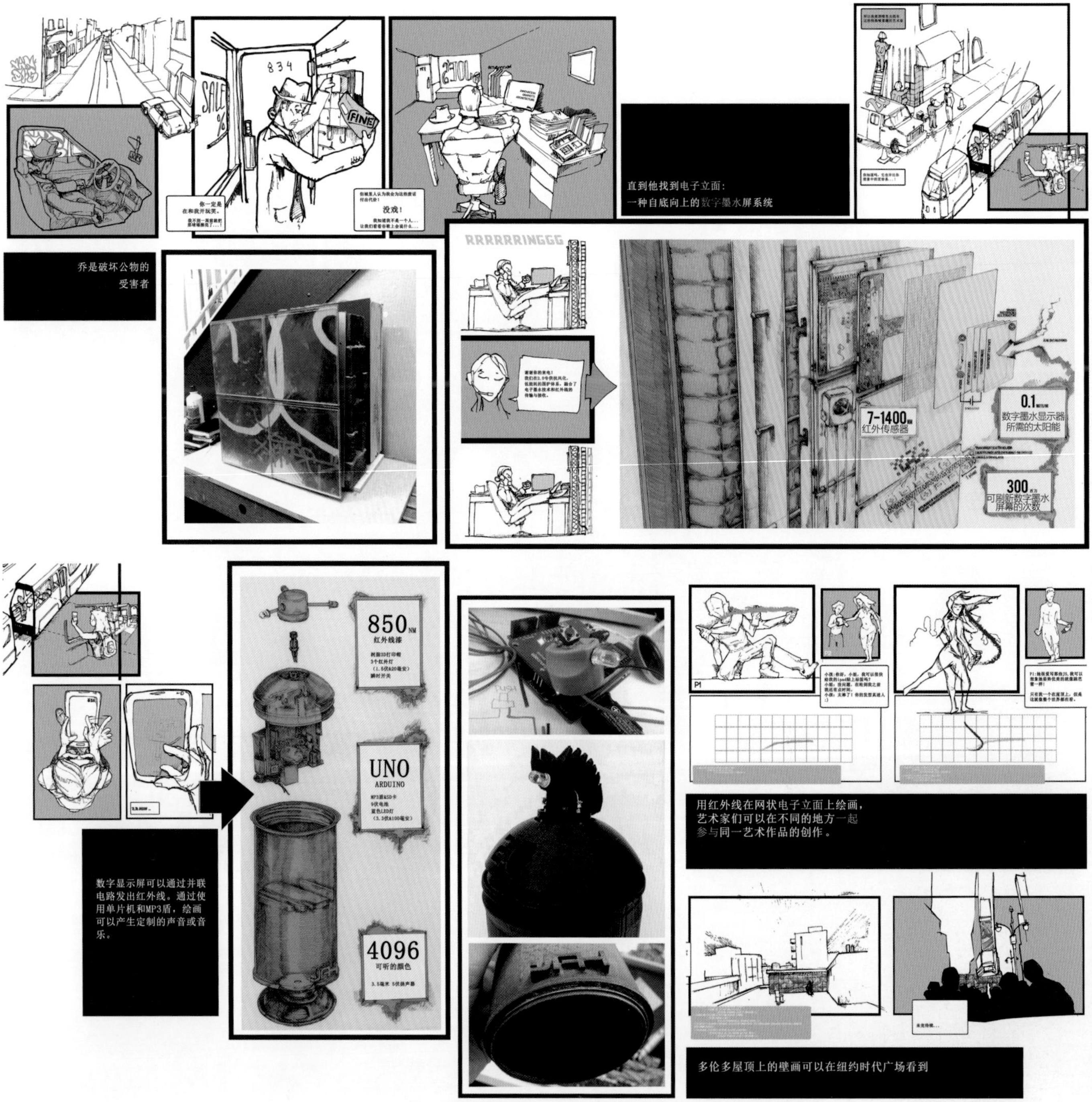
834
SALE
FINE
你一定是
在和我开玩笑。
没戏！
直到他找到电子立面：
一种自底向上的数字墨水屏系统
乔是破坏公物的
受害者
RRRRRRINGGG
7-1400NM
红外传感器
0.1
数字墨水显示器
所需的太阳能
300
可刷新数字墨水
屏幕的次数
850NM
红外线漆
树脂3D打印帽
3个红外灯
（1.5伏&20毫安）
瞬时开关
UNO
ARDUINO
MP3盾&SD卡
9伏电池
蓝色LED灯
（3.3伏&100毫安）
4096
可听的颜色
3.5毫米 5伏扬声器
数字显示屏可以通过并联电路发出红外线。通过使用单片机和MP3盾，绘画可以产生定制的声音或音乐。
P1
用红外线在网状电子立面上绘画，
艺术家们可以在不同的地方一起
参与同一艺术作品的创作。
多伦多屋顶上的壁画可以在纽约时代广场看到

现在，大众化的标准与标准的框架已经让位给了个性化定制，以识别各种各样的公众响应。与控制和监视相关的生物识别和定位技术被再次用于满足主观需求和愿望。下列项目探索了环境对于人类行为和特性的动态响应能力的新体验形式。在公共领域，连通性意味着扩大个体的影响范围，使参与者提高知名度和社交能力。在家庭环境中，建筑的响应能力支持不同的居住和生活方式。网络和电子媒介的环境使人们重新关注我们的感知和情感能力在构建群体和个体经验中所起到的作用，与其说是强调游乐性和戏剧性，倒不如说是为公众参与和私人居住提供设施和使用的便利。

Generic constraints and standardized frameworks give way to customized response, in recognition of a variegated public. Biometric and locative technologies associated with control and surveillance are re-appropriated in favour of subjective needs and desires. The following projects explore new forms of experience in the capacity for environments to dynamically respond to one's actions and attributes. In the public realm, the connectedness is meant to expand an individual's sphere of influence, projecting participants into situations of greater visibility and enhanced sociability. In the domestic environment the responsiveness of the architecture enables different scenarios of inhabitation and lifestyles. The networked and electronically mediated environments exercise a renewed interest in the role that our perceptual and affective faculties have in the construction of collective and individual experience, emphasising the ludic and theatrical, rather than the instrumental or practical facilitation of public participation and private dwelling.

6

个体感应

INDIVIDUATED RESPONSE

6.1 胡氏人体模型 Hu-mannequins

散发气味的人体模型变成了快闪店。

NMinusOne 工作室

Scent emitting mannequins transform a pop-up store.

Studio NMinusOne

Hu-mannequins is a pop-up store proposal for a fashion designer. The proposal attempts to enrich the experience of viewing by engaging the olfactory sense.

The installation is composed of scent-emitting mannequins that release a variety of smells into the environment throughout the day. Seats are placed within the store to harness the scent of sitting individuals. The scent of those occupying the seats is transported through a system of ducts, where the smell is released into the air via perforated mannequins. The person occupying the seat metaphysically wears the clothing through smell. The mannequin, typically a lifeless object, is now animated by human qualities through the production of smells — becoming a Hu-mannequin.

胡氏人体模型是为时装设计师设计的一种快闪店。它的设计概念是通过迷人的嗅觉来丰富视觉体验。

这个装置是由能够一直释放各种气味的人体模型组成的。商店中的座椅很好地利用了个人的独特气味。座椅上的气味通过一个管道系统输送到人体模型中，并由模型上的小孔释放出来。坐在椅子上人通过气味感觉像穿着店里的服装一样。人体模型是没有生命的物体，现在因为通过散发气味而感觉像是变成了活生生的人，这就是胡氏人体模型。

人们坐在这个椅子上的时候，简直就像穿着这件衣服一样，并且给予了衣服以这个人的独特气味。这就创建了一种全新的动态模式。与在更衣间试穿衣服不同，伙伴、朋友甚至是陌生人都能够通过嗅到衣服上虚拟穿戴者的气味想象人们穿上这件衣服的感觉。

While sitting on the furniture, a person literally inhabits the clothing, giving an outfit that person's unique smell. This sets up a new dynamic. Unlike the act of physically trying on a suit in a dressing room, partners, friends, and strangers can visualise an outfit by smelling the virtual wearer's body on the clothing.

图 6.1-1

图 6.1-1 一个穿着这件西装的不知情旁观者
An innocent bystander wearing the suit

图 6.1-2

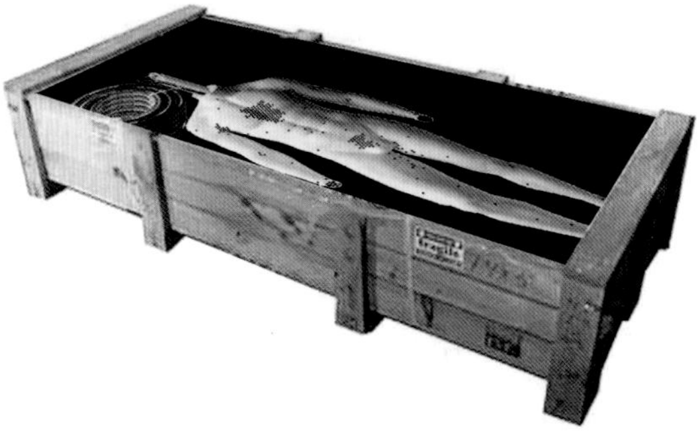

图 6.1-2 胡氏人体模型打包准备装运
The Hu-mannequin ready for shipping

图 6.1-3

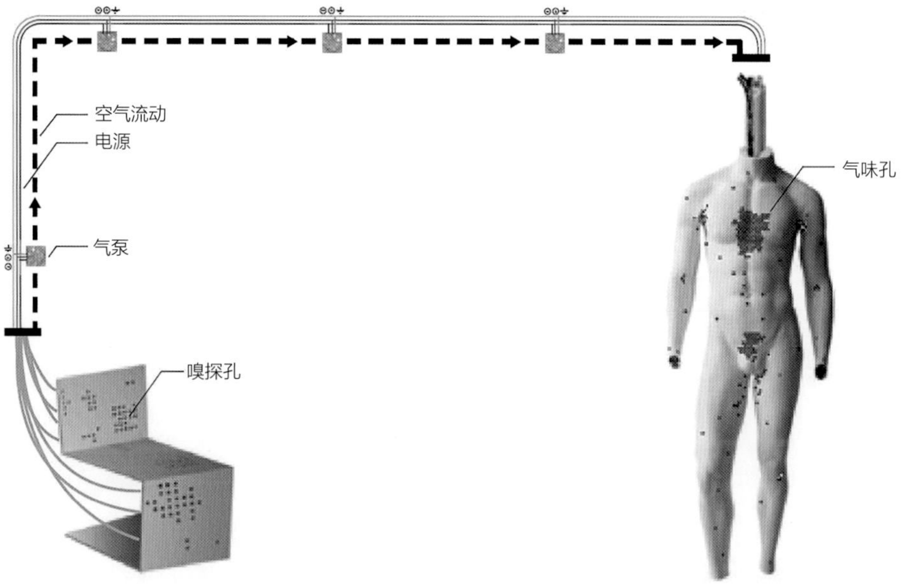

图 6.1-3 工作原理：椅子上的孔洞吸入空气，并通过管道传递到人体模型上，在相应的区域释放气味
How it works: air is sucked in through the perforations in the chair, relayed through ducts to the mannequin's body where scents are released in their corresponding areas

图 6.1-4 在纽约高线公园下的户外快闪店
Exterior of pop-up store underneath New York City's High-Line

图 6.1-5 不知道为什么，我就是喜欢这件套装
I don't know why, but I love this suit

图 6.1-4

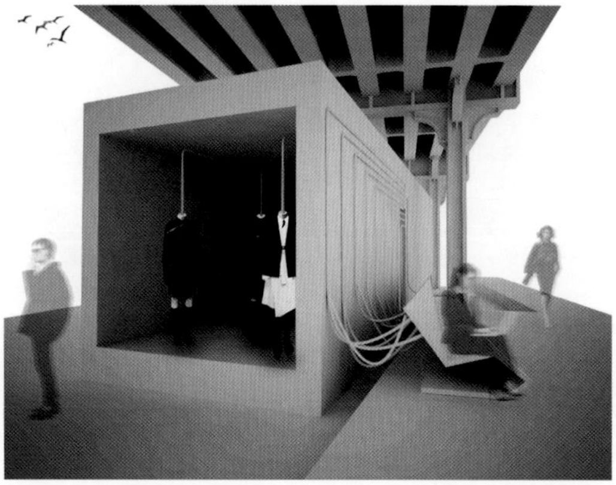

图 6.1-5

6.2 实时遮阳 Spot Shade

垂直遮阳是通过跟踪太阳的移动来调节光线。[1]

梅丽莎 · 刘

Vertical shades modulate the light by tracking the movement of the sun.[1]

Melissa Lui

“实时遮阳”是一个交互式遮光装置，利用寻光机制不断跟踪太阳的位置，通过调节和投射外部光线进入室内空间来提供理想的自然采光状况。

“实时遮阳”被安装在玻璃幕墙的后面，通过数字增强技术来调节自然光，为室内专门的阅读空间创造理想的采光条件。项目研究了为特定的室内空间位置进行专门定制，检验了利用数字技术为已占用的空间实现最终定制化空间体验的可能性。

装置原型由 5 个悬挂在玻璃幕墙后面雕刻出来的格栅构成。每个格栅都由一个与 Arduino 板相连的伺服电机单独控制。这个装置的程序设计是根据太阳的位置去旋转格栅，为特定的阅读点

‘Spot Shade’ is an interactive shading device that constantly tracks the location of sun in the sky with a light-seeking mechanism, regulating and projecting exterior light into the interior space to provide desirable, natural lighting conditions.

‘Spot Shade’ is installed behind a glass wall and is digitally enhanced to regulate natural light, creating desirable lighting conditions for a particular reading spot inside. Uniquely customised for a specific internal location and actuated if that location is occupied, the project examines the possibility of using digital technology to achieve the ultimate customisation of the spatial experience.

The prototype consists of five hanging sculptural bars installed behind a glass wall. Each bar is individually controlled by a servo motor connected to an Arduino board. The device is programmed so that, depending on the registered location of the sun, the bars are rotated to ward off the direct sunlight from a particular reading spot while letting

❶ 由鲁道夫 · 埃尔 – 库利与纳希德 · 纳比安指导。

❶ Guided by Rodolphe el-Khoury and Nashid Nabian.

in ambient light. Five light sensors are installed within the gap between the shading device and the glass wall. Computing the differential of light intensity detected by these sensors, the Arduino board pin-points the direction of the sunlight and actuates the shading device accordingly. To make the digitally enhanced experience a hyper-customised one, the device is activated only when the reading spot is actually occupied. A single chair placed at the reading spot is equipped with a pressure sensor that detects when someone is sitting in it. This information is communicated to the Arduino board to activate the shading device.

遮挡直射光线，同时保证周围的光线进入。在遮光装置与玻璃墙之间的间隙内安装了 5 个光传感器。通过计算这些传感器检测到的光强差值，Arduino 板精准定位阳光的方向，并相应启动遮光装置。为了使数字增强的体验成为“超定制”的体验，只有当阅读点真的被使用的时候遮阳装置才会被启动。阅读点放置一把装有压力传感器的椅子，当有人坐到椅子上的时候，该信息就被传送到 Arduino 板以启动遮光装置。

图 6.2-1

图 6.2-2

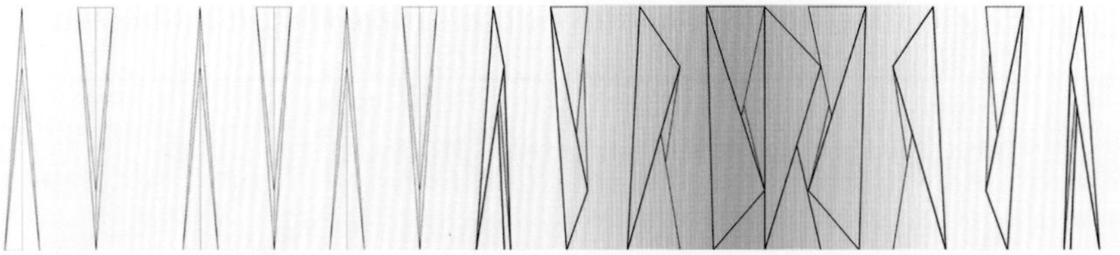

图 6.2-1 从室外观看遮阳系统
View of shading system from outside

图 6.2-2 局部阳光直射受阻示意
Diagram showing local blocking of direct sunlight

图 6.2-3

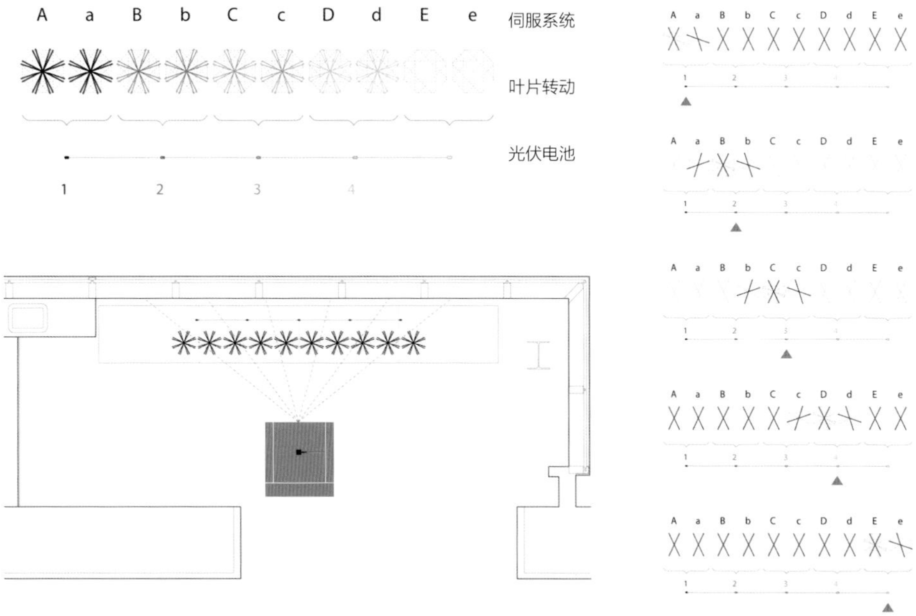

图 6.2-3 丹尼尔斯建筑与景观设计学院图书馆阅览室平面图，以及屏幕激活序列
Floor plan of reading alcove in the library at the Daniels Faculty of Architecture, Landscape and Design and sequence for screen activation

图 6.2-4

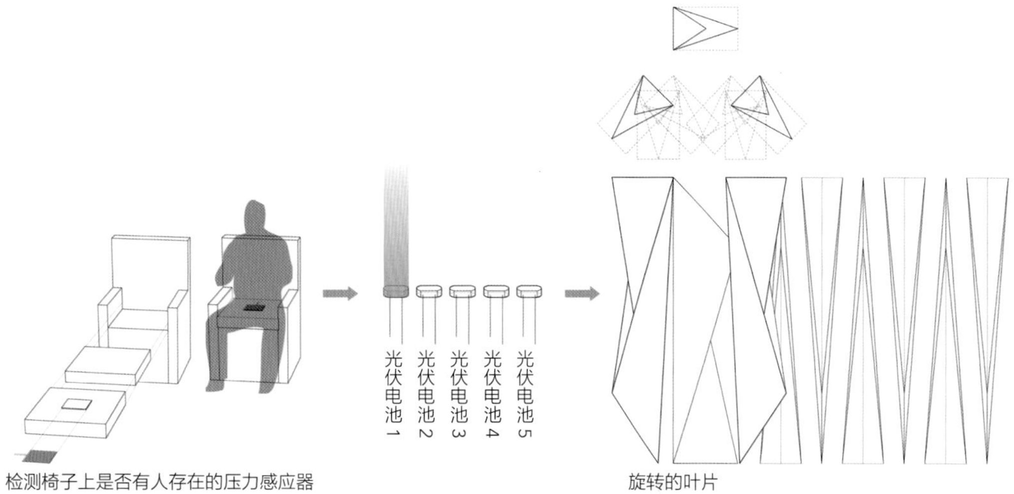

图 6.2-4 从建筑外部看屏幕
Screen as viewed from exterior of the building

图 6.2-5

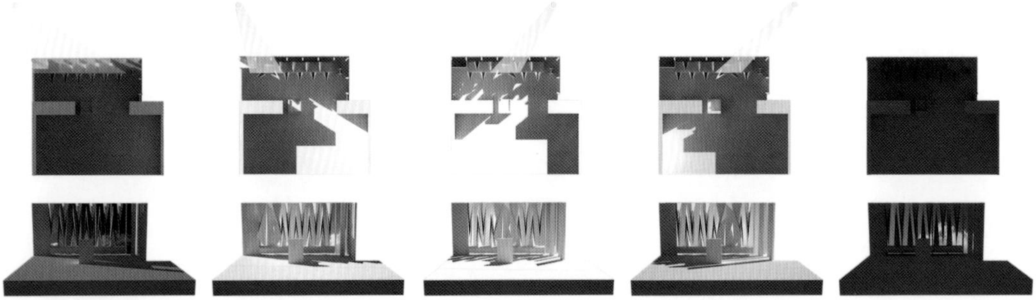

图 6.2-6

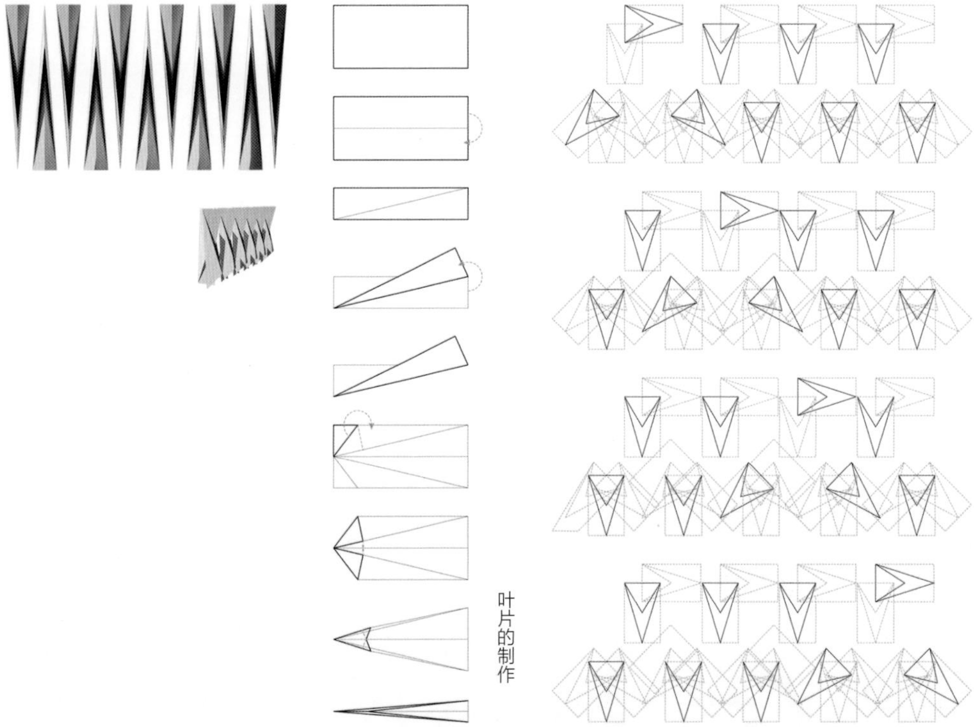

图 6.2-5 叶片从早上六点到下午六点之间 12 个小时的运动情况
Movement of fins over the course of a 12 hour period from 6am to 6pm

图 6.2-6 阴影移动的平面图
Plan view of rotating shades

6.3 光晕室 Auratic Chamber

> **该环境用灯光和颜色来对人们的心理状态进行拓展。❶**
>
> 吉米·陈
>
> **This environment uses light and colour to become an extension of one's Psychological state.❶**
>
> Jimmy Tran

“光晕室”营造了一种并非通常意义上的墙、地板或顶棚的空间体验，相反是用光线根据住户的心理状态来营造这种体验。

房间的照明与住户穿戴的神经脉冲传感器检测到的生理和心理状态产生互动作用。房间根据人们的行为和性情产生光晕，住户可以有意识地与光晕互动，把它作为身体或者精神的拓展。

当越来越多的人进入房间，每个人都创造了自己的光晕，光晕随着他们穿越空间，提供了一种全新形式的生物反馈和自我表达。光晕和色彩成了一种新的身体语言形式，赋予每个人更高的自我意识，将每个人的情绪和心理状态展示给其他人。

‘Auratic Chamber’ creates a variety of spatial experiences without the usual arrangements of walls, floors or ceilings. Instead, light is used to create experiences which vary depending on the inhabitant’s psychological state.

The lighting of the room interacts with the inhabitant’s physiological and psychological state-measured by a neural impulse sensor worn by the inhabitant. The room creates a halo of light according to the person’s behaviour and temperament. The inhabitant can consciously interact with this halo, treating it as an extra limb of his or her body, or an extension of his or her psyche.

As more people enter the room, each person creates their own halo. Each person’s halo follows them as they traverse the space, providing a new form of biofeedback and self expression. The halo of light and colour becomes a new form of body language, giving each person a heightened state of self-awareness, and sending visual cues to others about each person’s emotional and psychological state.

---

❶ 由克里斯托·马尔科普洛斯监制。

❶ Supervised by Christos Marcopoulos.

图 6.3-1

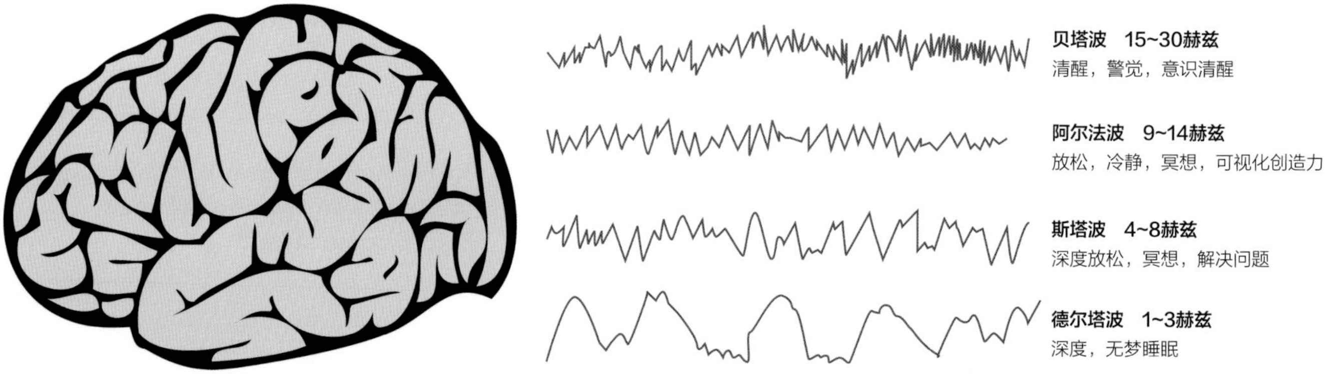

图 6.3-2

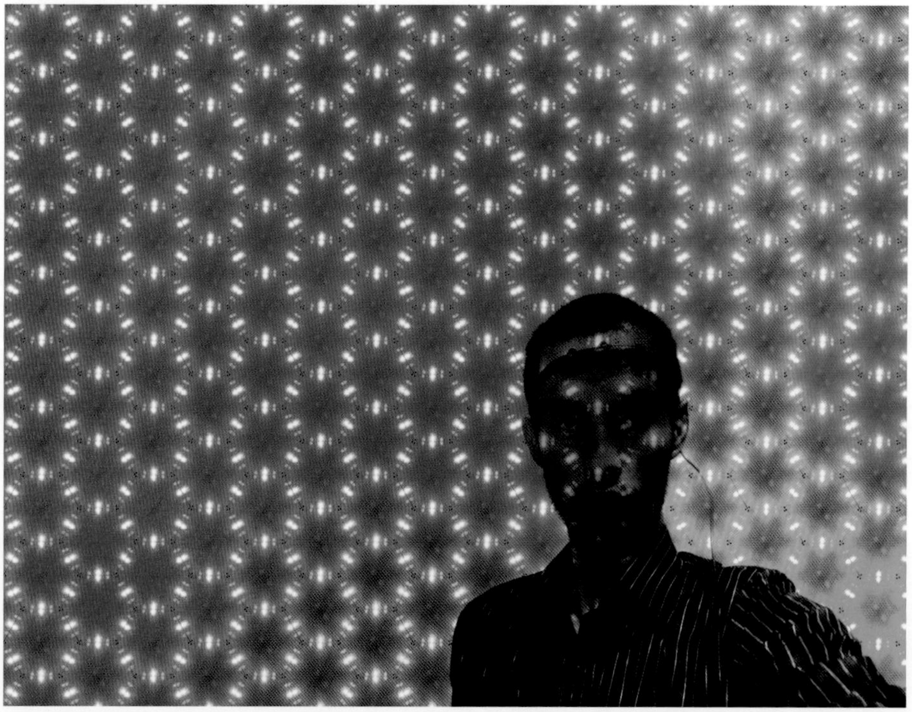

图 6.3-1 神经脉冲
Neural impulses

图 6.3-2 光晕室由排列在空间内表面的联锁砖组成
The Auratic Chamber consists of interlocking tiles that line the interior surfaces of a space

图 6.3-3 这组图片显示了不同社会环境下的空间状态，有些人照亮了房间，而另一些人则起到了相反的效果
This sequence shows the space under different social circumstances. Some people light up a room, while others have the opposite effect

图 6.3-3

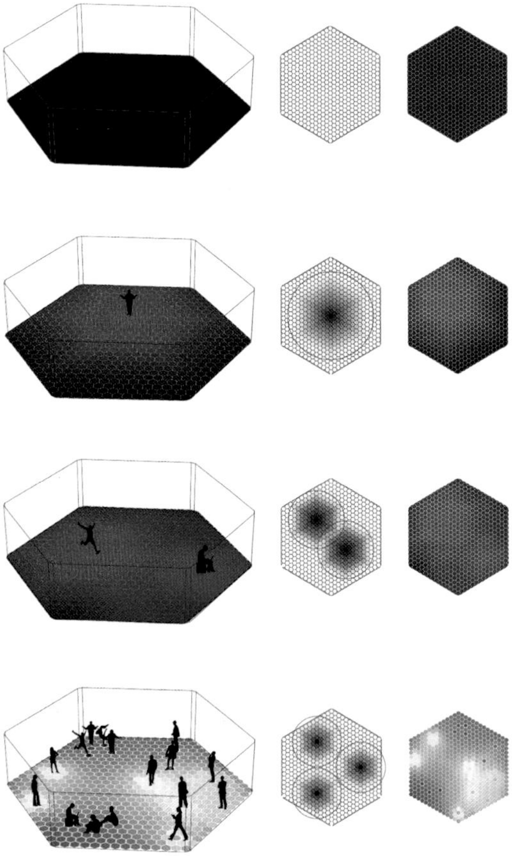

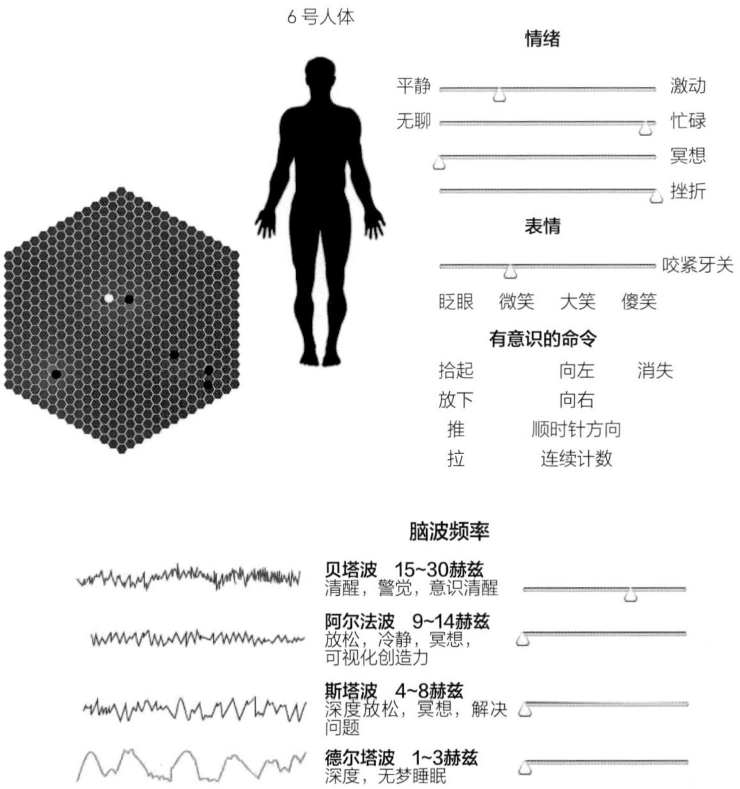

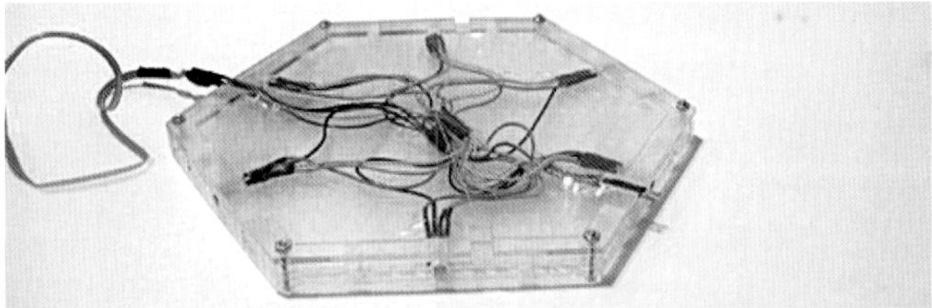

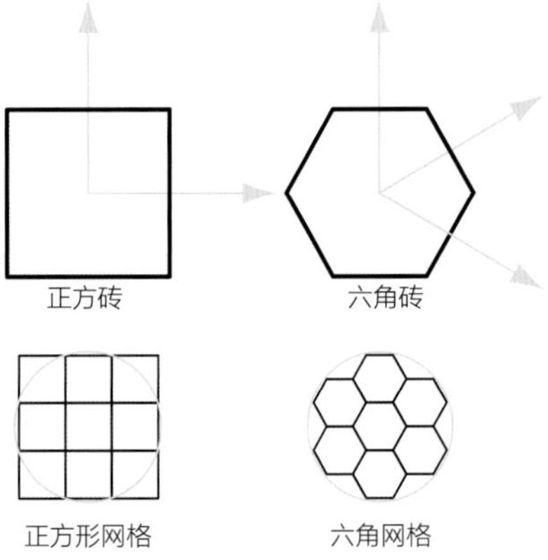

六角砖提供砖与径向网格之间的 3 个轴向的运动

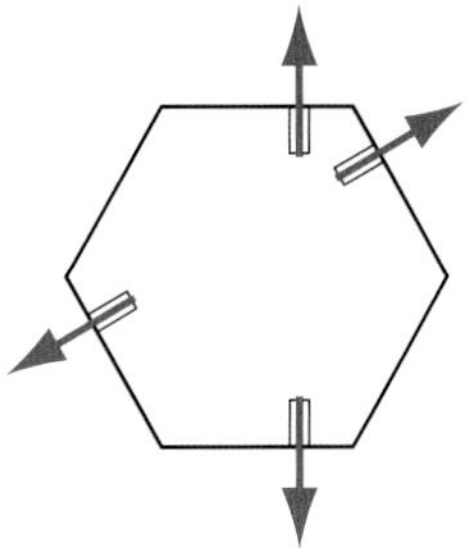

六角砖用 4 个连接点来组装六角网格

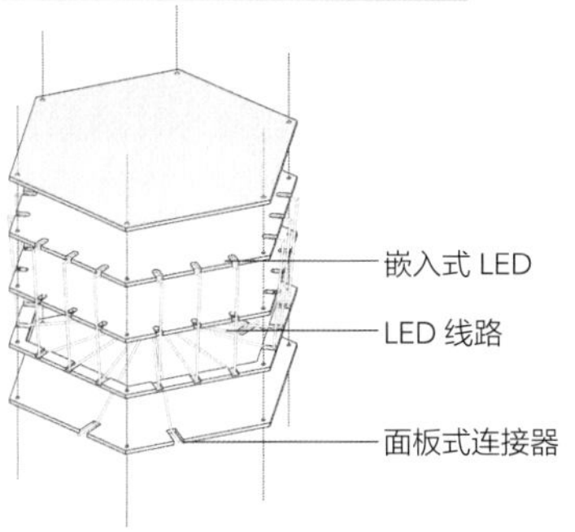

图 6.3-4

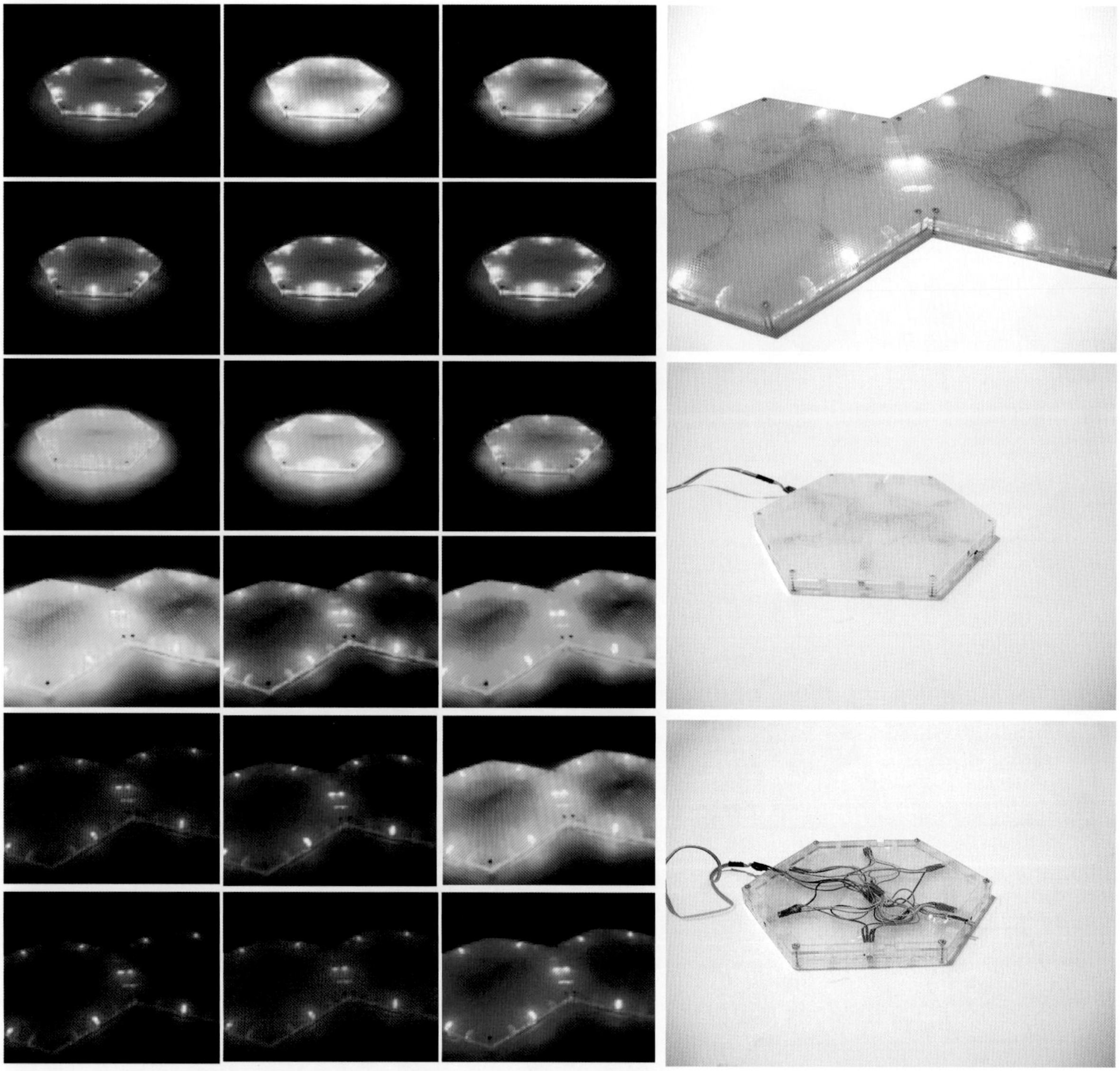

图 6.3-4 六角砖单体和联锁砖的原型
Prototypes of individual and interlocking tiles

6.4 气味筛选器 Scentisizer

气味引擎和可触界面是气味柱的重要组成部分和执行组件。

KLF 事务所，纳希德 · 纳比安

Scent Engine and Tangible Interface for Composing and Performing Scentscapes.

Khoury Levit Fong with Nashid Nabian

“气味筛选器”通过对动态构成与扩散的气味进行精确控制来进行复杂气味的调配和传递。

一个有形界面结合触觉和图形特征，管理一个阵列的 64 个气味分配器，描绘了一个广泛的嗅觉范围。用户可以像操作调音器一样独立控制气味箱，以直接和直观的方式去创作和调校气味。也可以通过数字图形用户界面用更复杂和高精度的方法来控制气味箱，让它能够以多种方式来排列和调制气味的基调和动态。模拟控制以触觉装置的形式提供可触摸及可视化的嗅觉体验，与此同时，在数码设备上添加的实用功能和网络连接功能，让“气味筛选器”成为实体的或可无线访问的数字化气味商场。

利用触觉界面，用户可以在规定的程序段操作一系列的调配器来控制香气成份的表现方式。

‘Scentisizer’ allows for the orchestration and delivery of complex fragrances by giving precise control over the constitution and diffusion of scent accords in dynamic compositions.

A tangible interface combines haptic and graphic features for managing an array of sixty-four scent dispensers that maps a wide-ranging olfactory field. Users can manipulate individual scent containers, much like organ stops, to compose and time an accord in a direct and intuitive fashion. They can also control them with greater complexity and precision by means of a digital graphic user interface that allows for multi-channelled sequencing and modulation of scent tones and dynamics. The analogue controls provide the haptic means for tangibly sculpting and visualising olfactory phenomena while the digital features add functionality and web connectivity, allowing Scentisizer to perform digitised olfactory scores stored locally or accessed wirelessly.

With the haptic interface users can control the behaviour of their scent-compositions over a prescribed period – programmed by manipulating the array of dispensers. The slightest elevation from the default low position activates a dispenser while varying elevations trigger different pre-programmed diffusion cycles. Once the cycle is launched, the motorised tubes are automatically animated, sinking gradually into their default rest position while transcoding the status of the unfolding olfactory phenomenon into a topographic image at any given time to build a tangible and instantly graspable 3-d diagram of the evolving scent structure.

从默认的低位以最细微的高度激活分配器，随着高度的变化触动各个阶段的程序去展开不同的循环周期。循环周期一旦启动，机动试管就会自动激活，逐渐下沉到默认的静止位置，而嗅觉现象则将在特定的时间将气味体系的演化展现为一个直观的三维演示图。

The graphic user interface enables greater control over the scent-dispensing array, providing user-adjustable parameters for shaping the dynamics of the individual constituents of the fragrance and the overall effect over time.

图形用户界面增强了对于气味调配系统的控制，通过可调参数去调整随着时间变化的香气成份的动态变化和整体效果。

图 6.4-1

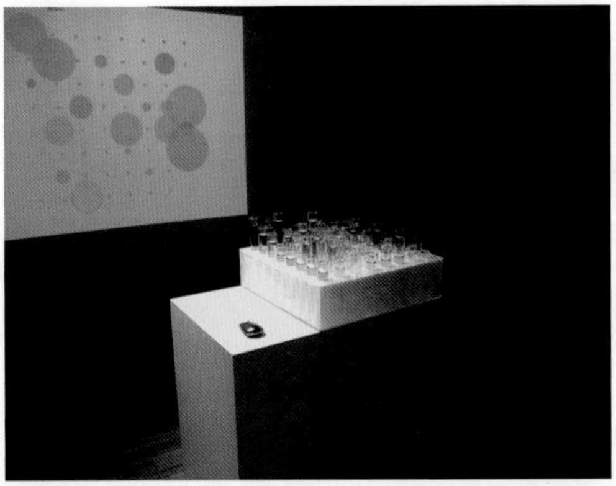

图 6.4-2

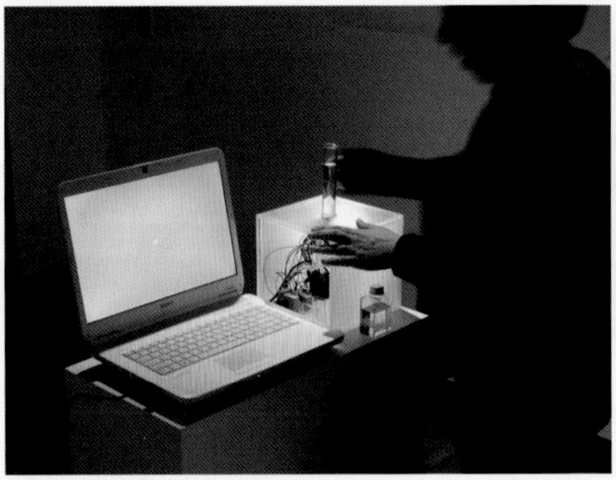

图 6.4-1 位于多伦多的气味筛选器在执行“渗透”模式
Scentisizer in “perform” mode at InterAccess, Toronto

图 6.4-2 触感用户界面及其原型
Demonstration and prototype of haptic user interface

图 6.4-3

图 6.4-3 气味图谱。根据香调轮的变化，每个调配器被分配了一个气味，这是由香水行业开发的一种分类方法，对不同但有亲缘关系的香味进行了用户友好型分类，如花香、清香、东方调、木香和馥奇香调

Scent mapping. The scents assigned to each dispenser are mapped according to a variation on the Fragrance Wheel, a taxonomy developed by the fragrance industry for a user-friendly classification of scents in distinct but related families: Floral, Fresh, Oriental, Woody, and Aromatic Fougère

图 6.4-4

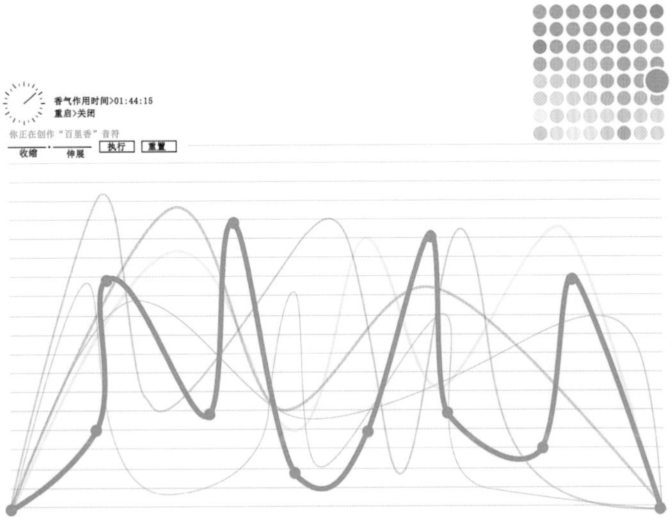

图 6.4-6

图 6.4-5

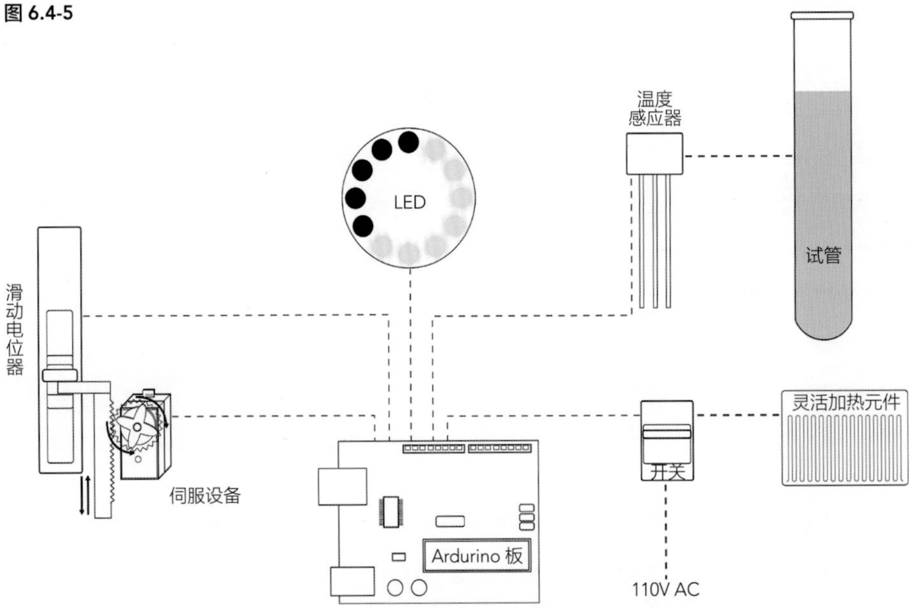

图 6.4-4 用于创作和执行的图形用户界面
Graphic user interface for composing and performing scentsapes

图 6.4-5 触觉用户界面的电路图
Circuit diagram of haptic user interface

图 6.4-6 64 个气味分配器阵列
An array of sixty-four scent dispensers

6.5 变色龙休息厅 Chameleon Lounge

安装有传感器的椅子模仿使用者服装的颜色，还可以模仿与他们接触的物体的色彩。

KLF 事务所，纳希德 · 纳比安

Sensor equipped chairs imitate the colour of users' clothing and objects with which they come in contact.

Khoury Levit Fong with Nashid Nabian

“变色龙休息厅”是一个由含有颜色响应增强器的塑模制椅子组成的集合。嵌入式技术使得每把椅子能够识别使用者的颜色，并且像变色龙一样变换自己的色彩。

“变色龙休息厅”属于利用信息技术来增强人体感官及其时空表征作为个人的社交轨迹的建筑联盟成员——KLF 事务所的一系列研究项目的一部分。它探讨了通过适度的技术措施来体现环境对于主体行为和特性的密切且精准的响应。个性化的连通性从表面上看放大了主体的魅力光环，从而扩大了其公众影响范围。

电子器件由色彩传感器、Arduino 微控制器以及高亮度三原色 LED 光源组成。Arduino 微控制器可以根据不同的场景进行编程。在新奥尔良，当椅子不使用时，LED 灯光会渐渐变暗，并且嵌

‘Chameleon Lounge’ is a collection of plastic moulded chairs that have been electronically enhanced to respond to colour. Embedded technology allows each chair to pick up the colour of the person who uses it and to imitate it in a chameleon-like fashion.

‘Chameleon Lounge’ belongs to a series of projects by Khoury Levit Fong which aligns architecture with information technology to empower the sensate body and its spatio-temporal envelope as the intimate locus of social connection. It explores, in a modest application, the capacity for environments to respond precisely and intimately to a subject’s actions and characteristics. The individuated connectedness seemingly amplifies the subject’s charismatic aura, thus expanding his or her sphere of influence in public.

The electronics are comprised of a colour sensor, an Arduino microcontroller and high-intensity RGB LEDs. The Arduino is programmable for a variety of scenarios. In the New Orleans version, the LEDs gradually turn off when the chair is not in use and the embedded sensor does not pick up a new colour. In Casablanca, the LEDs are programmed to cycle through a sequence of colours when the chairs are unoccupied, simulating spontaneous swarm behaviour when left undisturbed.

入式传感器也不会识别新颜色。在卡萨布兰卡，当椅子无人使用时，LED 被编程设计为通过一系列的颜色循环，模拟不受干扰时自发的群体行为。

图 6.5-1

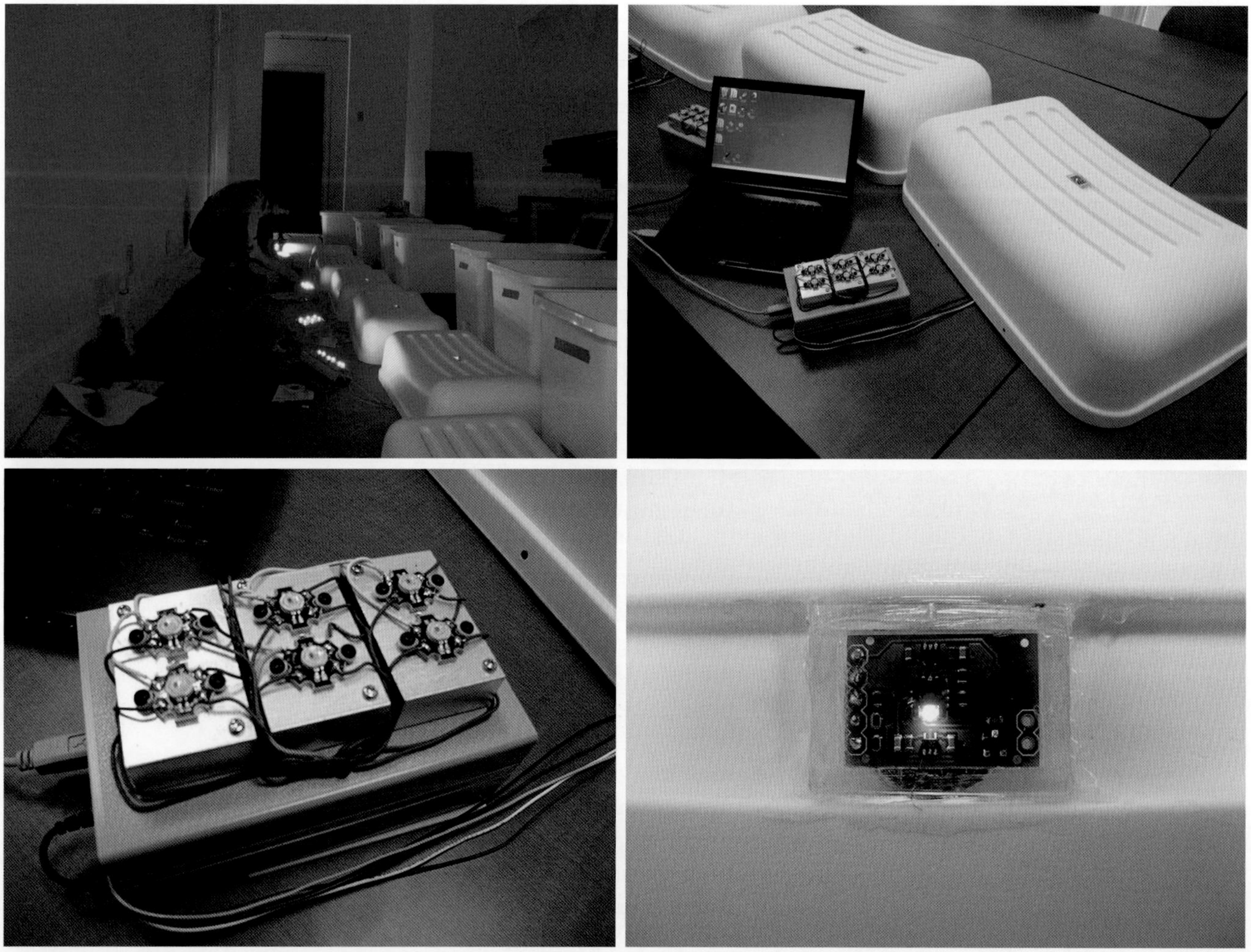

图 6.5-1 电子元件集成到从宜家采购的凳子上
Electronic components integrated into IKEA sourced stool

图 6.5-2

图 6.5-3

图 6.5-2 在新奥尔良 2008 年的友邦保险节上的装置
Installation in operation at New Orleans AIA DesCours Festival 2008

图 6.5-3 在 2009 年卡萨布兰卡艺术节上的装置
Installation in operation at Casablanca art festival in 2009

6.6 空气中的建筑 The Architecture of Air

反思建筑入口。

奈格尔·桑耶德菲斯

Rethinking the architectural threshold.

Negar Seyedfathi

Exploring the architectural interventions of using air as an invisible environmental separator at a monumental scale of an outdoor urban plaza allows for an emergence of a new typology: a ‘conditioned’ outdoor public room. In this typology, air curtains located on the periphery of the site form a microclimate where all the dividing physical barriers between inside and outside, between the conditioned and unconditioned are dissolved. Also, the floor tiles as the only active surface on the urban square create dynamically distinguished microclimatic zones expressed in gradient threshold conditions similar to the primal environment of the campfire.

Although psychologically internalized, the square is left as an agora, a forum and a theatre for the city, an environmentally-controlled landscape of social interactions that attracts various activities all year round.

在一个具有纪念意义的城市户外广场，用空气这种介质作为一种无形的环境分离手段是新出现的一种探索建筑介入的类型：一个“有条件的”户外公共空间。在这种类型中，位于场地周围的空气幕形成了一个微气候，所有内部和外部、有条件和无条件之间的物理分隔都消失了。同样，地面铺装作为城市广场唯一的活力表面，创造了一个动态、可识别的微气候区域，类似于以篝火的原始环境的梯度阈值条件来表达。

虽然广场在人们的心理上被内化了，但它仍然是城市的集会场所、辩论场所和城市剧院，是一个一年四季都能吸引各种活动的开展社会交往，并进行环境调节的景观。

图 6.6-1

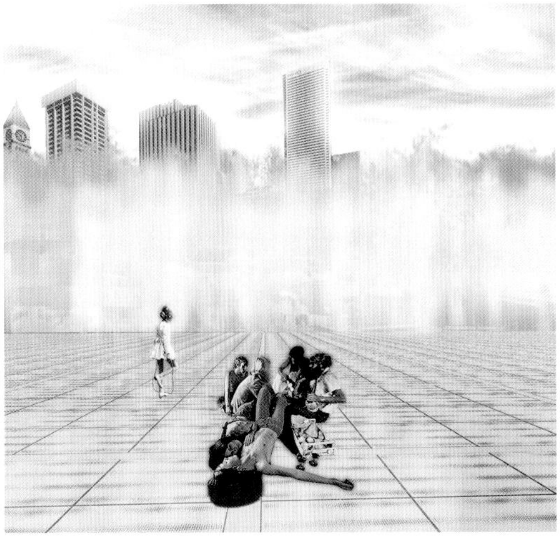

图 6.6-1 概念效果图
Render-Concept

图 6.6-2 人体热力学：除了辐射温度、相关湿度和风速外，空气温度是影响人类舒适度的最主要因素。通过控制这些因素，许多在特定季节里无法开展的户外活动都变得可能
Thermodynamics of human body：air temperature is the most dominant environmental factor in human comfort followed by radiant temperature, relative humidity and air speed. By controlling these factors, outdoor activities that are particularly impossible during certain seasons are expanded

图 6.6-2

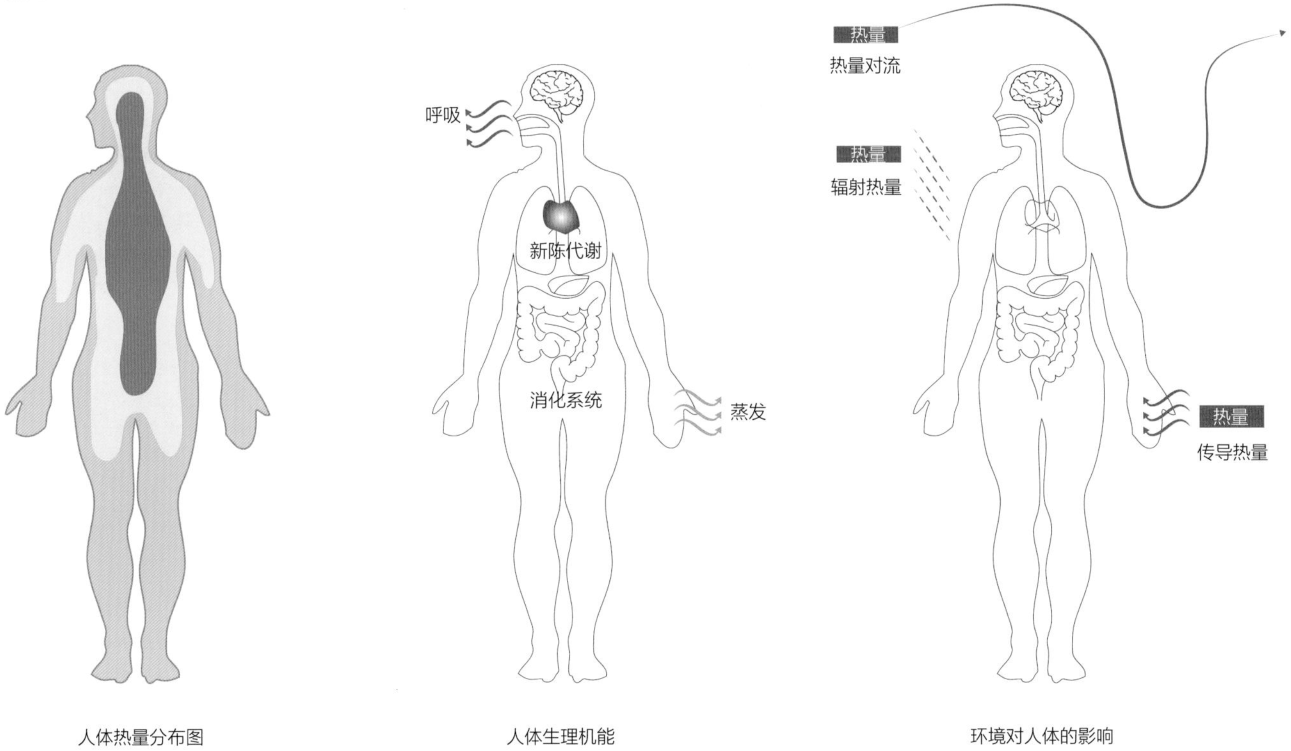

图 6.6-3

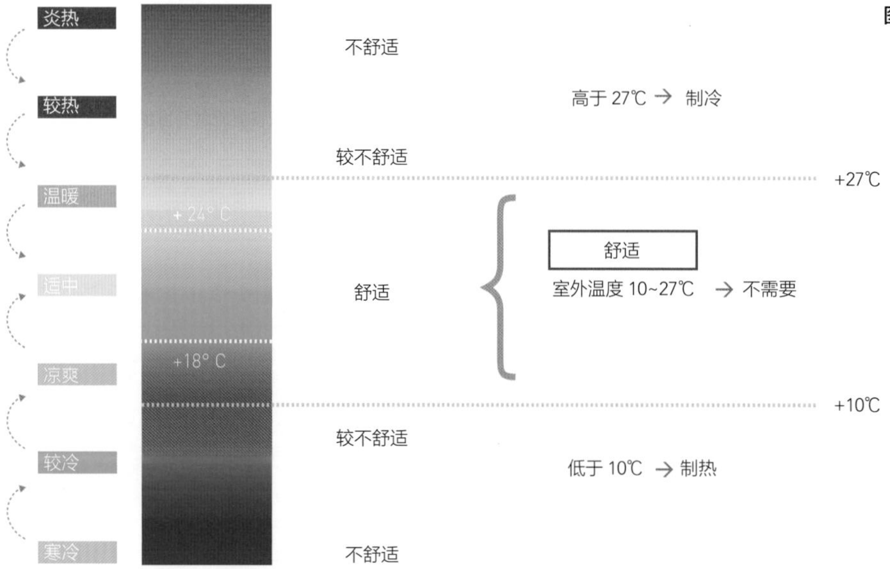

图 6.6-3 热舒适范围：拟建的系统可以将温度从一个范围调节到下一个范围。如果室外凉爽，系统可以将人体的热感觉从凉爽调节到适中。然而，如果室外非常冷，系统将轻微地调节温度，将温度从冰点提高到寒冷，而不是剧烈地将温度从冰点改变到适中

Thermal comfort range: the proposed system takes the temperature from one range to the next. For instance, if it is cool outside the system changes the body's thermal sensation from cool to neutral. However, if it is freezing outside, the proposed system would slightly moderate the temperature and raise it from freezing to cold instead of drastically changing it from freezing to natural

图 6.6-4

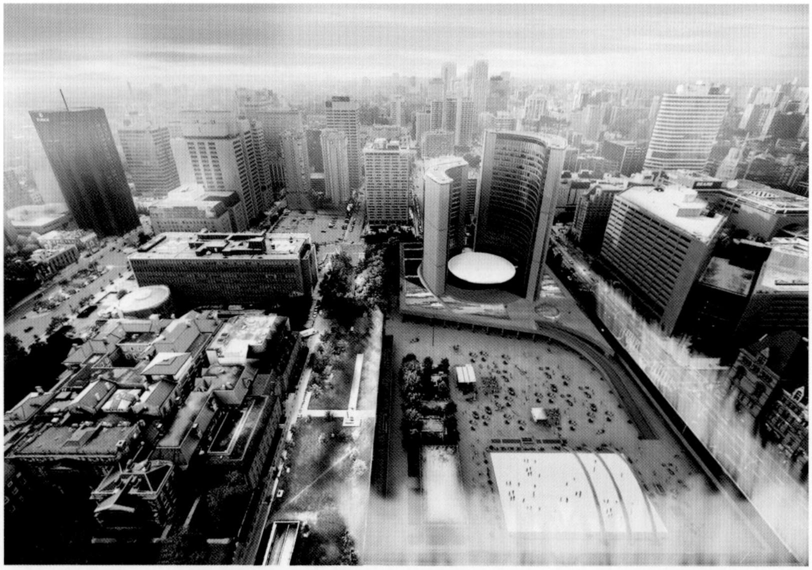

图 6.6-4 冬季架空效果图：营造一个热舒适度适宜的户外公共空间，以吸引全年各种不同活动

Render- winter- aerial: a thermally comfortable outdoor public room is created that attracts various activities all year round

图 6.6-5

图 6.6-7

图 6.6-6

图 6.6-5 典型冬日白天效果图
Render-Typical cold winter day

图 6.6-6 典型夏日白天效果图
Render-Typical hot summer day

图 6.6-7 入口效果图：添加雾层，将气幕转换为交互表面
Render-Threshold condition：adding a mist layer transforms the air curtain to an interactive surface

图 6.6-8

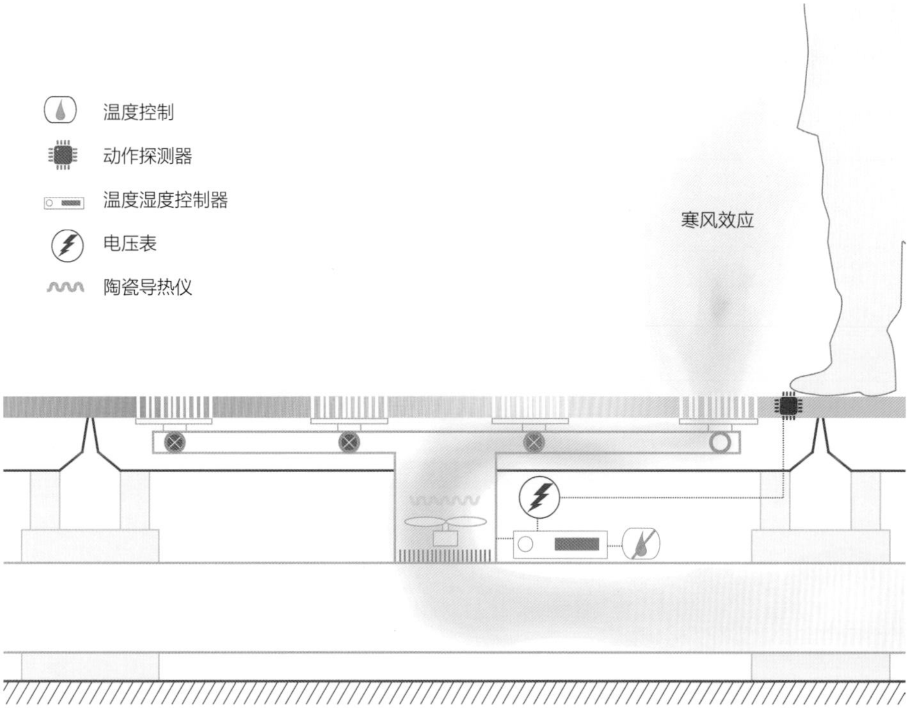

图 6.6-8 夏季瓷砖制冷系统：通过使用电扇这种简单的方式，整个系统可以制造出一种“寒风效应”，使空气更容易蒸发皮肤上的汗水，从而降温。当有人靠近的时候，传感器就会激活瓷砖。风扇的速度和湿度可以被一个能实时检测外部空气状况的控制器所控制

Tile cooling system in Summer: simply by using fan, the system creates 'wind chill effect' making it easier for the air to evaporate sweat from the skin and eliminating body heat. When approached by a person, the sensors activate the tile. The speed of the fans and the level of humidity are controlled by a controller that constantly monitors the outside air conditions

图 6.6-9

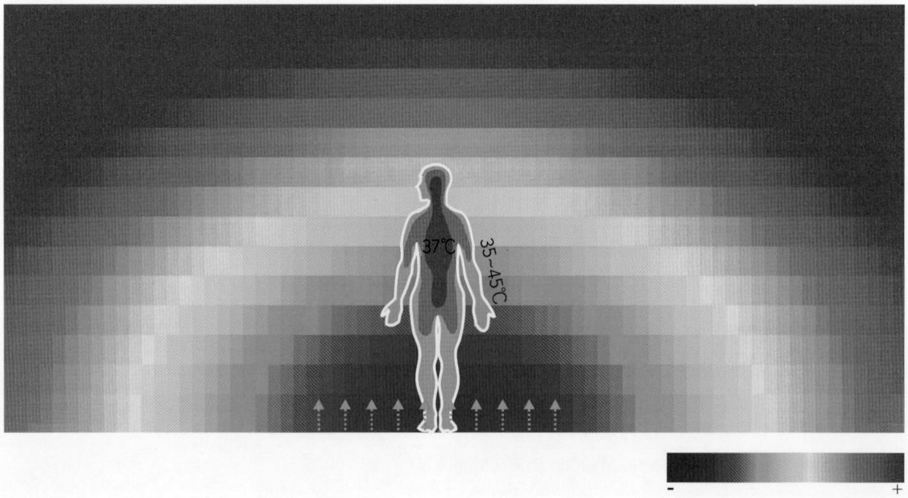

图 6.6-9 夏季热量图：当夏季冷空气消散时，有空调区域的热量图

Summer – thermograph: thermal map of the conditioned zone when cold air is dissipated in summer

图 6.6-10 冬季瓷砖制热系统：同样的风扇系统可配备陶瓷加热元件，可以在热量消散之前制热空气。在寒冷天气时，当有人靠近瓷砖时，加热元件根据恒温器监控的外部温度开始制热。该系统主要通过对流制热和制热瓦的辐射制热来调节环境
Tile Heating system in Winter: the same fan system is equipped with ceramic heating elements that warm the air before being dissipated. On a cold day, when the tile is approached by a person, the heating element starts to get warmer depending on external temperature being monitored by a thermostat. The system mainly conditions the environment through convective heating as well as some radiation heating from the heated tiles

图 6.6-10

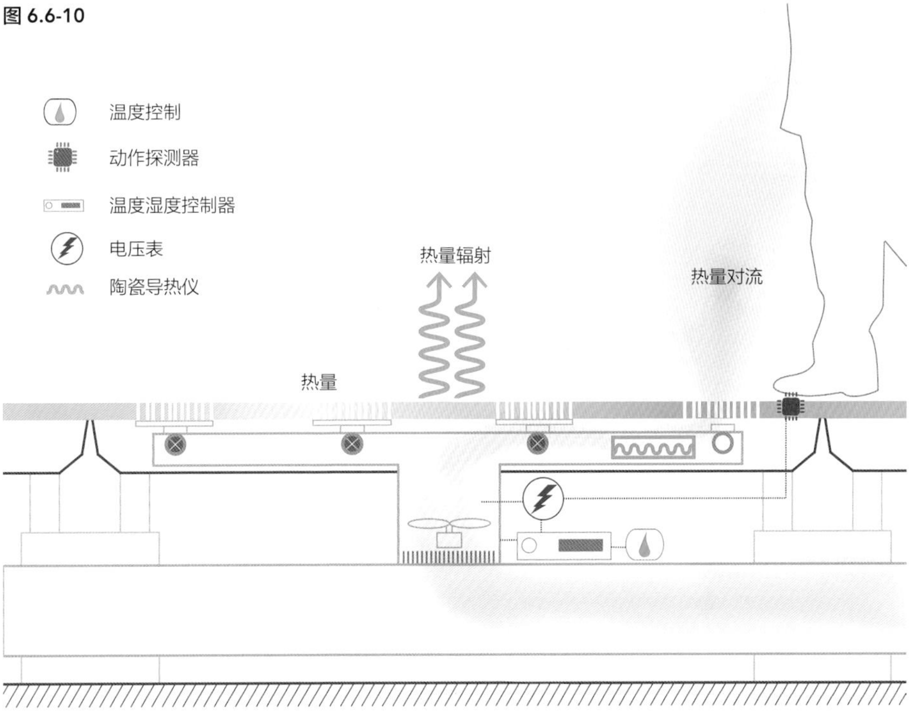

图 6.6-11 冬季热量图：当冬季热空气消散时，有空调区域的热量图
Winter – thermograph: thermal map of the conditioned zone when hot air is dissipated in winter

图 6.6-11

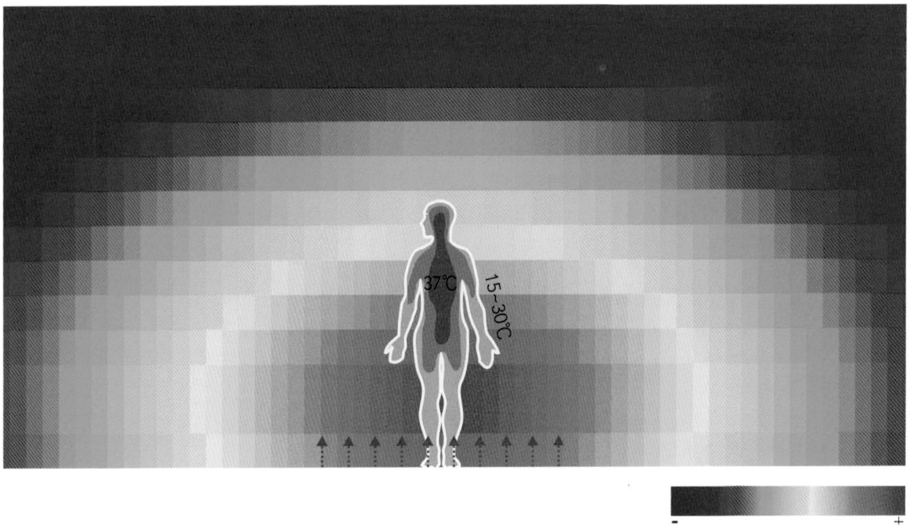

图 6.6-12

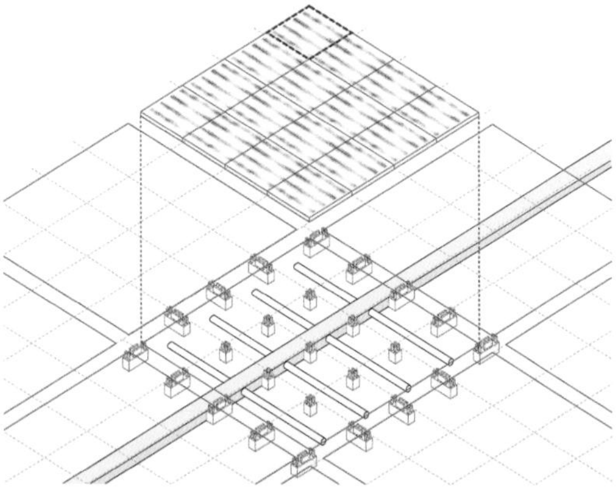

图 6.6-13

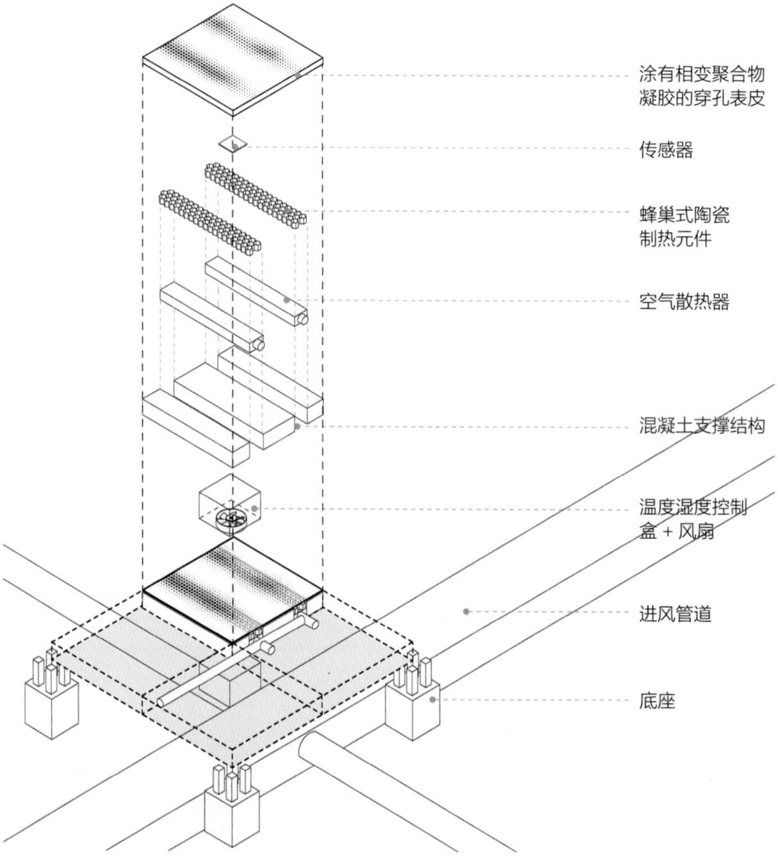

图 6.6-12 瓷砖构造：所有调节环境所需的技术都被置于瓷砖下面的隐蔽空间，所以并不会对行人的正常行走造成障碍

Tile infrastructure: all the required technology for conditioning the environment is located in the plenum space under the tiles hidden from the view so that it does not add any physical obstacle interfering with the pedestrian movements

图 6.6-13 瓷砖细部：瓷砖由一个风扇、温度湿度控制盒、混凝土支撑结构、空气散热器、陶瓷制热元件和传感器组成。调节后的空气从顶部涂有相变聚合物凝胶的穿孔表面被排出

Tile detail: the tiles are composed of a fan, temperature and humidity control box, concrete structural support, air diffusers, ceramic heating elements, and a sensor. The conditioned air is dissipated from a perforated surface on the top coated with phase-change polymer gel-filled layer

图 6.6-14

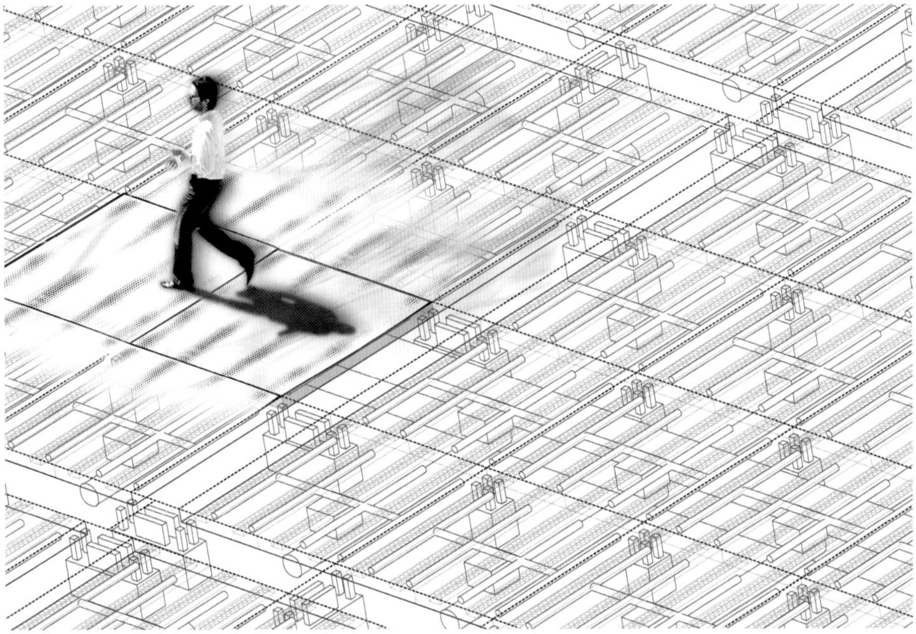

图 6.6-14 瓷砖系统效果图：该系统以一种优化的方式工作，而不是调节整个广场的面积。嵌在地砖里的传感器可以探测到人的位置，因此它只对人所在的位置进行调节。每一块瓷砖独立工作，且可调节各自的温度，从而创造一个更高效迅速的系统

Render-Tile system：instead of conditioning the entire area of the square, the system works in an optimized way. The sensors embedded in the tiles detect the location of the people so that it only conditions where it is occupied. Each tile work independently and regulate its own temperature creating a more efficient and much faster system

图 6.6-15

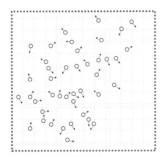

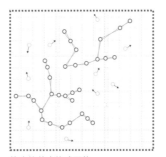

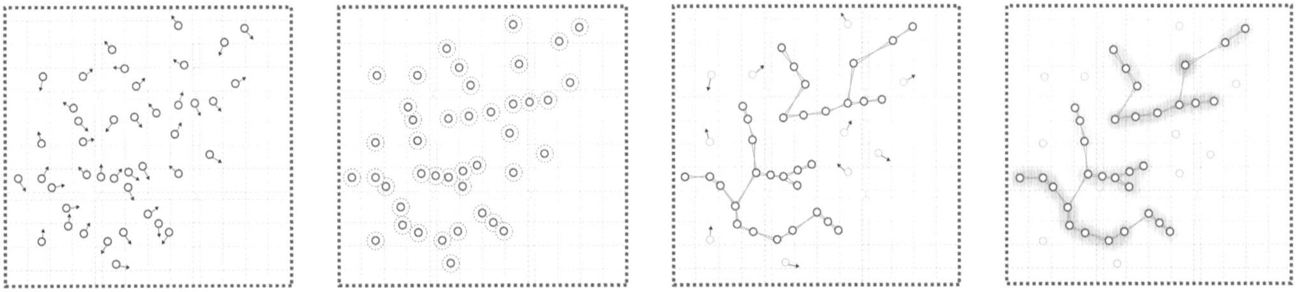

人们在广场上随意漫步

每个人都被嵌入地砖中的传感器作为节点检测到

静止的节点构成网络

温度变化的集群是根据节点的数量和它们在一个地方停留的时间长短而形成的

图 6.6-15 聚集系统

Tile aggregation system

作者简介
About the Authors

Rodolphe el-Khoury

Rodolphe el-Khoury is Dean of the University of Miami School of Architecture. Trained as both a historian and a designer, he divides his time between scholarship, research, and practice. He is the author of numerous books on architecture and urbanism, including See Through Ledoux; Architecture Theatre, and the Pursuit of Transparency; Monolithic Architecture; and Figures: Essays on Contemporary Architecture. el-Khoury's leads RAD-UM, a research lab for embedded technology and robotics aiming at enhancing responsiveness and resilience in buildings and smart cities. The work of his firm, Khoury Levit Fong (KLF), has won international awards.

el-Khoury's work has been featured in national and international media outlets that cut across disciplines ranging from WIRED Magazine to the Wall Street Journal to the Space Channel to BBC World. He has shared his work through teaching, visiting professorships and lectures at dozens of institutions in the U.S. and abroad.

鲁道夫 · 埃尔 – 库利

鲁道夫 · 埃尔 – 库利是美国迈阿密大学建筑学院院长。作为一名历史学家和设计师，他将主要精力放在学术、研究和实践上。他已经完成了许多关于建筑和城市主义的书，包括《透视勒杜》《建筑剧场》《追求透明性》《整体建构》《人物：当代建筑随笔》。库利主持的 RAD–UM 是一家研究嵌入式技术和机器人的实验室，旨在提高建筑和智能城市的感应式和弹性能力。其公司 KFL 事务所的作品获得了不少国际性奖项。

库利的作品在美国与国际媒体上都有特别报道，包括从《连线》杂志到《华尔街日报》，再到《太空频道》《BBC World》的各个领域的媒体。他通过教学、担任客座教授和在美国及海外的几十家机构的讲座分享了自己的研究与实践工作。

卡罗尔·穆海贝尔

卡罗尔·穆海贝尔是多伦多大学丹尼尔斯建筑、景观和设计学院的副教授和城市设计主任。她也是 NMinusOne 工作室的联合创始人和合伙人。在家庭尺度上，她的工作重心是将家庭作为一个沉浸式的环境—— 一个能够产生新的身体和情感体验的环境——通过增强或重新校准基础设施与方案。她是《狂野城市主义，重新设计加利福尼亚州》(CCA 2006) 的联合编辑。她的设计作品被旧金山现代艺术博物馆收藏，并在学术与主流媒体广泛发表自己的设计作品，包括《纽约时报》杂志和《建筑实践杂志》。她是加拿大创新基金会的资助人，为新成立的 RAD 提供支持。NMinusOne 工作室已经在 2012 年入选纽约建筑联盟“新兴之声”系列讲座。

Carol Moukheiber

Carol Moukheiber is Assistant Professor and Director of Urban Design at the University of Toronto's Daniels Faculty of Architecture, Landscape, and Design. She is also co-founder and partner in the architecture practice Studio NMinusOne (n-1). At the domestic scale, her work has focused on the home as an immersive environment — one capable of generating new physical and emotional experiences — through the enhancement or recalibration of its infrastructure, or set of programmes. She is the co-editor of Wild Wild Urbanism, Redesigning California (CCA 2006). Her design work has been acquired by the San Francisco Museum of Modern Art, and published widely in academic and mainstream media, including The New York Times Magazine and Praxis Journal of Architecture. She is the recipient of a Canada Foundation for Innovation grant providing support for the newly launched RAD. Studio NMinusOne has been selected by the New York Architectural League for their 'Emerging Voices' lecture series, 2012.

克里斯托·马尔科普洛斯

克里斯托·马尔科普洛斯是多伦多大学丹尼尔斯建筑、景观和设计学院和 NMinusOne 工作室的副教授。马尔科普洛斯曾经在由雷姆·库哈斯、斯基德莫尔、奥文斯和美林在旧金山主持的 OMA 事务所 (鹿特丹大都会建筑事务所) 工作过。他的技术专长在 NMinusOne 工作室的项目开发中发挥了重要作用。他有关家庭环境的作品已被旧金山现代艺术博物馆收藏。他还是《狂野城市主义，重新设计加利福尼亚州》(CCA 2006) 的联合编辑。他的建筑作品被广泛发表在学术和主流媒体上，包括《纽约时报》杂志和《建筑实践杂志》。NMinusOne 工作室已经在 2012 年入选纽约建筑联盟“新兴之声”系列讲座。

Christos Marcopoulos

Christos Marcopoulos is Assistant Professor at the University of Toronto's Daniels Faculty of Architecture, Landscape, and practice Studio NMinusOne (n-1). Marcopoulos has extensive professional working experience having worked in the offices of OMA, (Office for Metropolitan Architecture, Rotterdam), led by Rem Koolhaas, and Skidmore, Owings and Merrill, San Francisco. His technical expertise has been instrumental in the development of Studio NMinusOne's projects. His work on the domestic environment has been acquired by the San Francisco Museum of Modern Art. He is the co-editor of Wild Wild Urbanism, Redesigning California (CCA 2006). His built work has been published widely in academic and mainstream media, including The New York Times Magazine and Praxis Journal of Architecture. Studio NMinusOne has been selected by the New York Architectural League for their 'Emerging Voices' lecture series, 2012.

作品目录
List of Projects

Studio NMinusOne Projects Information:
NMinusOne 工作室项目信息：

–
“IM 感应毯”
2011 年 / 产品原型
展　览：《缝线》，拉里 · 韦恩 · 理查兹策展，多伦多工作室画廊
负责人：克里斯托 · 马尔科普洛斯、卡罗尔 · 穆海贝尔与鲁道夫 · 埃尔 – 库利
团　队：瓦伦蒂娜 · 梅勒、塞巴斯蒂安 · 萨瓦内、乔纳 · 罗斯 – 马尔斯、叶平西、萨马尔 · 萨比、迪娜 · 萨比
‘IM Blanky’
2011 / Product Prototype
Exhibition: Stitches, curated by Larry Wayne Richards, WORKShop Gallery, Toronto
Principals in Charge: Christos Marcopoulos and Carol Moukheiber in collaboration with Rodolphe el-Khoury
Team: Valentina Mele, Sebastian Savone, Jonah Ross-Marrs, Yie Ping See, Samar Sabie, Dina Sabie

–
“范斯沃斯帘幕”
2011 年 / 室内设计展装置
地　点：多伦多
负责人：克里斯托 · 马尔科普洛斯、卡罗尔 · 穆海贝尔与鲁道夫 · 埃尔 – 库利
团　队：金敏宇，方安丽
“Farnsworth Curtain”
2011 / Interior Design Show Installation Proposal

Location: Toronto
Principals in Charge: Christos Marcopoulos and Carol Moukheiber with Rodolphe el-Khoury
Team: Min Woo Kim, Andria Fong

–
“胡氏人体模型”
2010 年
竞　赛：美国知名建筑网站 Architizer，纽约 Siki Im 时装屋
负责人：克里斯托 · 马尔科普洛斯与卡罗尔 · 穆海贝尔
团　队：瓦伦蒂娜 · 梅勒，迈克尔 · 斯帕塔福拉
“Hu-mannequins”
2010
Competition: Architizer, House of Fashion for Siki Im, New York
Principals in Charge: Christos Marcopoulos and Carol Moukheiber
Team: Valentina Mele, Michael Spatafora

–
“室外的房间”
2009 年
展　　览：《缺水》，里埃特 · 玛格丽丝和艾丽莎 · 朝尼策展，多伦多 LWR 画廊
负 责 人：克里斯托 · 马尔科普洛斯、卡罗尔 · 穆海贝尔与唐 · 施力博格
景观顾问：罗伯 · 怀特
“The Out-House”
2009
Exhibition: Out of Water, curated by Liat Margolis and Aziza Chaouni, LWR Gallery,Toronto
Principals in Charge: Christos Marcopoulos and Carol Moukheiber in collaboration with Don Shillingburg
Landscape Consultant: Rob Wright

–
“RGB 花园”
2009 年
比赛项目：魁北克第 11 届国际梅蒂斯花园节
负 责 人：克里斯托 · 马尔科普洛斯、卡罗尔 · 穆海贝尔与纳希德 · 纳比安
团　　队：梅丽莎 · 刘，瓦伦蒂娜 · 梅勒，克莱奥 · 巴斯特
“RGB Garden”
2009
Competition: The 11th International Metis Garden Festival, Quebec
Principals in Charge: Christos Marcopoulos, Carol Moukheiber in collaboration with Nashid Nabian
Team: Melissa Lui, Valentina Mele, Cleo Buster

–
“双重状态房间”
2008 年 / 产品原型
负责人：克里斯托 · 马尔科普洛斯与卡罗尔 · 穆海贝尔
团　队：迈克尔 · 德 · 容，梅丽莎 · 刘
“Dual State Space”
2008 / Product Prototype
Principals in Charge: Christos Marcopoulos and Carol Moukheiber
Team: Michael de Jong, Melissa Lui

–
“数字化窗户”
2008 年 / 产品原型
负责人：克里斯托 · 马尔科普洛斯与卡罗尔 · 穆海贝尔
“Digital Window”
2008 / Product Prototype
Principals in Charge: Christos Marcopoulos and Carol Moukheiber

–
“范斯沃斯墙”
2008 年 / 产品原型
负责人：克里斯托 · 马尔科普洛斯，卡罗尔 · 穆海贝尔与曼妮 · 曼妮
“Farnsworth Wall”
2008 / Product Prototype
Principals in Charge: Christos Marcopoulos and Carol Moukheiber in collaboration with Mani Mani

–
“天空之宅”
2004 年
负责人：克里斯托 · 马尔科普洛斯与卡罗尔 · 穆海贝尔
“Sky House”
2004
Principals in Charge: Christos Marcopoulos and Carol Moukheiber

–
“看不见的住宅”
2004 年 / 产品原型
负责人：克里斯托 · 马尔科普洛斯与卡罗尔 · 穆海贝尔

团　队：帕特里克·蒂尔尼，迈克尔·德·容
“Invisible House”
2004 / Product Prototype
Principals in Charge: Christos Marcopoulos and Carol Moukheiber
Team: Patrick Tierney, Michael de Jong

KLF事务所项目信息：
KLF Projects Information:

–
“规则”
2011 年 / 装置
地　点：麻省理工学院
负责人：鲁道夫·埃尔－库利和罗伯特·莱维特
团　队：撒迦利亚·格伦农，林赛·霍克曼，蕾妮·梁，乌尔坦·伯恩
“Rule”
2011 / Installation
Location: MIT School of Architecture
Principals in Charge: Rodolphe el-Khoury and Robert Levit
Team: Zachariah Glennon, Lindsay Hochman, Renee Leung, Ultan Byrne

–
“自然 2.0”
2011 年 / 装置
地　点：安大略省多伦多城市美术馆
负责人：鲁道夫·埃尔－库利和罗伯特·莱维特
团　队：纳里曼·穆萨维和撒迦利亚·格伦农，费萨尔·巴希尔，伊琳娜·索罗
‘Nature 2.0’
2011 / Installation
Location: Harborfront Gallery,Toronto, Ontario
Principals in Charge: Rodolphe el-Khoury and Robert Levit
Team: Nariman Mousavi and Zachariah Glennon, Faisal Bashir and Irina Solop

–
“光之隧道”
2011 年 / 室内设计展装置方案
地　点：安大略省多伦多市

负责人：鲁道夫 · 埃尔 – 库利和罗伯特 · 莱维特
组　员：撒迦利亚 · 格伦农，纳里曼 · 穆萨维
“Tunnel of Light”
2011 / Interior Design Show Installation Proposal
Location: Toronto, Ontario
Principals in Charge: Rodolphe el-Khoury and Robert Levit
Team: Zachariah Glennon, Nariman Mousavi

–
“超级块”
2011 年 / 产品原型
负责人：鲁道夫 · 埃尔 – 库利与纳希德 · 纳比安
团　队：扎克 · 格伦农，纳里曼 · 穆萨维
“Superblock”
2011 / Product Prototype
Principal in Charge: Rodolphe el-Khoury in collaboration with Nashid Nabian
Team: Zack Glennon, Nariman Moussavi

–
“闪烁”
2010 年
竞　赛：“大地艺术发电机计划”
负责人：鲁道夫 · 埃尔 – 库利和罗伯特 · 莱维特
团　队：林赛 · 霍克曼，哈立德 · 纳赛尔，梅丽莎 · 刘
“Blink”
2010
Competition: Land Art Generator Initiative
Principals in Charge: Rodolphe el-Khoury and Robert Levit
Team: Lindsay Hochman, Khaled Al-Nasser. Melissa Lui

–
“纪念碑 + 比特”
2009 年 / 展览
地　点：多伦多 Eric Arthur 画廊，安娜堡 CMKY 画廊
负责人：鲁道夫 · 埃尔 – 库利和罗伯特 · 莱维特
团　队：詹姆斯 · 迪克森，迈克 · 冯和梁锦松
“Monuments + Bits”
2009 / Exhibition

Location: Eric Arthur Gallery,Toronto; CMKY Gallery, Ann Arbor
Principals in Charge: Rodolphe el-Khoury and Robert Levit
Team: James Dixon, Mike Fung, and Renée Leung

–
“可连接的遮阳”
2009 年
竞　赛：“凤凰城给我庇护”
负责人：鲁道夫 · 埃尔 – 库利，罗伯特 · 莱维特，库特 · 拉尼尔建筑事务所
团　队：迈克 · 冯
“Linking Shade”
2009
Competition: Gimme Shelter, Phoenix
Awards: Finalist
Principals in Charge: Rodolphe el-Khoury and Robert Levit in collaoration with Kuth Ranier Architects
Team: Mike Fung

–
“我的城市”
2009 年 / 装置
地　点：多伦多市便利画廊
负责人：鲁道夫 · 埃尔 – 库利
编　程：马克 · 琼斯（新兴软件）
团　队：詹姆斯 · 迪克森
“MyCity”
2009 / Installation
Location: Convenience Gallery, Toronto
Principal in Charge: Rodolphe el-Khoury
Programming: Mark Jones (Emergent Software)
Team: James Dixon

–
“无穷集合”
2009 年 / 装置
地　点：多伦多白夜
负责人：鲁道夫 · 埃尔 – 库利与纳希德 · 纳比安
团　队：迈克 · 冯
“Aleph TO”

2009 / Installation
Location: Nuit Blanche, Toronto
Principal in Charge: Rodolphe el-Khoury in collaboration with Nashid Nabian
Team: Mike Fung

–
“气味筛选器”
2009 年 / 装置
地　点：多伦多 InterAccess
负责人：鲁道夫 · 埃尔 – 库利与纳希德 · 纳比安
团　队：阿里 · 亚伯赫特，史蒂文 · 阿维斯，马汉 · 贾瓦迪
“Scentisizer”
2009 / Installation
Location: InterAccess, Toronto
Principal in Charge: Rodolphe el-Khoury in collaboration with Nashid Nabian
Team: Ali Yarbakht, Steven Avis, Mahan Javadi

–
“变色龙休息厅”
2008 年 / 装置
地　点：路易斯安那州新奥尔良市 DesCours 艺术节
负责人：鲁道夫 · 埃尔 – 库利与纳希德 · 纳比安
团　队：罗伯特 · 莱维特，纳希德 · 纳比安，迈克 · 冯，詹姆斯 · 迪克森
“Chameleon Lounge”
2008 / Installation
Location: DesCours Arts Festival, New Orleans, Louisiana
Principal in Charge: Rodolphe el-Khoury in collaboration with Nashid Nabian
Team: Robert Levit, Nashid Nabian, Mike Fung, James Dixon

–
“窗户 2.0”
2008 年 / 装置
地　点：多伦多 LWR 画廊
负责人：鲁道夫 · 埃尔 – 库利
团　队：詹姆斯 · 迪克森
“Window 2.0”
2008 / Installation
Location: LWR Gallery, Toronto

Principal in Charge: Rodolphe el-Khoury
Team: James Dixon

–
“深圳当代艺术与城市规划馆”
2007 年 / 竞赛
地　　　点：深圳
获 奖 情 况：入围（四分之一比赛的第一场）AIA 建筑优秀奖
负　责　人：KFL 事务所
景观设计师：艾丽莎 · 彼得 · 诺斯，NDO
团　　　队：塔拉尔 · 拉赫迈赫，李山、拉吉 · 帕特尔，阿德里安 · 菲弗，卢克 · 斯特恩，阿齐亚 · 达马尼，尤瑟夫 · 弗雷泽，迈克 · 冯，尼克 · 马鲁利斯
“MOCAPE”
2007 / Competition
Location: Shenzhen
Awards: Competition Finalist (First of four / initial competition) AIA Merit award for unbuilt architecture
Principals in Charge: Khoury Levit Fong
Landscape Architects: Alissa Peter North, NDO
Team: Talal Rahmeh, Shan Li, Raj Patel, Adrian Phiffer, Luke Stern, Azia Dhamani, Yusef Frasier, Mike Fung, Nick Maroulis

–
“斯特拉特福德广场”
2006 年
竞　赛：斯特拉特福德市集广场
地　点：安大略省斯特拉特福德
奖　项：公开赛荣誉奖
负责人：鲁道夫 · 埃尔 – 库利
团　队：刘玲晨，王成龙
“Stratford Square”
2006
Competition: Market Square, Stratford
Location: Stratford, Ontario
Awards: Honourable Mention in Open Competition
Principal in Charge: Rodolphe el-Khoury
Team: Lingchen Liu and Chenglong Wang

致谢
Acknowledgments

The projects documented in this book have benefitted tremendously from the resources, ideas and research at the academic programs, cultural institutions, and professional offices where they developed; we thank the following for their support: California College of the Arts, Convenience Gallery, AIA New Orleans, HarbourFront Center, InterAccess, Khoury Levit Fong, Massachusetts Institute of Technology, Studio NMinusOne, University of Hong Kong, University of Michigan, University of Toronto, DIAC, and WORKshop.

With contributions by Drew Adams, Yuliya Bentcheva, Cleo Buster, Justin Cheng, Maya Desai, James Dixon, Holly Jordan, Samuel Ko, David Long, Melissa Lui, Ricky Luk , Mani Mani, Fadi Massoud, Taraneh Meshkani, Azadeh Mohammadi, Nashid Nabian, Maggie Nelson, Otto Ng, Ada Juwah, Shadi Ramos, Julianna Sassaman, Mae Shaban, Don Shillingburg, Rick Sole, Lailee Soleimani, Studio NMinusOne, Elaine Tong, Jimmy Tran, Arthur Tseng, Nadya Volicer, and Ali Yarbakhti

本书所记录的项目很大程度上受益于一些学术项目、文化机构和他们所发展的专业研究室的资源、理念和研究。我们感谢以下机构的支持：加州艺术学院、便利画廊、美国友邦新奥尔良保险公司、港湾中心、InterAccess、KLF 事务所、麻省理工学院、迈纳斯工作室、香港大学、密歇根大学、多伦多大学、DIAC 及工作坊。

为本书作出贡献的还有：朱·亚当斯、尤利娅·本茨瓦、克莱奥·巴斯特、贾斯汀·陈、玛雅·德赛、詹姆斯·狄克逊、霍利·乔丹、塞缪尔·高、戴维·朗、梅丽莎·刘、瑞奇·鲁克、曼妮·曼妮、法迪·马苏德、塔兰涅·米什卡尼、阿扎德·穆罕默迪、纳希德·纳比安、麦琪·纳尔逊、奥托·吴、艾达·尤瓦、沙迪·拉莫斯、朱利安纳·萨斯曼、曼·沙巴、唐·施力博格、瑞克·索莱、莱利·萨利曼尼、NMinusOne 工作室、伊莲·童、吉米·陈、亚瑟·曾、娜迪亚·沃勒斯和埃利·雅巴哈提。

几位同行、同事和朋友为我们贡献了批判性的见解和专业知识，我们特别感谢：汤姆·贝赛、阿德里安·布莱克威尔、布莱恩·拜尔根、大卫·埃德曼、史蒂文·冯、乔丹·盖革、埃利·古尔德、安妮塔·曼瑟夫斯·郝特、米切尔·约阿希姆、希拉·肯尼迪、泰德·坎斯克、罗伯特·莱维特、大卫·利伯曼、安德鲁·佩恩、蒂姆·普波、拉里·韦恩·理查兹、安德里亚·桑普森、爱丽丝·雪莱、约翰·夏纳、理查德·萨默、纳德·泰若妮、汤姆·沃普斯、梅森·怀特、肖恩·威廉姆森和尹弥京。

还要感谢著作的出版团队，尤其是多伦多大学的瓦伦蒂娜·梅勒和安德鲁·佩斯基。

感谢上海大学上海美术学院的魏秦副教授与华诚博远建筑设计集团的张昕先生将此书带到中国。

Among several peers, colleagues and friends who have contributed critical insights and expert knowledge to the work we especially recognize and thank Tom Bessai, Adrian Blackwell, Brian Boigon, David Erdman, Steven Fong, Jordan Geiger, Arleen Gould, Anita Matusevics-Halter, Mitchell Joachim, Sheila Kennedy, Ted Kesik, Robert Levit, David Lieberman, Andrew Payne, Tim Poupore, Larry Wayne Richards, Andrea Sampson, Elise Shelley, John Shnier, Richard Sommer, Nader Tehrani, Tom Verbes, Mason White, Shane Williamson and Meejin Yoon.

We thank the book's production team, in particular Valentina Mele and Andrew Piotrowski in Toronto.

We are grateful to Professor Wei Qin of Shanghai University and Mr. Zhang Xin of Huachengboyuan Engineering Technology Group for bringing this book to China.

译后记
Postscript

With the development of information revolution and digital technology, our global village seems to be getting smaller. Quite different from the previous ones, the temporal and dimensional concepts have changed the life style of human beings because the boundaries between virtuality and reality, man and machine, art and science become fuzzier. As a result, intelligence and interaction will evolve into a new branch of learning in the near future. As AI and information technology are widely used in the human life, the various forms of 'cloud life' have penetrated into our daily life. In addition, diversification of the client ends ranges from mobile phones, sensors, electronic readers, AI automobiles and smart houses…The digitalized information interchanges give rise to the fundamental revolution in human interactions, urban management and operation and human environment.

When we worked in the international digital workshop for the theme 'Designing for Internet of Things' initiated by Shanghai Academy of Fine Arts in 2013. Jointly launched by Prof. Li Qiansheng, chairman of Digital Art Department, and me, our workshop invited Prof. Rodolphe el-Khoury, dean of Architectural Engineering School, University of Toronto

随着信息革命与数字技术的发展，地球的距离在逐渐缩小，时间与维度的概念也和从前大为不同，扩展到人类的生活方式，虚拟与现实、人与机器、艺术与科学之间的界限也越来越模糊，智能与互动成为我们当下生活日趋显现的发展趋势。随着智能化与信息技术的普及，越来越多样化的“云端”生活已渗透到人们最日常和触手可及的生活层面，用户端也呈现多元化：从手机、传感器、电子阅读器，到数字轿车、智能家居……数字化的信息交换对人的行为、城市运行与建成环境产生了根本性的变革。

本书能够引入中国，源自 2013 年在上海大学美术学院举行的《为物联网而设计》的国际数字空间工作营。工作营由我和数码艺术系的李谦升主任共同发起，华诚博远设计集团的张昕先生特别引荐他的好友——时任多伦多大学建筑学院院长鲁道夫·埃尔－库利教授来到中国，亲自指导在当时对我们还非常陌生的“物联网”技术的设计实践营。来自建筑学、城乡规划与数字媒体等不同专业的学生利用物联网的核心技术——传感器技术、射频识别标签、嵌入式系统，对未来家居、交通、保健、物流、购物等几大主题进行了脑洞大开的想象与空间构思，同学们活跃的思维与耳目一新的想法，确实出乎我们的意料，取得了意想不到的教学效果。如在家居中植入传感器技术，指导“土豆先生”日常的家庭医疗保健；嵌入式技术可以使“一个杯子”通过与物体的触碰方式完成相识、交友、互动等行为，呈现一种云上的社交生活模式……种种智能城市下的生活愿景在当时看似是学生的天马行空，但是在当下已经越来越变成即将实现的生活方式。上海大学美术学院的汪大伟院长非常支持跨科学工作营的开展，他富有前瞻性的思维与注重跨学科人才培养的思想，以及工作营的成功举办都激发了我和张昕对智能城市与未来建筑领域的兴趣，与本书引入中国的想法。感谢中国建筑工业出版社李东禧主任的信任与举荐，为本书在中国建筑工业出版社翻译出版奠定了基础。

鲁道夫·埃尔－库利教授与他的研究伙伴在世界各地完成了大量设计实践作品，在国际上相关领域具有较高的知名度，从 2004 开始，他们专注于物联网与人工智能在城市可持续发展领域的研究，尤其是机器人、嵌入式技术在建成环境的应用、互动性与感应式环境的原型等研究，并结合建筑与城市设计的实践项目，引领了智能城市与未来建筑发展的新方向。

随着智能化与云技术的普及，物品与互联网

at the time, to our academy to render direction for the technical design of ‘Internet of Things’ alien to us with assistance of Mr. Zhang Xin who works in Huacheng Boyuan Design Group. Coming from the faculties such as architecture, urban-rural planning and digital media, our students freed their imaginative designs and innovations in the coming housing, traffic, logistics and shopping with the core technologies of Internet of Things, including sensor technology, radio frequency identification tag and embedded system, during which the innovative ideas and designs made us surprised and we achieved the teaching results beyond our expectation. For example, the designs such as inversion of sensors into houses, direction of ‘Mr. Potato’ in household daily medical care, embedded technology which enabled ‘a cup’ to help us to complete the behaviors of making acquaintance, friends and interactions through the touches between the objects… Various living prospects and visions seem to be the fantastic ideas of the students, but they are being turned into the modes of our life in the intelligent cities. Prof. Wang Dawei, dean of Shanghai Academy of Fine Arts, showed his full support for the inter-disciplinary research in the workshop, whose prospective insight and focus on cultivating the students of inter-disciplines, and the success of the workshop inspired and encouraged the interest in intelligent cities and the future architecture of the translators. The monograph *Buildings Alive Architecture for the Internet of Things* caught our attention and thus we decided to translate it into Chinese. We own our indebtedness to Mr. Li Dongxi, director of China Architecture and Building Press for his recommendation of this foreign monograph and his support for the publication of the Chinese-English bilingual version.

Prof. Rodolphe el-Khoury and his partners are internationally well-known for their numerous designs and completed works. From 2004, his team was engaged in the studies of Internet of Things and urban sustainable development, particularly the research in the application of robot and embedded technology in the built environment, the prototypes of interactive and responsive environments. In integrating the practical architectural project with the urban design, the team leads the new developing trend of smart city and future architectural development.

With the popularization of intelligence and cloud technology, the connection between goods and the Internet has realized the real-time information exchange, intelligent identification, positioning and monitoring, etc. Intelligent terminal products characterized by embedded system can be available everywhere, ranging from the electronic devices to aerospace satellite system, which will be embedded in building envelope, furniture, household wares, and which will offer such services of intelligent household lives, health care, home shopping, etc. The trend will be bound to break our traditional thinking on the building space functions and spatial scales, make the interaction process between the people and the architecture present a new activated state. As a result, a building will become a life entity with perceptible, breathing, thinking actions to meet the requirements of the residents.

The Chinese-English bilingual edition of *Buildings Alive Architecture for the Internet of Things* is an expanded edition of the original English monograph, adding some latest case study and materials on the basis of the original book. The book covers design practices and prototypes of multiple disciplines, amalgamating architecture, urban and rural planning, landscape, art design, computer science, communication engineering, automation and some other disciplines. The contents in the book cover six aspects, such as artificial nature, immersive space, kinetic components, variable cladding, surface as interface and Individuated response. From overall urban layout to the architectural details at different levels, this book systematically introduces the research on the prototype of the sentient architecture, puts forth the theory on the life characteristics and adaptive ability of buildings. The book deals with the synthetic sentience of intelligent cities and provides the new ideas and new orientations for future architecture and the urban development.

This book enlightens the architecture in the following two respects.

1. To construct a sharing platform of multidisciplinary collaboration

The projects listed in the book involve the questions about dynamic living systems, with the aim to develop a digitally enhanced architecture from collaborative experimentation to the buildings and urban space which are adaptable to the development model of smart cities. At

的联结实现了信息的实时交换、智能化识别、定位与监控等。以嵌入式系统为特征的智能终端产品随处可见；小到人们身边的电子设备，大到航空航天的卫星系统，将之植入建筑的围护结构、家具、生活物件中，给人们提供智能化的家居生活、保健服务、居家购物等，这势必打破我们对建筑空间功能与空间尺度的传统思维，使人与建筑之间的交互过程呈现一种新的激活状态：建筑成为一个可感知、可呼吸、会思考、可行动，并满足居者个性需求的生命体。

《感应式建筑　物联网时代的建筑》中英双语版是英文原版书的扩充版，在原书基础上补充了一些最新的案例与资料。该书内容横跨多学科领域的设计实践与设计原型，将建筑、城乡规划、景观，与艺术、计算机科学、通信工程、自动化等多学科融贯交叉。本书内容主要从人工自然、沉浸式空间、能动构件、可变覆盖面、表面作为接口与个体感应等六个方面，从城市整体到建筑构造细部不同层级，系统地介绍了感应式建筑的原型研究，从实践角度提出了建筑所具有的生命特征与自适应能力、建筑与城市具有的综合感知能力，并提供了未来建筑与城市发展的新思维与新方向。

本书对建筑学界的启发主要体现在以下两点：

1. 建构多学科协同的跨学科共享平台

书中的项目提出了与动态生命系统有关的问题，其主旨是研发一种数字增强的建筑，从协同实验到适应智慧城市发展模式下的建筑与城市空间。目前，国内关于交互性的研究多局限于产品设计与数字媒体技术方面，将物联网技术与城市、景观与建筑空间多维一体结合的研究成果在国内仍是全新领域，尚属起步阶段，本书为相关领域的研究提供了大量可借鉴的参考案例。

2. 人工智能时代城市与建筑的新范式

技术革新推动了城市与建成环境根本性的空间变革，也推动了一种从机械化范式走向生物学模型的转变，对城市空间与建筑的塑造提出了新的范式。从复合智能化的建筑材料到由人、物体、空间与环境景观构成的巨大网络都可以通过信息技术手段实现虚拟自然，使建成环境具有类似于生命系统的自适应能力，依赖互联网与信号处理系统，在建筑、环境与居住者之间建立反馈机制，使建筑具有感知、思考、行动与交流能力，适应居住者对建筑的形态适变、适宜环境、空间氛围、个体感知等环境需求。

人工智能时代的建筑与城市体现了技术发展对人们生活与行为的巨大影响与变革。美国《连线》杂志创始人凯文·凯利在《失控》一书中指出了未来技术发展所遵循的“蜂群效应”，即像生命系统那样的自组织性，无论是生物系统还是技术系统，技术逻辑无法实现对复杂生命系统的管理，人的生命系统将始终参与到人类进化与技术的进步中，未来科技系统将逐渐以模仿自然系统的方式而发展。因而，人类可以驾驭技术并成为技术的主导者，也可以在技术支撑下和谐而诗意地生活，相信未来的城市与建筑将更加体现人、生态与技术的和谐共生。希望此书的翻译出版能

present, the domestic research on interactivity is mostly confined to product design and digital media technology, while the studies on integrating Internet of Things technology with urban, landscape and architectural space are still in the initial stage in China. This book provides a large number of reference cases for research in related fields.

2. A new paradigm for urban architecture in the time of artificial intelligence

Technological innovation promotes the fundamental spatial transformation of urban and the built environment, and also propels the transformation from mechanization paradigm to biological model, which brings forth a new paradigm for the re-shaping of urban space and architecture. Formed by the composite intelligent building materials and the environmental landscape of man, object and space, the gigantic network can virtualize the nature through information technology, making the built environment boasting the adaptive ability similar to the life system. By virtue of Internet and signal processing system, a feedback system is built among the buildings, the environment and the residents, enabling the architecture acquiring the perceiving, thinking, acting and communicating ability so as to meet the requirements of the residents for suitable typology, friendly environment, space atmosphere, individual perception.

The buildings and cities in the era of artificial intelligence mirror the tremendous influence and change of technological development on people's life and behavior. In his book *Out of Control*, Kevin Kelly, founder of *On-line* magazine, points out that the coming technological development follows ‘the swarm effect’, namely the self-organization like life system, in which both biological system and technology system, technology logic cannot achieve the management of complex life system, while life system will always be involved in human evolution and the advancement of the technology. Moreover, the future science and technology system will gradually evolve into the developing process in the form of imitation of natural systems. Therefore, human beings can control technology and still remain the masters of technology, who can also enjoy a harmonious and poetic life under the support of technology. It is believed that future cities and buildings will more represent the harmonious coexistence of human beings, ecology and technology. The translators expect that this book can offer beneficial inspiration for the scientific workers engaged in interdisciplinary collaborative research and cooperation in the fields of architecture, urban and rural planning, landscape, art design and other related disciplines in China.

As this book involves a wealth of knowledge in such fields as computer science, communication engineering and automation, even botany, which have posed huge challenge on the translators. There may be some oversights and mistakes in this bilingual version. All the critical opinions and proposals for improvement will be very much valued by the translators.

The foreword and Chapter One to Chapter Four are translated by Wei Qin, while Chapter Five and Chapter Six are co-translated by Zhang Xin and Wei Qin. The proofreading is completed by Wei Qin.

At the moment that this book will be printed, the two translators feel obliged to the teachers, students and friends who have offered us valuable assistance in the translation and publication of this book. We still feel indebted to the following students Li Jiayi, Zheng Ruirui, Kang Yilan, Zhang Jianhao, Zhang Zhuoyuan, for their translation and processing numerous pictures and diagrams in the book.

The translators are grateful to Prof. Rodolphe el-Khoury and Prof. Carol Moukheiber for their sincere support for book to be translated into Chinese and their selfless help during the translation.

Special acknowledgement are contributed to Li Dongxi and Wu Ling, directors of China Architecture & Building Press, Sun Shuyan and Chen Chang, editors of the press, for their painstaking efforts in the publication of the book. Without their care, concern and help, the publication of the book would have been impossible.

The translators earnestly expect that the book will be helpful to the professionals.

Wei Qin

September, 2020 in Shanghai

够为国内建筑、城乡规划、景观、艺术设计及其相关学科进行跨学科协同研究和合作提供有益的启发。

因为本书的跨学科背景，书中涉及大量计算机、通信工程与自动化，甚至是植物学等众多领域的知识，为翻译带来了很大挑战。限于译者的学识有限，书中难免会有错漏之处，还请读者包涵与指正。

本书序言、第 1 ~ 4 章由魏秦翻译，第 5 ~ 6 章由张昕与魏秦共同翻译，全书校对由魏秦完成。此书即将付梓之际，特别感谢对本书翻译出版做出贡献与提供帮助的老师、同学与朋友。书中案例图片数量很大，感谢在读和已毕业的学生参与了图片翻译与处理的大量工作，在此对李嘉漪、郑瑞瑞、康艺兰、张健浩、张卓源表示感谢！

衷心感谢鲁道夫 · 埃尔 − 库利教授、卡罗尔 · 穆海贝尔教授对译者的信任和支持，使本书得以引入中国。在翻译过程中，他们也给予耐心的指导与无私的帮助！

特别感谢中国建筑工业出版社的李东禧主任、吴绫主任、孙书妍编辑与陈畅编辑，为本书的出版倾注了大量心血，提供了细致周到的帮助，在此衷心感谢！

希望本书的翻译出版不负读者的期望！

魏秦

2020 年 9 月于上海